Water Disinfection and Natural Organic Matter

ACS SYMPOSIUM SERIES **649**

Water Disinfection and Natural Organic Matter

Characterization and Control

Roger A. Minear, EDITOR
University of Illinois

Gary L. Amy, EDITOR
University of Colorado

Developed from a symposium sponsored
by the Division of Environmental Chemistry, Inc.

American Chemical Society, Washington, DC

Library of Congress Cataloging-in-Publication Data

Water disinfection and natural organic matter: characterization and control / Roger A. Minear, editor, Gary L. Amy, editor.

p.   cm.—(ACS symposium series, ISSN 0097–6156; 649)

"Developed from a symposium sponsored by the Division of Environmental Chemistry, Inc., at the 210th National Meeting of the American Chemical Society, Chicago, Illinois, August 23–24, 1995."

Includes bibliographical references and indexes.

ISBN 0–8412–3464–7

1. Water—Purification—Disinfection—Congresses. 2. Water—Purification—Disinfection—By-products—Congresses. 3. Organic water pollutants—Congresses.

I. Minear, R. A.   II. Amy, Gary L.   III. American Chemical Society. Division of Environmental Chemistry.   IV. American Chemical Society. Meeting (210th: 1995: Chicago, Ill.)   V. Series.

TD459.W37   1996
628.1'662—dc20   96–36454
                                                             CIP

This book is printed on acid-free, recycled paper.

Foreword

THE ACS SYMPOSIUM SERIES was first published in 1974 to provide a mechanism for publishing symposia quickly in book form. The purpose of this series is to publish comprehensive books developed from symposia, which are usually "snapshots in time" of the current research being done on a topic, plus some review material on the topic. For this reason, it is necessary that the papers be published as quickly as possible.

Before a symposium-based book is put under contract, the proposed table of contents is reviewed for appropriateness to the topic and for comprehensiveness of the collection. Some papers are excluded at this point, and others are added to round out the scope of the volume. In addition, a draft of each paper is peer-reviewed prior to final acceptance or rejection. This anonymous review process is supervised by the organizer(s) of the symposium, who become the editor(s) of the book. The authors then revise their papers according to the recommendations of both the reviewers and the editors, prepare camera-ready copy, and submit the final papers to the editors, who check that all necessary revisions have been made.

As a rule, only original research papers and original review papers are included in the volumes. Verbatim reproductions of previously published papers are not accepted.

ACS BOOKS DEPARTMENT

Contents

NATURAL ORGANIC MATTER RELATIONSHIPS AND CHARACTERIZATION

INDEXES

Preface

A MAJOR CHALLENGE IN TREATING DRINKING WATER is how to reconcile the seemingly contradictory goals of achieving disinfection (inactivating pathogenic and indicator microorganisms) while minimizing the formation of disinfection by-products (DBPs). This paradox translates into the goal of minimizing overall risk from both waterborne pathogens and carcinogenic DBPs. In the 1970s and 1980s, chlorine was seen as the major contributor to the DBP problem largely because of the initial discovery of the linkage between trihalomethanes (THMs) and chlorinated drinking water. Alternative disinfectants, including ozone, chlorine dioxide, and chloramines, were embraced. However, this response ignored the contributions of chlorine toward the eradication of waterborne diseases. Moreover, alternative disinfectants were eventually found to harbor their own DBPs (e.g., bromate formation during ozonation). We have come to recognize that all of these disinfectants may play a role in disinfection, and that combinations of disinfectants may best contribute to overall risk minimization. Given recent episodes of *Cryptosporidium* in finished drinking water, disinfection cannot be compromised.

Initial attempts to control DBPs recognized natural organic matter (NOM) as an important precursor to THMs. However, the initial chemical identity of NOM was based on nonspecific total organic carbon measurements. We have now come to realize that the composition and reactive properties of NOM vary from source to source. Moreover, we now realize that there are actually two DBP precursors, an organic precursor (NOM) and an inorganic precursor (bromide ion, Br^-). Bromide influences DBP risk, which is a function not only of compound class but also of the degree of bromination versus chlorination.

This book attempts to add to our knowledge of the chemical identity of NOM and the interactions between NOM and Br^- in forming DBPs from both chlorine and alternative disinfectants. It is an outgrowth of a symposium presented at the 210th National Meeting of the American Chemical Society titled "Disinfection By-Products and NOM Precursors: Chemistry, Characterization, and Control," sponsored by the ACS Division of Environmental Chemistry, Inc., in Chicago, Illinois, from August 23–24, 1995.

ROGER A. MINEAR
University of Illinois
Urbana, IL 61801

GARY L. AMY
University of Colorado
Boulder, CO 80309

Chapter 1

Water Disinfection and Natural Organic Matter: History and Overview

Roger A. Minear[1] and Gary L. Amy[2]

[1]Institute for Environmental Studies, Department of Civil Engineering, University of Illinois, 1101 West Peabody Drive, Urbana, IL 61801–4723
[2]Department of Civil, Environmental, and Architectural Engineering, University of Colorado, Boulder, CO 80309

Historical Developments

With the introduction of chlorine in drinking waters in the early part of this century (Sawyer et al, 1994), a new era was started regarding public health protection and the evolution of trust in safe drinking waters. This did not preclude the need for maintenance of control for individual toxic chemicals and specific aesthetic features of the product water. This latter element was an integral component to public health considerations in that safe waters must also be palatable in order to prevent consumers from seeking alternative sources of more aesthetically pleasing water which might in fact not be biologically and chemically safe. Much of the early control and setting of standards related to a few select chemical and biological safety parameters plus a series of what were essentially aesthetic indices.

Natural organic matter (NOM) was one of these indices, indirectly, as it was reflected in the removal of color from product waters. Although iron could also impart a yellow color to waters, typically the undesirable color causing substances originated from naturally occurring humic substances. Much research among the engineers and water chemists of the 1950s to 1970s was directed at characterization and removal of these substance from drinking waters. Prominent and representative of these efforts was the work of A. P. Black and his students at the University of Florida (For example, Black and Christman, 1963a,b; Black and Willems, 1961). In the late 1950s to the early to mid 1960s, the fundamental nature of aquatic humic substances was the subject of debate among researchers (Christman, 1968; Shapiro, 1957,1958). At the end of that decade, it was generally accepted that natural water color was principally the result of humic substances and that these humic substances were complex polyfunctional organic molecules of varying molecular weight, ranging up to relatively large macromolecules, and this characterization continued into the 1970s (Christman

0097–6156/96/0649–0001$15.00/0

and Ghassemi, 1966, Gjessing, 1965, 1970, 1976). Continued characterization relied upon newly evolving instrumental techniques. The book by Thurman (1985) and extensive work by the U.S. Geological Survey on Suwannee River humic substances (USGS 1989) have collected much of the current understanding based on the research efforts to that time.

Among the common practices for removing this undesirable color for domestic drinking waters was bleaching out the color by use of the strong oxidizing power of free chlorine (HOCl and OCl⁻); however, the common disinfection practice of the time was to ensure that a residual of free chlorine was achieved in the disinfection process so this frequently led to oxidation of nitrogen and organic species in the process. It wasn't until the early 1970s when the work of the Dutch water chemist, Johannes Rook (1974) demonstrated that the chlorination of Amsterdam drinking water produced chloroform and other trihalomethane species in the finished waters. US studies completed in the mid to late 1970s (Westrick, 1990) also confirmed the Dutch findings and touched off a flurry of Trihalomethane (THM) related research in the Drinking Water field around the world.

At this time, a scientist at Oak Ridge National Laboratory was completing a Ph.D. in which radio-labeled chlorine was used to study products produced upon chlorination of domestic waste waters and model compounds. Associated with his continuing interest in this research area and the evolving concerns relative to human health implications, Robert Jolley, along with assistance from the US EPA and other colleagues, initiated in 1975, what was to become a series of 6 "Chlorination Conferences" (Jolley et al. 1978, 1980, 1983a, 1983b, 1985, 1990). These conferences became a forum for presentation of the active research related in one way or another to water disinfection. As the implications of the presence of THMs in drinking waters were pondered by the regulatory agencies in the United States and other countries, the linkage between the science and evolving regulations became in integral part of the "Chlorination Conferences." Phil Singer has portrayed this interrelation ship in a series of excellent review publications (Singer, 1992, 1993, 1994).

Regulatory evolution focused initially on THM control and means of reducing the quantities produced. Three targets existed: the organic precursor (NOM), the disinfectant type and dose, and the removal of THM compounds themselves. The branching web of research directions spawned by these foci included engineering processes, biological studies relating microbial disinfection, human health implications and the basic chemistry associated with all these activities. The original focus on THMs has evolved in several areas:

* expanded interest in disinfection by-products (DBPs) associated with other (alternative) disinfectants,

* effectiveness of disinfection with alternative disinfectants,

* relative risk factors of products and or processes and use of advances in fields like molecular biology,

* continuing interest in the role of NOM and how its characteristics can be related to DBP formation and control,

* analytical chemistry advances and the development of better analysis procedures, and

* fundamental understanding of the chemistry involved in disinfection and DBP formation processes.

Overview of DBP and NOM Issues

To expand on these themes, it is important to realize that while free chlorine came under much earlier scrutiny (Rook, 1974; Bellar, 1974) in the formation of disinfection by-products (DBPs), alternative chemical disinfectants such as ozone (O_3), chlorine dioxide (ClO_2), and chloramines (e.g., NH_2Cl, monochloroamine) each has since been shown to have its own host of DBPs. Trihalomethanes (THMs) and halogenated acetic acids (HAAs) represent the most important groups of chlorination by-products, formed in the presence of natural organic matter (NOM), serving as the organic precursor, and bromide ion (Br^-), playing the role of the inorganic precursor. While measurement of Br^- is straightforward, NOM measurement is achieved through surrogates (Edzwald, 1985): the amount of NOM is quantified through total (TOC) or dissolved (DOC) organic carbon, while NOM character can be deduced by UV absorbance at 254 nm and by specific UV absorbance (SUVA, UVA_{254}/DOC), an index of the humic/aromatic character of the NOM.

Both NOM properties (e.g., molecular weight) and the relative amount of Br^- affect the magnitude and species distribution of THMs and HAAs. In addition to chlorine dose, THM and HAA formation is also influenced by pH and temperature. Chlorinated drinking water has also been shown to contain haloacetonitriles (HANs), haloketones, and miscellaneous chloro-organic compounds such as chloral hydrate, chloropicrin, cyanogen chloride, and 2,4,6 trichlorophenol (Singer, 1994; Oliver, 1983; Uden, 1983). In the presence of bromide, brominated THMs, brominated HAAs, brominated HANs, cyanogen bromide, and bromopicrin can be formed. Specific compounds constitute only about half of the overall pool of chlorinated organic material (Singer, 1994), as measured by total organic halide (TOX). THMs are the dominant class of chlorination DBPs, followed by the HAAs; reasonable source-specific correlations have been established between THMs and HAAs (Singer and Chang,

1989). In a nation-wide survey, the median concentrations of THMs and HAAs were found to be 36 and 17 ug/L, respectively (Krasner, 1989); the proposed DBP regulations will limit THMs and HAAs to 80 and 60 ug/L, respectively. While the health effects of these individual compounds vary (Bull, 1991), MX [3-chloro-4-(dichloromethyl)-5hydroxy-2(5H)-furanone] has been shown to significantly contribute to the mutagenicity of chlorinated drinking water.

In the presence of NOM, ozonation produces several groups of organic by-products, including aldehydes, ketoacids, and carboxylic acids (Xie, 1992; Weinberg, 1993), most of which are relatively biodegradable. In the presence of Br^-, bromate (BrO_3^-) is formed during ozonation (Haag and Hoigne, 1983) with influential factors being NOM, ozone dose, Br^-, pH, and temperature (von Gunten, 1992). Bromate can form through either a molecular ozone or a hydroxyl radical pathway, with the latter predominant in the presence of NOM (Siddiqui, 1995). Proposed regulations will limit BrO_3^- to 10 ug/L. In the presence of both NOM and Br^-, organo-bromine (TOBr) compounds such as bromoform, bromoacetic acids, and bromoacetonitriles can form.

The major by-products associated with chlorine dioxide (Gordon, 1990) include chlorate (ClO_3^-) and chlorite (ClO_2^-); the U.S. EPA has recommended that the combined residuals of ClO_2, ClO_2^-, and ClO_3^- be less than 1 mg/L. Chloramination significantly reduces but does not eliminate THM formation (Jensen, 1985); cyanogen chloride and TOX represent the major DBP issues with respect to chloramines.

Since DBPs are formed by all of the above chemical disinfectants, the adoption of alternative disinfectants for DBP control often means only a tradeoff between one group of DBPs versus another. The most effective DBP control strategy is organic precursor removal through enhanced coagulation, biofiltration, granular (or biological) activated carbon (GAC or BAC), or membranes. There has been little success in Br^- removal. Other DBP control options include water quality modifications; for example, acid or ammonia addition for bromate minimization.

NOM removal is strongly influenced by NOM properties embodying the size, structure, and functionality of this heterogeneous mixture. NOM consists of a mixture of humic substances (humic and fulvic acids) and non-humic (hydrophilic) material. It is the humic substances that are more reactive with chlorine (Reckhow, 1990; Christman, 1983; Collins, 1985) and ozone, both in terms of oxidant/disinfectant demand and DBP formation. Processes such as coagulation, adsorption, and membranes are separation processes which remove NOM intact, while ozonation transforms part of the NOM into biodegradable organic matter (BOM), potentially removable by biofiltration/BAC. Coagulation preferentially removes humic/higher molecular weight NOM; the selectivity of membranes for NOM removal is largely dictated by the molecular weight cutoff of the (nanofiltration or ultrafiltration) membrane; the use of GAC requires significant empty bed contact times; biofiltration can only remove the rapidly degradable fraction of the BOM. Other than removal at the treatment plant, source (watershed) control represents another control option; for example, the control of algal-derived NOM in water supply reservoirs (Hoehn, 1980).

Purpose of the Symposium and Its Organization

Even though the official "Chlorination Conferences" did not continue beyond 1989, research has continued as the issues have become increasingly complex and important. Traditional forums for research presentation continue to be used and this book evolves from one of these, a thematic symposium at a national American Chemical Society meeting. This particular symposium was organized around the theme of DPBs and natural organic matter (NOM), either as a direct precursor or an influential factor in the overall solution chemistry involving DBP formation.

The chapters that follow are grouped to reflect general areas of research in the following fashion:

Regulatory analysis sets the stage for the driving force behind much of the current research even though the researchers would prefer to argue to the contrary. To this end, it was appropriate to include a focus on the regulatory environment. While pure science is the currency of the researcher, practicality and application play a role in much of the environmental field especially since research costs money. Krasner et al. have provided this linkage.

From this point on, the book organization follows the original symposium organization which was in the thematic groupings, Chlorination/Chloramination Products and Reactions; NOM Relationships and Characterization; and Ozone and Other Processes. The ordering is not arbitrary. Chlorine based disinfection is still the dominant process used in the United States whether it be free chlorination or intentional formation of chloramines. Variations in processes and specific configurations with the resultant impact on both DBP formation and disinfection efficiency continue to be the subject of active research.

How the characteristics of NOM relate to the overall processes makes research into NOM properties and reactions with both disinfection chemicals and NOM removal processes an area of continuing importance to the water treatment industry. However, improved knowledge in this area reaches beyond only drinking water treatment and it was not the intent to restrict presentations to direct linkage with water treatment.

As was pointed out in the overview section above, many other disinfection processes are being examined as either replacements for or adjuncts to chlorination processes as means of reducing DBP formation in drinking waters. While ozone is prominent in this effort, it is not exclusive. The collected papers in this third section are a forum for the active research into these alternatives and the potential problems that may be associated with their application.

REFERENCES

Black, A.P. and Christman, R.F., "Characteristics of Colored Surface Waters", Journal AWWA, 55:753, (1963a).

Black, A.P. and Christman, R.F., "Chemical Characteristics of Fulvic Acids", Journal AWWA, 55:897, (1963b).

Black, A.P. and Willems, D.G., "Electrophoretic Studies of Coagulation for the Removal of Organic Color", Journal AWWA, 53:589, (1961).

Bellar, T., et al., The Occurrence of Organohalides in Chlorinated Drinking Water", Journal AWWA, 66:703 (1974).

Bull, R., and Kopfler, F., "Health Effects of Disinfectants and Disinfection By-Products", AWWA Research Foundation (1991).

Christman, R.F., "Chemical Structures of Color Producing Organic Substances in Water" In Symposium on Organic Matter in Natural Waters, D.W. Hood, ed. University of Alaska, pp181-198, (1968).

Christman, R.F., "Identity and Yields of Major Halogenated Products of Aquatic Fulvic Acid Chlorination", Environ. Sci. & Technol., 17:10:625 (1983).

Christman, R.F. and Ghassemi, M., "Chemical Nature of Organic Color in Water", Journal AWWA, 58:723, (1966).

Collins, M., et al., "Molecular Weight Distribution, Carboxylic Acidity, and Humic Substances Content of Aquatic Organic Matter: Implications for Removal During Water Treatment", Environ. Sci. & Technol., 20:10:1028 (1986).

Edzwald, J., "Surrogate Parameters for Monitoring Organic Matter and Trihalomethane Precursors in Water Treatment", Journal AWWA, 77:4:122 (1985).

Gjessing, E.T., "Use of 'Sephadex' Gel for the Estimation of Molecular Weight of Humic Substances in Natural Water" Nature, 208: 1091, (1965).

Gjessing, E.T., "Ultrafiltration of Aquatic Humus", Environ. Sci. & Technol., 4:437, (1970).

Gjessing, E.T., Physical and Chemical Characteristics of Aquatic Humus, Ann Arbor Science, (1976).

Gordon, G., et al., "Minimizing Chlorite Ion and Chlorate Ion in Water Treated with Chlorine Dioxide", Journal AWWA, 82:4:160 (1990).

Haag, W., and Hoigné, J., "Ozonation of Bromide-Containing Waters: Kinetics of Formation of Hypobromous Acid and Bromate", Environ. Sci. & Technol., 17:5:261 (1983).

Hoehn, R., et al., "Algae as Sources of Trihalomethane Precursors", Journal AWWA, 72:6:344 (1980).

Krasner, S., et al., "The Occurrence of Disinfection By-Products in U.S. Drinking Water", Journal AWWA, 81:8:41 (1989).

Jensen, J., et al., "Effect of Monochloramine on Isolated Fulvic Acid" Org. Geochem., 8:1:71 (1985).

Jolley, R.L., "Water Chlorination: Chemistry, Environmental Impact and Health Effects", Volume 1, Ann Arbor Science Publishers INC. (1978).

Jolley, R.L., Gorchev, H., and Hamilton, D.H., "Water Chlorination: Chemistry, Environmental Impact and Health Effects", Volume 2, Ann Arbor Science Publishers INC. (1978).

Jolley, R.L., Brungs, W.A., Cumming, R.B., and Jacobs, V.A., "Water Chlorination: Chemistry, Environmental Impact and Health Effects", Volume 3, Ann Arbor Science Publishers INC. (1980).

Jolley, R.L., Brungs, W.A., Cotruvo, J.A., Cumming, R.B., Mattice, J.S., and Jacobs, V.A., "Water Chlorination: Chemistry, Environmental Impact and Health Effects - Chemistry and Water Treatment", Volume 4 (1), Ann Arbor Science Publishers INC. (1983a).

Jolley, R.L., Brungs, W.A., Cotruvo, J.A., Cumming, R.B., Mattice, J.S., and Jacobs, V.A., "Water Chlorination: Chemistry, Environmental Impact and Health Effects - Environment, Health, and Risk", Volume 4 (2), Ann Arbor Science Publishers INC. (1983b).

Jolley, R.L., Bull, R.J., Davis, W.P., Katz, S., Roberts, M.H., and Jacobs, V.A., "Water Chlorination: Chemistry, Environmental Impact and Health Effects", Volume 5, Lewis Publishers (1985).

Jolley, R.L., Condie, L.W., Johnson, J.D., Katz, S., Minear, R.A., Mattice, J.S., and Jacobs, V.A., "Water Chlorination: Chemistry, Environmental Impact and Health Effects", Volume 6, Lewis Publishers (1990).

Oliver, B., "Dihaloacetonitriles in Drinking Water: Algae and Fulvic Acid as Precursors", Environ. Sci. & Technol., 17:2:80 (1983).

Reckhow, D., et al., "Chlorination of Humic Materials: By-Product Formation and Chemical Interpretations", Environ. Sci. & Technol., 24:11:1655 (1990).

Rook, J., "Formation of Haloforms during Chlorination of Natural Waters", Water Treat. Exam. 23:234 (1974).

Shapiro, J., "Chemical and Biological Studies on the Yellow Acids of Lake Water" Limnol Oceanog., 2:161, (1957).

Shapiro, J., "Yellow Acid-Cation Complexes in Lake Water" Science 127:702, (1958).

Siddiqui, M. et al., "Bromate Ion Formation: A Critical Review", Journal AWWA, 89:10:58 (1995).

Singer, P.C., "Formation and Characterization of Disinfection By-Products", Paper from the First International Conference on the Safety of Water Disinfection: Balancing Chemical and Microbial Risks, International Life Sciences Institute, Washington, DC, (1992).

Singer, P.C., "Trihalomethanes and Other By-Products Formed by Chlorination of Drinking Water" in Keeping Pace with Science and Engineering, National Academy Press, pp 141-164, (1993).

Singer, P.C., "Control of Disinfection By-Products in Drinking Water", ASCE Journal of Environmental Engineering, 120:4:727 (1994).

Singer, P.C., and Chang, D., "Correlations Between Trihalomethanes and Total Organic Halides Formed During Drinking Water Treatment", Journal AWWA, 81:8:61 (1989).

Thurman, E.M., Organic Geochemistry of Natural Waters, Nijhoff/Junk, (1985).

Uden, P. and Miller, J., "Chlorinated Acids and Chloral in Drinking Water", Journal AWWA, 75:10:525 (1983).

U.S.G.S., Humic Substances in the Suwannee River, Georgia: Interactions, Properties, and Proposed Structures, Open - File report 87-557, (1989).

von Gunten, U., and Hoigne, J., "Factors Affecting the Formation of Bromate During the Ozonation of Bromide-Containing Waters", Aqua, 41:5:299 (1992).

Weinburg, H., et al., "Formation and Removal of Aldehydes in Plants that Use Ozone", Journal AWWA, 85:5:72 (1993).

Westrick, J.J., "National Surveys of Volatile Organic Compounds in Ground and Surface Waters", in Significance and Treatment of Volatile Organic Compounds, Christman, et al, eds. Lewis Publishers, 1990.

Xie, Y., and Reckhow, D., "A New Class of Ozonation By-Products: the Ketoacids", Proceedings, AWWA Conference (1992).

Chapter 2

Regulatory Impact Analysis of the Disinfectants–Disinfection By-Products Rule

Stuart W. Krasner[1], D. M. Owen[2], and J. E. Cromwell III[3]

[1]Water Quality Division, Metropolitan Water District of Southern California, 700 Moreno Avenue, La Verne, CA 91750–3399
[2]Malcolm Pirnie, Inc., 703 Palomar Airport Road, Suite 150, Carlsbad, CA 92009
[3]Apogee Research, Inc., 4350 East-West Highway, Suite 600 Bethesda, MD 20814

As part of the rule-making process, regulatory impact analyses (RIAs) are performed to determine the potential effects of implementing different regulatory options. An RIA was prepared for the Disinfectants/ Disinfection By-Products (D/DBP) Rule. DBPs can be controlled by using either precursor removal technologies or alternative disinfectants. In the RIA, total organic carbon (TOC) was used as a surrogate for DBP precursors. To represent the occurrence and control of chlorination by-products, trihalomethanes (THMs) and haloacetic acids (HAAs) were evaluated. It was predicted that reductions in DBPs would parallel the TOC reductions anticipated for the precursor removal strategies. In addition, it was predicted that ozone/chloramines would produce THM levels as low as those produced by granular activated carbon and HAA levels comparable to those produced by optimized coagulation. The RIAs were used during the rule-making process to evaluate the efficacy of various technologies in controlling DBPs and to develop the nationwide and household costs of different regulatory scenarios.

As part of the rule-making process, the U.S. Environmental Protection Agency (USEPA) performs regulatory impact analyses (RIAs) to determine the potential effects of implementing different regulatory options. RIAs include the following elements: (1) the extent to which utilities might need to change from present treatment practices to other treatment alternatives in order to meet various regulatory options under consideration; (2) the nationwide occurrence of the contaminants of interest before and after the treatment changes have been implemented; (3) the benefits to society accruing from reduced exposure to the contaminants under regulation; and (4) the capital and annual costs to utilities--as well as increases in the annual household cost of water--resulting from implementation of the treatment changes.

For the Disinfectants/Disinfection By-Products (D/DBP) Rule, the USEPA's most significant concern was to develop regulations for DBPs while also continuing to require adequate treatment for the control of microbiological contaminants (*1,2*). The USEPA decided to utilize a negotiated rule-making process to help the stakeholders understand the complexities of the microbial/DBP risk/risk tradeoff issue and reach a consensus on the most appropriate regulation to address both DBP and microbial concerns (*2-4*). An Advisory Committee was formed from representatives of the drinking water industry, environmental and consumer groups, public health officials, and federal and state regulators. In addition, a Technology Workgroup was made up of technical experts--representing the same diversity of interests as the Advisory Committee--to assist the USEPA and its contractors in developing RIAs for the D/DBP Rule.

Methodology

Overview of D/DBP RIA. Detailed RIAs were prepared for surface-water systems that use conventional coagulation/filtration (nonsoftening systems) and serve 10,000 people or more (*4,5*). Systems in this category serve water to 116 million people in the United States, or 72 percent of the population for whom surface water is the drinking-water source. Also, because surface-water systems will need to meet additional disinfection requirements to control microbial pathogens (*6*), chemical and microbial risks were carefully balanced in evaluating technologies for these systems. RIAs were also developed for the other water categories in order to assess the full national implications of the D/DBP Rule (*4,5*).

In general, most DBPs can be controlled by using either precursor removal technologies--i.e., optimized or enhanced coagulation (*7,8*), granular activated carbon (GAC) (*9,10*), or membranes (*11,12*)--or alternative disinfectants--e.g., chloramines (NH$_2$Cl) (*13,14*), ozone (O$_3$) (*7,14-15*). To represent the occurrence and control of chlorination by-products, the two major classes of identified DBPs--trihalomethanes (THMs) and haloacetic acids (HAAs) (*16*)--were evaluated in the RIA. The analyses primarily focused on total THMs (TTHMs) and the sum of five HAAs (HAA5), mono-chloro-, dichloro-, trichloro-, monobromo-, and dibromoacetic acid, which are the HAAs for which data and predictive equations existed at the time of the rule-making (*17*). In addition, total organic carbon (TOC) was used as a surrogate for DBP precursors. Reckhow and Singer (*7*) have shown that the precursors for THMs and HAAs, as well as for total organic halogen (TOX) (a surrogate for halogen-containing organic compounds), are removed as well or better than TOC during coagulation.

For preliminary analyses, optimized coagulation was defined as an enhancement of the existing coagulation process achieved by the addition of 40 mg/L alum over the system's baseline dose. The median alum dose--in the American Water Works Association (AWWA)/AWWA Research Foundation (AWWARF) Water Industry Database (WIDB)--was 10-15 mg/L; therefore, enhanced coagulation would require ~50 mg/L alum (on average) (*4*). For example, Cheng and co-workers (*18*) found that enhanced coagulation in two source waters that currently use 5-10 mg/L alum (and have raw-water TOC concentrations of ~3-4 mg/L) would require 40 mg/L alum. Likewise, Reckhow and Singer (*7*) observed optimal removal of TOC and DBP precursors from a moderately colored raw water (containing 6.8 mg/L TOC) with alum doses of ~30-50 mg/L. However, in some waters (especially those with very high concentrations of raw-water TOC), higher amounts of coagulant may be needed.

The feasible range of GAC usage was bounded by defining the operation (i.e., the removal of a synthetic organic contaminant versus the removal of DBP precursors). One GAC usage was established to fit the existing definition of "feasible" GAC technology for the removal of 90-99 percent of trichloroethylene (19); it operates with a 10-min empty-bed contact time (EBCT) and a regeneration frequency of 180 days and is referred to as GAC10. An alternative usage was created to correspond to a GAC operation resulting in 90-percent TTHM/HAA5 reduction, where the GAC follows conventional treatment; it operates with a 20-min EBCT and a 60-day regeneration frequency and is referred to as GAC20. These operating parameters were based partially on bench- and pilot-scale studies (9,10) and partially on predictions made using a DBP Regulatory Assessment Model (DBPRAM) (see below).

Membrane treatment was based on the use of nanofiltration. Membrane technology--in particular, nanofiltration--has been evaluated for the removal of DBP precursors in both groundwaters (11) and surface waters (20). For example, Taylor and colleagues (11) found that membranes with molecular-weight cutoffs (MWCs) of 400 daltons or less, under normal operating conditions, were required for the control of THM precursors. Typically, higher MWC membranes were found to be ineffective for the removal of substantial concentrations of precursor material.

Ozone/chloramine treatment was set up to include preozonation, biologically active filtration to remove the assimilable organic carbon (AOC) (e.g., aldehydes) produced by ozonation (21,22), some free-chlorine contact time (to inactivate the heterotrophic-plate-count bacteria that can be found in the effluent of biologically active filters), and postchloramination. Such a disinfection strategy for ozonation--in order to control DBPs and microbials, as well as to produce biologically stable water-- is being explored by a number of systems (23).

RIA Predictive Model. The DBPRAM that was used as part of the D/DBP rule-making process included predictive equations to estimate DBP concentrations during water treatment (24). However, because reliable equations for predicting individual DBP formation in a wide range of waters (e.g., those containing high bromide levels) were not available at the time the DBPRAM was developed, the RIAs emphasized TTHM (25) and HAA5 (17) formation. The mixed bromine- and chlorine-containing HAAs were not commercially available when DBPs were measured during the development of the HAA predictive equations. However, for a low-bromide water, the error introduced by not including mixed bromochloro HAA species was believed to be low. For waters with moderate or high levels of bromide, TTHMs (where all of the bromine-containing species are measured) are probably the class of DBPs that will drive most systems to make treatment modifications so as to comply with the proposed rule (18).

The DBPRAM predicted the removal of TOC during alum coagulation, GAC adsorption, and nanofiltration (24,26). These equations were developed based on a number of bench-, pilot-, and full-scale studies. The removal of TOC during precipitative softening has not been modeled to date, although systems that soften represent only a small percentage (~10 percent) of the surface-water treatment plants. The

DBPRAM also predicted the alkalinity and pH changes resulting from chemical addition (24), as well as the decay of residual chlorine and chloramines in the plant and the distribution system (27).

In developing the RIA, the first step was to estimate the occurrence of relevant source-water parameters (28). TOC data from the AWWA/AWWARF WIDB (29) and bromide data from a nationwide bromide survey (30) formed the basis for determining the DBP precursor levels (31). The WIDB was designed to emphasize breadth rather than depth in its coverage (32). For example, the TOC data in the WIDB were limited because only a small percentage of systems were monitoring for TOC (i.e., there were 703 individual surface-water treatment plants in the WIDB, but only 154 provided influent TOC data). In the nationwide bromide survey, a smaller number of systems (a total of 100 utilities) were surveyed as compared to the number represented in the WIDB, but all the source waters were analyzed for the same parameters. In addition, the majority of waters selected for the bromide survey were picked at random, with geographical representation as a selection criterion. For the RIA, actual water quality data were used to simulate predicted occurrence values based on a statistical function such as log-normal distribution (28,31).

In running the DBPRAM, the production of DBPs was restricted to surface-water plants that filtered but did not soften. Surface waters typically have higher disinfection (i.e., *CT*) criteria--and thus a greater likelihood of producing more DBPs--than do groundwaters (i.e., *Giardia* in surface waters is more difficult to inactivate than viruses in groundwaters). As noted above, an equation to predict TOC removal during softening was not available. However, there was a mechanism for accounting for DBP occurrence in other water systems during the rule-making (see below). Part of the RIA for the surface-water plants that filter but do not soften was the creation of 100 representative waters through a Monte Carlo simulation.

The second step of the RIA was to predict a probability distribution of nationwide THM and HAA occurrence if all surface-water plants that filter but do not soften--based on the 100 simulated waters--utilized a particular technology for DBP control (i.e., enhanced coagulation, GAC, nanofiltration, or alternative disinfectants). Although individual utilities will consider a range of technologies to meet disinfection and D/DBP rules, the RIA initially examined one technology at a time. Subsequently, a decision-making process was employed to examine the predicted compliance choices that systems will make (33). As part of the DBPRAM, compliance with the Surface Water Treatment Rule (SWTR), a potential enhanced SWTR, the Total Coliform Rule, and the Lead and Copper Rule were modeled. Although nationwide DBP studies typically measured DBP occurrence prior to implementation of these new microbial and corrosion rules (16), the DBPRAM made it possible to assess the effects of meeting a multitude of rules simultaneously.

During the D/DBP negotiated regulation, a Technology Workgroup of engineers and scientists was formed (4). The workgroup reviewed the DBPRAM and RIAs and provided input to ensure that the predicted output was consistent with real-world data. Prior validation of the DBPRAM in Southern California--where bromide occurrence was relatively high--indicated that the central tendency was to underpredict TTHMs by 20-30 percent (24). In addition, evaluation of the model in low-bromide North Carolina waters also revealed that the model tended to underpredict both THM and HAA concentrations and resulted in absolute median deviations of approximately 25-30 percent (*34*). Therefore, the Technology Workgroup adjusted the DBPRAM output to correct for the underpredictions (i.e., it divided the output by 0.7). In addition, the resultant ("corrected") data were confirmed against full-scale data that included geographically diverse systems throughout the United States (*35*).

Prior validation of the alum coagulation part of the model was performed in several eastern states as well as in Southern California. The overall central tendency was to overpredict TOC removal by 5-10 percent. The Technology Workgroup felt that utilities would need an overdesign factor to ensure that precursor removal technologies could consistently meet water quality objectives. The use of a 15-percent overdesign factor for TOC removal compensated for a typical overprediction in TOC removal by alum. For plants that do not filter or filter with softening, case studies on a number of systems throughout the nation were used to assess compliance choices and predicted water qualities. For groundwaters, data from the WIDB (29) and the Ground Water Supply Survey (*36*) on TOC and THM levels were used in developing RIAs for those systems.

Results and Discussion

Effects of Alternative Treatments on Control of DBPs and TOC. For those surface-water systems modeled in the DBPRAM that filter but do not soften (i.e., the 100 simulated waters), the median and 90th-percentile TOC concentrations in raw water were 3.9 and 8.4 mg/L, respectively. In these analyses, conventional treatment was capable of reducing the TOC of the water based on the use of coagulant dosages necessary to remove turbidity (Figure 1). The use of alternative disinfectants does not reduce the TOC level (mineralization of TOC with ozonation is minimal under drinking water conditions). It was predicted that optimized coagulation and GAC10 would result in similar reductions in TOC (median and 90th-percentile TOC ranges of 1.6-1.8 and 2.7-3.1 mg/L, respectively). Membranes provided somewhat better TOC reduction (median and 90th-percentile values of 0.5 and 0.6 mg/L, respectively) than did GAC20 (median and 90th-percentile values of 0.6 and 1.0 mg/L, respectively).

As a baseline, it was predicted that for those systems using conventional treatment and chlorine as the only disinfectant, the median and 90th-percentile TTHM concentrations were 46 and 90 μg/L, respectively. In comparison, for those systems treating surface water (or a mixture of surface and ground waters) that were included in the AWWA/AWWARF WIDB, the median and 90th-percentile TTHM values were 43 and 74 μg/L, respectively. The lower value for the 90th-percentile TTHM occurrence probably resulted, in part, from the use of alternative disinfectants (e.g., chloramines) and other technologies to comply with the existing TTHM maximum contaminant level (MCL) of 100 μg/L (*37*). As part of the baseline analysis, the predicted median and 90th-percentile concentrations for HAA5 were 28 and 65 μg/L, respectively. In comparison, for systems that treat surface water (or a mixture of surface and ground waters)--primarily with conventional treatment and chlorine disinfection--in a database of actual DBP occurrence (*35*), the median and 90th-percentile HAA5 values were 28 and 73 μg/L, respectively. Because there was no MCL for

HAAs, specific strategies to control HAAs (beyond what was done to meet the TTHM MCL) were not practiced.

In the RIA, the reductions in DBPs (i.e., TTHMs and HAA5) (Figures 2 and 3) paralleled the TOC reductions observed for the precursor removal strategies. It was predicted that optimized coagulation and GAC10 would lower the median and 90th-percentile TTHM ranges to 29-30 and 58-61 µg/L, respectively, and that GAC20 and membranes would lower the median and 90th-percentile TTHM ranges to 11-14 and 23-28 µg/L, respectively. Likewise, it was predicted that optimized coagulation would lower the HAA5 median and 90th-percentile concentrations to 17 and 37 µg/L, respectively, and that GAC20 would achieve median and 90th-percentile HAA5 levels of 11 and 24 µg/L, respectively. In addition, it was predicted that ozone/chloramines would produce TTHM levels as low as those produced by GAC20 and would produce HAA5 levels comparable to those produced by optimized coagulation. Furthermore, it was predicted that addition of postchloramination to conventional treatment (primary disinfection with free chlorine) would lower the 90th-percentile TTHM concentration to 69 µg/L, which is close to the WIDB 90th-percentile TTHM value of 74 µg/L (12 percent of the systems in the WIDB were using chloramines).

Best Available Technologies (BATs) and MCLs. BATs are selected as part of the rule-making process. A specific technology (or group of technologies) is chosen that represents a cost-effective and feasible control measure. Other control strategies may be used to comply with a rule, as long as they can meet the same water quality standards as the BAT. MCLs are set as close to the (health-based) MCL goals (MCLGs) as feasible, with treatment techniques, costs, and analytical detection limits as additional considerations.

As an example of the costs of these DBP control strategies, the Technology Workgroup used a central-tendency cost model (*38*) to estimate the additional cost per household per year ($/HH/yr) for each technology, where the current nationwide cost of drinking water for large systems (according to the WIDB) was ~$166/HH/yr (on average). For an "average-to-treat" water, a range of additional costs--where the higher end of the range was usually for small systems that would not benefit from an economy of scales for the installation of expensive equipment and processes, as the large systems would--was $6-$14/HH/yr for optimized coagulation; $12-$526/HH/yr for ozone; $35-$600/HH/yr for GAC with a 15-min EBCT and $50-$800/HH/yr for GAC with a 30-min EBCT; and $70-$262/HH/yr for nanofiltration.

Because of concerns over known and unknown DBPs associated with alternative disinfectants (*22,23,39-42*), as well as the cost of advanced precursor removal technologies (*10*), enhanced coagulation (or GAC10)--with chlorine as both the primary and secondary disinfectant--was chosen as the BAT for the first stage of the D/DBP Rule (*43*). The use of enhanced coagulation would allow systems to utilize existing treatment processes to achieve initial reductions in DBP exposure. In addition, the cost of optimizing an existing coagulation process is lower than that of installing a new unit process such as GAC.

The proposed rule also includes a second stage in which either (1) enhanced coagulation followed by GAC10 or (2) GAC20 was selected as the BAT. Analysis by the Technology Workgroup demonstrated that enhanced coagulation installed to

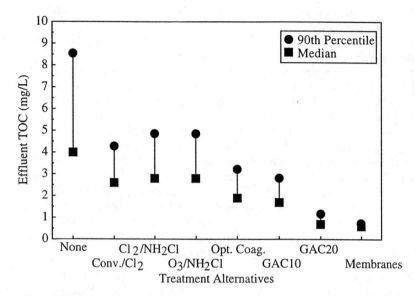

Figure 1. Effect of Alternative Treatments on TOC for Large Surface-Water Systems That Filter But Do Not Soften. (Reproduced, with permission, from reference 4; copyright 1995, American Water Works Association.)

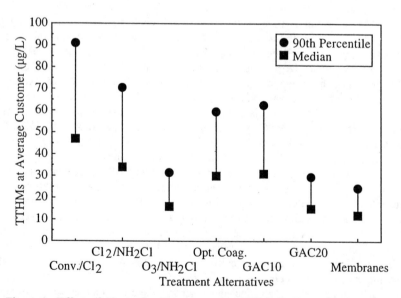

Figure 2. Effect of Alternative Treatments on TTHMs for Large Surface-Water Systems That Filter But Do Not Soften. (Reproduced, with permission, from reference 4; copyright 1995, American Water Works Association.)

comply with Stage 1 of the D/DBP Rule could be used to reduce the carbon usage rate, thus enabling systems installing a GAC10 operation in Stage 2 of the rule to achieve the same TOC reductions that could have been accomplished by conventional treatment and GAC20. Membranes were not selected as BAT because the cost differential between membranes and GAC20 in large surface-water systems is unwarranted for TOC removal. BAT is established as that technology which is affordable by large metropolitan water systems (compare the low end of the cost ranges for GAC [$35-$50/HH/yr] and membranes [$70/HH/yr]). However, for small systems, membranes may be more cost-effective than GAC (compare the high end of the cost ranges for GAC [$600-$800/HH/yr] and membranes [$262/HH/yr]). In addition, groundwater utilities with decentralized distribution systems may be better able to afford the use of membranes than they would the construction of centralized GAC treatment facilities.

In developing technical and economic data to support potential MCLs, the Technology Workgroup considered two issues: (1) a typical system should be able to comply with a regulation by installing the BAT, and (2) a utility must design a safety factor into treatment operations to ensure compliance over time. It was decided to define the BAT as applicable to 90 percent of the waters, with the others being more difficult to treat; for this remaining 10 percent of the waters, systems would need to install other technologies, which could be more expensive than the BAT, to ensure compliance. For all systems, a 20-percent safety factor was utilized.

Thus, for Stage 1, 80 µg/L TTHMs and 60 µg/L HAA5 were ultimately proposed as the MCLs (*43*), in which compliance with a safety factor would require the BAT to achieve <64 µg/L TTHMs and <48 µg/L HAA5. The 90th-percentile values of TTHMs and HAA5 for optimized coagulation are 58 and 37 µg/L, respectively (Figure 2), which meet the specified criteria. Likewise, for Stage 2, 40 µg/L TTHMs and 30 µg/L HAA5 are the proposed MCLs (*43*), in which compliance with a safety factor would require the BAT to achieve <32 µg/L TTHMs and <24 µg/L HAA5. The 90th-percentile values of TTHMs and HAA5 for GAC20 are 28 and 24 µg/L, respectively (Figure 2).

In addition, Stage 1 of the proposed D/DBP Rule requires surface-water systems--including those that treat groundwater under the influence of surface water-- that use conventional treatment (with sedimentation basins) to enhance their coagulation (or softening) process in order to control DBP precursors (*8,43*). The DBP precursor control requirement will be imposed in addition to the TTHM and HAA5 MCLs. Enhanced coagulation (or softening) can be used to remove the precursors for other DBPs as well as those associated with THMs and HAAs. Reckhow and Singer (*7*) demonstrated that the formation potential (a measure of precursor levels) of THMs, HAAs, dichloroacetonitrile, 1,1,1-trichloropropanone, and TOX could all be reduced with enhanced coagulation. Thus, enhanced coagulation can be used to control other DBPs as well, although they are not part of the proposed D/DBP Rule.

Compliance Forecasts. In addition to the BAT(s), a utility can use alternative technologies (e.g., alternative disinfectants) to comply with the MCLs. Many combinations of the above technologies--with mixed precursor removal strategies and/or alternative disinfectants (e.g., enhanced coagulation followed by chlorine for primary disinfection and chloramines for secondary disinfection)--were explored. A least-cost sorting routine was established that predicted the percentage of systems that would use each type of technology combination to comply with the proposed rule (in particular, with Stage 2) (Table I). (Although chlorine dioxide was not specifically examined in this list of compliance choices, it is expected that some of the systems utilizing alternative disinfectants will choose chlorine dioxide.)

Table I. Compliance Forecast for Large Surface Water Systems[a]

| | Compliance Forecast (%) | |
Treatment Technology	Stage 1	Stage 2
Current	28	21
Cl_2/NH_2Cl	3	6
Enhanced coagulation	43	15
Enhanced coagulation + Cl_2/NH_2Cl	10	23
O_3/NH_2Cl	5	5
Enhanced coagulation + O_3/NH_2Cl	6	17
Enhanced coagulation + GAC10	3	0
Enhanced coagulation + GAC10 + Cl_2/NH_2Cl	0	1
Enhanced coagulation + GAC10 + O_3/NH_2Cl	0	6
Enhanced coagulation + GAC20	3	4
Enhanced coagulation + GAC20 + O_3/NH_2Cl	0	2
NH_2Cl	0	1
O_3/Cl_2	0	0.1
O_3/diatomaceous-earth filtration	0	0.1
O_3/direct filtration/NH_2Cl	0	0.1

[a]Includes surface-water systems that soften, as well as systems that currently do not filter.

It was predicted that 28 and 21 percent of the large surface systems--including systems that soften, as well as systems that currently do not filter--would be able to comply with the proposed requirements for Stages 1 and 2, respectively, with their current treatment. In general, these systems treat low-TOC waters with low levels of DBP precursors. Most surface-water systems would be required to implement enhanced coagulation with or without alternative disinfectants. For Stage 1, 43 percent of the systems would be able to use enhanced coagulation with free chlorine for disinfection in order to comply with the proposed Stage 1 MCLs, whereas in Stage 2 only 15 percent of the systems would be able to use enhanced coagulation with chlorine. In Stage 2, the other systems that had implemented enhanced coagulation in Stage 1 would be able to comply with the proposed Stage 2 MCLs by either switching to alternative disinfectants or adding GAC (in addition to maintaining enhanced coagulation from Stage 1). Because of this wide variety of treatment options, it was predicted that large surface-water systems would not need to install membranes, whereas it was predicted that membranes would be used by a certain percentage of small systems and groundwater systems in order to comply with the proposed rule (*43*).

For surface-water systems that currently do not filter, some additional least-cost alternatives (to the construction of a full conventional treatment plant) that they might consider would be the use of preozonation/postchlorination, ozonation with diatomaceous-earth filtration, and ozonation with direct filtration and postchloramination (Table I). Because many of these systems treat low-TOC waters, such options would probably not be required until the promulgation of the Stage 2 MCLs.

Because of the limited scientific data available on the occurrence, health effects, and treatability of DBPs, a two-stage D/DBP Rule was developed. Furthermore, the cost impacts of Stage 2 of the D/DBP Rule are substantial (*4*). Consequently, large utilities will be required to collect data via an Information Collection Rule (ICR) (*44*) for use in characterizing occurrence and treatment data that could be used to evaluate various regulatory options for controlling DBP contaminants. Data collected under the ICR will be used in a second negotiated rule-making process to resolve Stage 2 D/DBP standards.

Standards for Other DBPs and the Disinfectants. RIAs were prepared for other DBPs (i.e., bromate and chlorite ions) and the disinfectants themselves (i.e., chlorine, chloramines, and chlorine dioxide) (*43*). Two brief examples of how other RIAs were addressed are presented below.

Chloramine Standard. In the AWWA/AWWARF WIDB, the range of distribution-system average chloramine residuals was 0.3 to 4.8 mg/L, with a median value of 1.6 mg/L (*35*). The proposed maximum residual disinfectant level (MRDL) for chloramines is 4.0 mg/L, based on a (health-based) MRDL goal of 4.0 mg/L (*43*). The 75th- and 90th-percentile occurrence values in the WIDB were 2.2 and 3.0 mg/L, respectively. In order to reliably stay below the proposed MRDL, as many as 10 percent of the systems using chloramines will need to change their treatment practices in some manner. Typically, high chloramine residuals are associated with waters containing high concentrations of TOC. Thus, reducing TOC levels to meet the precursor control requirements should make it possible to use lower disinfectant residuals in systems treating high-TOC water.

Chlorite Ion MCL. When chlorine dioxide is used as an alternative disinfectant, the inorganic by-products chlorite ion and chlorate ion are produced. During water treatment, approximately 50-70 percent of the chlorine dioxide dosed will immediately appear as chlorite ion and the remainder as chlorate ion (*45*). The residual chlorite ion continues to degrade in the water distribution system in reactions with oxidizable material in the finished water or in the distribution system.

Data were obtained for up to 17 utilities in USEPA Region VI (Texas, Oklahoma, New Mexico, Arkansas, and Louisiana) that employ chlorine dioxide (Bubnis, B., Novatek, unpublished data). In many instances, data were available for chlorite ion occurrence on a monthly basis. As an example, for June 1992, chlorite ion concentrations were 0.4-1.2 mg/L, whereas in January 1993 chlorite ion levels were at values of 0.2-0.8 mg/L. The higher values in June may have been partially a result of the need for more chlorine dioxide in warmer months to meet the oxidant demand of the water. When the data were examined quarterly (e.g., quarter 1 is January through March), Figure 4 shows that the highest occurrence for chlorite ion in the systems sampled in USEPA Region VI was during the spring and summer seasons (*43*).

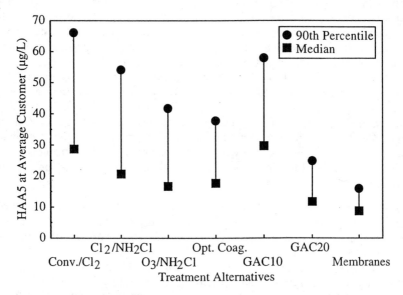

Figure 3. Effect of Alternative Treatments on HAA5 for Large Surface-Water Systems That Filter But Do Not Soften. (Reproduced, with permission, from reference 4; copyright 1995, American Water Works Association.)

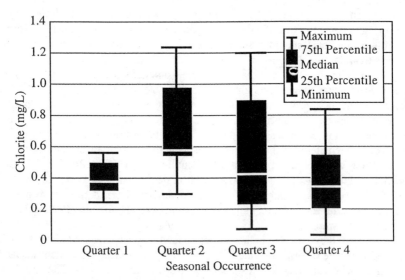

Figure 4. Chlorite Occurrence for 13 Utilities in USEPA Region VI in 1992.

In the proposed D/DBP Rule, the USEPA listed the MCLG for chlorite ion as 0.08 mg/L (*43*). If an MCL were set at the MCLG value, such a strict standard would probably preclude the use of this alternative disinfectant. However, the proposed MCLG includes uncertainty factors based on data gaps in the health effects research on this DBP. New health effects research to resolve these data gaps is under way (Romano, R. R., Chemical Manufacturers Assoc., personal communication, 1995).

Although research is being conducted on ways of reducing chlorite ion residuals at the treatment plant (e.g., using ferrous iron) (*46,47*), additional work is required. At this time, the only BAT for chlorite ion is control of the treatment and disinfection processes to minimize chlorite ion formation in the treatment plant (*43*). Because most of the chlorine dioxide dosed will be converted to chlorite ion, the chlorine dioxide dose must be controlled. However, depending on the oxidant demand of the water, which can increase in warmer months, a certain amount of chlorine dioxide must be added to be effective. Because many systems that use chlorine dioxide have been able to reduce THM formation (*40*), setting a chlorite ion MCL that might preclude the use of chlorine dioxide could lead to systems switching back to free chlorine, which would result in an increase in THM and HAA production.

Therefore, an MCL was proposed that was believed to represent the lowest level that could be reliably achieved on a monthly basis for typical utilities employing chlorine dioxide. (Compliance must be achieved each month, as chlorite ion health effects are acute rather than chronic.) The Advisory Committee agreed to propose 1.0 mg/L as the MCL for chlorite ion, with certain qualifications and reservations (*43*). If the current health effects study indicates that a chlorite ion level of 1.0 mg/L is safe, the MCL shall remain at that level. In addition, if the new data suggest that the MCLG for chlorite ion is 1.0 mg/L or higher, the MCL could be set at the revised MCLG value. However, if the new health effects study indicates that the chlorite ion at <1.0 mg/L is of concern, the chlorite ion MCL will have to be reassessed. It will be necessary to reexamine the tradeoffs and regulatory impact of a lower chlorite ion MCL in light of the positive aspects of chlorine dioxide disinfection.

Conclusions

The USEPA proposed the D/DBP Rule on July 29, 1994. The proposal includes (1) a lower MCL for TTHMs; (2) new MCLs for other DBPs (i.e., HAA5, bromate, and chlorite); (3) establishment of MRDLs for chlorine, chloramines, and chlorine dioxide; and (4) requirements for the removal of DBP precursors, with TOC removal as an indicator of precursor control performance. RIAs were used during the rule-making process to evaluate the efficacy of utilizing different technologies to control DBPs and to develop the nationwide and household costs of different regulatory scenarios.

References

1. Gelderloos, A. B.; Harrington, G. W.; Schaefer, J. K.; Regli, S. In *Advances in Water Analysis and Treatment; American Water Works Association Proceedings: 1991 Water Quality Technology Conference*; AWWA: Denver, CO, 1992; Part II, pp 1407-1440.
2. Means, E. G., III; Krasner, S. W. *J. AWWA* **1993**, *85 (2)*, 68-73.

3.   USEPA. *Fed. Reg.* **1992**, *57 (179)*, 42533-42536.
4.   Roberson, J. A.; Cromwell, J. E., III; Krasner, S. W.; McGuire, M. J.; Owen, D. M.; Regli, S.; Summers, R. S. *J. AWWA* **1995**, *87 (10)*, 46-57.
5.   Wade Miller Associates, Inc. *Regulatory Impact Analysis for the National Primary Drinking Water Regulations: Disinfectants/Disinfection By-Products Rule;* USEPA Contract 68-C3-0368, Work Assignment No. 3; Wade Miller Associates, Inc.: Arlington, VA, 1994.
6.   USEPA. *Fed. Reg.* **1994**, *59 (145)*, 38832-38858.
7.   Reckhow, D. A.; Singer, P. C. *J. AWWA* **1990**, *82 (4)*, 173-180.
8.   Krasner, S. W.; Amy, G. *J. AWWA* **1995**, *87 (10)*, 93-107.
9.   Lykins, B. W., Jr.; Clark, R. M.; Adams, J. Q. *J. AWWA* **1988**, *80 (5)*, 85-92.
10.   McGuire, M. J.; Davis, M. K.; Tate, C. H.; Aieta, E. M.; Howe, E. W.; Crittenden, J. C. *J. AWWA* **1991**, *83 (1)*, 38-48.
11.   Taylor, J. S.; Thompson, D. M.; Carswell, J. K. *J. AWWA* **1987**, *79 (8)*, 72-82.
12.   Fu, P.; Ruiz, H.; Thompson, K.; Spangenberg, C. *J. AWWA* **1994**, *86 (12)*, 55-72.
13.   Norman, T. S.; Harms, L. L.; Looyenga, R. W. *J. AWWA* **1980**, *72 (3)*, 176-180.
14.   Jacangelo, J. G.; Patania, N. L.; Reagan, K. M.; Aieta, E. M.; Krasner, S. W.; McGuire, M. J. *J. AWWA* **1989**, *81 (8)*, 74-84.
15.   Ferguson, D. W.; Gramith, J. T.; McGuire, M. J. *J. AWWA* **1991**, *83 (5)*, 32-39.
16.   Krasner, S. W.; McGuire, M. J.; Jacangelo, J. G.; Patania, N. L.; Reagan, K. M.; Aieta, E. M. *J. AWWA* **1989**, *81 (8)*, 41-53.
17.   Mallon, K.; Najm, I.; Gramith, K; Jacangelo, J.; Krasner, S. In *American Water Works Association Proceedings: 1992 Water Quality Technology Conference*; AWWA: Denver, CO, 1993; Part II, pp 1801-1830.
18.   Cheng, R. C.; Krasner, S. W.; Green, J. F.; Wattier, K. L. *J. AWWA* **1995**, *87 (2)*, 91-103.
19.   Durenberger, D.; et al. *Congressional Record—Senate* **1986** (May 21), pp 11655-11673.
20.   Jacangelo, J. G.; Laîné, J.-M. *Evaluation of Ultrafiltration Membrane Pretreatment and Nanofiltration of Surface Waters*; AWWARF: Denver, CO, 1993.
21.   Huck, P. M.; Fedorak, P. M.; Anderson, W. B. *J. AWWA* **1991**, *83 (12)*, 69-80.
22.   Weinberg, H. S.; Glaze, W. H.; Krasner, S. W.; Sclimenti, M. J. *J. AWWA* **1993**, *85 (5)*, 72-85.
23.   Glaze, W. H.; Weinberg, H. S. *Identification and Occurrence of Ozonation By-Products in Drinking Water;* AWWARF and AWWA: Denver, CO, 1993.
24.   Harrington, G. W.; Chowdhury, Z. K.; Owen, D. M. *J. AWWA* **1992**, *84 (11)*, 78-87.
25.   Amy, G. L.; Chadik, P. A.; Chowdhury, Z. *J. AWWA* **1987**, *79 (7)*, 89-97.
26.   Harrington, G. W.; Chowdhury, Z. K.; Owen, D. M. In *Water Quality for the New Decade; American Water Works Association Proceedings: 1991 Annual Conference*; AWWA: Denver, CO, 1991; pp 589-624.
27.   Dharmarajah, H.; Patania, N. L.; Jacangelo, J. G.; Aieta, E. M. In *Water Quality for the New Decade; American Water Works Association Proceedings: 1991 Annual Conference*; AWWA: Denver, CO, 1991; pp 569-588.

28. Letkiewicz, F. J.; Grubbs, W.; Lustik, M.; Cromwell, J.; Mosher, J.; Zhang, X.; Regli, S. In *Water Quality; American Water Works Association Proceedings: 1992 Annual Conference*; AWWA: Denver, CO, 1992; pp 1-48.
29. AWWARF; AWWA. *Water Industry Database*; AWWARF and AWWA: Denver, CO, 1991.
30. Amy, G.; Siddiqui, M.; Zhai, W.; DeBroux, J.; Odem, W. *Survey of Bromide in Drinking Water and Impacts on DBP Formation;* AWWARF and AWWA: Denver, CO, 1994.
31. Wade Miller Associates, Inc. *Technical Memorandum for Disinfectant/ Disinfection By-Product Regulatory Impact Analysis: Sensitivity of Raw Water Quality and Treatment Parameters;* Report to USEPA, Office of Groundwater and Drinking Water; Wade Miller Associates, Inc.: Arlington, VA, 1992.
32. Cromwell, J. E., III; Lee, R. G.; Kawczynski, E. In *American Water Works Association Proceedings: 1990 Annual Conference*; AWWA: Denver, CO, 1990; pp 1029-1049.
33. Gelderloos, A. B.; Harrington, G. W.; Owen, D. M.; Regli, S.; Schaefer, J. K.; Cromwell, J. E., III; Zhang, X. In *Water Quality; American Water Works Association Proceedings: 1992 Annual Conference*; AWWA: Denver, CO, 1992; pp 49-79.
34. Grenier, A. D.; Obolensky, A.; Singer, P. C. *J. AWWA* **1992**, *84 (11)*, 99-102.
35. Gramith, K.; Krasner, S.; Jacangelo, J.; Means, E. In *American Water Works Association Proceedings: 1993 Water Quality Technology Conference*; AWWA: Denver, CO, 1994; Part II, pp 677-693.
36. Westrick, J. J.; Mello, J. W.; Thomas, R. F. *The Ground Water Supply Survey Summary of Volatile Organic Contaminant Occurrence Data;* USEPA, Technical Support Division: Cincinnati, OH, 1983.
37. McGuire, M. J.; Meadow, R. G. *J. AWWA* **1988**, *80 (1)*, 61-68.
38. Malcolm Pirnie, Inc. *Technologies and Costs for Control of Disinfection By-Products;* Report to USEPA, Office of Science and Technology; Malcolm Pirnie, Inc.: White Plains, NY, 1992.
39. Krasner, S. W.; Glaze, W. H.; Weinberg, H. S.; Daniel, P. A.; Najm, I. N. *J. AWWA* **1993**, *85 (1)*, 73-81.
40. Aieta, E. M.; Berg, J. D. *J. AWWA* **1986**, *78 (6)*, 62-72.
41. Richardson, S. D.; Truston, A. D., Jr.; Collette, T. W.; Patterson, K. S.; Lykins, B. W., Jr.; Majetich, G.; Zhang, Y. *Environ. Sci. Technol.* **1994**, *28 (4)*, 592-599.
42. Diehl, A. C.; Speitel, G. E., Jr.; Symons, J. M.; Krasner, S. W., Hwang, C. J. In *Water Quality; American Water Works Association Proceedings: 1995 Annual Conference*; AWWA: Denver, CO, 1995; pp 535-546.
43. USEPA. *Fed. Reg.* **1994**, *59 (145)*, 38668-38829.
44. USEPA. *Fed. Reg.* **1994**, *59 (28)*, 6332-6444.
45. Aieta, E. M.; Roberts, P. V.; Hernandez, M. *J. AWWA* **1984**, *76 (1)*, 64-70.
46. Griese, M. H.; Kaczur, J. J.; Gordon, G. *J. AWWA* **1992**, *84 (11)*, 69-77.
47. Iatrou, A.; Knocke, W. R. *J. AWWA* **1992**, *84 (11)*, 63-68.

CHLORINATION–CHLORAMINATION PRODUCTS AND REACTIONS

Chapter 3

Empirical Models for Chlorination By-Products: Four Years of Pilot Experience in Southern Connecticut

John N. McClellan[1], David A. Reckhow[1], John E. Tobiason[1], James K. Edzwald[1], and Alan F. Hess[2]

[1]Department of Civil and Environmental Engineering, 18 Marston Hall, University of Massachusetts, Amherst, MA 01003
[2]South Central Connecticut Regional Water Authority, 99 Sargent Drive, New Haven, CT 06511

Site-specific empirical models of trihalomethanes (THMs), haloacetic acids (HAAs), and the organic precursors of these compounds were developed based on pilot study data. Either chlorination by-product precursors or chlorination by-products themselves were treated as dependent variables, and either water quality parameters or process parameters comprised the independent variables. Ultraviolet absorbance was found to be the most effective conventional water quality parameter for predicting precursor removal in treatment processes. A first-order model was developed to predict the removal of THM and HAA precursors in waters treated with ozone and subsequent granular activated carbon filtration. In a separate analysis, haloacetic acid concentrations were predicted as a function of THM concentrations. Among other findings, these models have suggested a negative impact of filter aids on chlorination by-product precursor removal at a direct filtration plant.

Trihalomethanes (THMs) and haloacetic acids (HAAs) are by-products of the reactions between chlorine and natural organic matter in drinking water treatment. New rules will lower existing trihalomethane standards for drinking water and impose a standard on haloacetic acids for the first time (1). Measurements of the concentrations of THMs and HAAs and their precursors are necessary to control and assess treatment process performance, but direct measurements are time consuming and require sophisticated equipment. Empirical relationships which allow the estimation of THM and HAA concentrations as functions of other more easily measured parameters are therefore of interest.

Empirical models of chlorine disinfection by-products (DBPs) and their precursors can be used in several ways. First, models can be used for process control. Precursor concentrations during treatment and THM and HAA concentrations in distribution systems can be estimated using easily measured parameters. Information needed for process control can thus be obtained faster and with less expense using models than by direct measurements. Empirical models can also be used to evaluate the relative impact of individual parameters on process performance. Sensitivity analysis employing empirical models can be used for preliminary evaluation of process modification options. Finally, empirical models can provide a convenient way of summarizing large data sets and of comparing the treatability of different waters.

The South Central Connecticut Regional Water Authority (RWA) and the University of Massachusetts (UMass) conducted studies at pilot plants owned and operated by the RWA in which various process modifications aimed at improving the removal of the organic precursors of HAAs and THMs were investigated. A database that includes measurements of THMs, HAAs, and other water quality parameters was developed using results from the pilot studies. This database was used to calibrate and test empirical models for DBP precursors and for the formation of THMs and HAAs.

The objective of this paper is to present the results of these modeling efforts, and to provide a broad discussion of site-specific empirical modeling for control of DBPs.

Background

Previous Chlorination By-product Modeling Efforts. In view of the need to control THM and HAA concentrations and the expense and time required to measure them directly, there has been considerable interest in developing predictive models for these substances. It was recognized early on that chlorine concentration, the concentration of organic precursor compounds, pH, temperature, and contact time all affect THM formation (2,3). Engerholm and Amy (4) proposed a model that predicts chloroform concentration as a function of total organic carbon (TOC), time, and a parameter defined as chlorine dose normalized by TOC concentration. Urano et al. (5) proposed a model which is similar to the Engerholm and Amy (4) model, but which incorporates the effects of pH and temperature. Morrow and Minear (6) explored the effect of bromide on THM formation, and calibrated a model which incorporated bromide concentration.

The three models described above use TOC alone as a surrogate for THM precursor concentration. This is a limitation because TOC-THMFP correlations tend to be site specific and are often poor. Ultraviolet (UV) absorbance at 254 nm is a good surrogate for THMFP (7). Amy et al. (8) found that a parameter defined as the product of TOC (mg/L) and UV absorbance (cm^{-1}) correlated more closely with THMFP than did either parameter individually in experiments with a variety of natural waters. Amy et al. (8) presented a model for THM formation which employs UV absorbance, TOC, chlorine dose, pH, and bromide concentration as predictor variables.

The database used for calibration contains over 1000 data points and was developed using nine natural waters in bench scale experiments. The Montgomery Watson *(9)* models, which are similar in form to the Amy et al. *(8)* THM models, provide separate equations for each of the THM and HAA species. These models were calibrated using a data set which includes the Amy et al. *(8)* database, two bench-scale databases developed by Montgomery Watson, and a bench-scale data set provided by the Metropolitan Water District of Southern California.

A difficulty with general empirical models is that they can be unreliable if the value of any independent variable is outside the range found in the calibration data. In addition, the dependent variable could be sensitive to parameters not included in the model. These factors may be less problematic in site-calibrated models. However, general models (e.g. the Montgomery Watson *(9)* models) have the advantage of large calibration databases. It is unclear which of these modeling approaches is superior in a given situation, or what level of accuracy can be expected from either approach.

The models mentioned above, which can be classified as formation models, predict DBP formation based on reaction conditions and reactant concentrations. Another class of models can be used to predict the concentration of organic DBP precursors. Examples of precursor models from the literature are the equations for THMFP as functions of UV absorbance and TOC given by Edzwald et al. *(10),* the equations presented by Harrington et al. *(12)* that predict TOC and UV absorbance removal as functions of alum dose and pH, and the model proposed by Huck et al. *(11)* for the removal of THMFP in biologically active granular activated carbon (GAC). The Huck et al. *(11)* model predicts the removal of precursors in a treatment process as a function of process parameters (in this case, empty bed contact time [EBCT]) and influent concentration. Although modeling of precursor removal versus process parameters could be very useful in the design and operation of water treatment facilities, a relatively small amount of work has been done in this area to date.

RWA/UMass Pilot Studies. Pilot studies were conducted at the RWA's West River Treatment Plant (WRTP) in 1992, at the Lake Saltonstall Treatment Plant (LSTP) in 1993 and 1994, and at the Lake Gaillard Treatment Plant (LGTP) in 1993, 1994 and 1995. The effects of various configurations of ozone and granular activated carbon (GAC) on organics removal were examined in these studies. Results from the 1992 WRTP and 1993 LGTP studies were published previously *(13)*. In all these studies, measurements were made of THM and HAA precursor concentrations, UV absorbance, dissolved organic carbon (DOC), and turbidity in samples taken from raw waters, intermediate points in the pilot plants, and full- and pilot-scale treated waters. The data generated in these studies was organized and used for the development and testing of empirical models.

Experimental Methods

Pilot Plants. Lake Gaillard Water Treatment Plant (LGTP) is a direct filtration plant. The process includes coagulant addition (alum and cationic polymer), rapid mix, three-stage flocculation (30 minutes total contact time), and anthracite/sand (A/S) filtration at a hydraulic loading rate of 2.5-5 gpm/sq ft. Pilot Train #1 at LGTP consisted of a direct filtration process which simulated the full scale process, followed by a counter-current ozone contactor (10 minute contact time) and four granular activated carbon (GAC) contactors configured in series. The GAC contactors each contained 20 inches of Calgon Filtrasorb 300 (8 x 30 mesh) media over a gravel support layer and were typically operated at a hydraulic loading rate of 2.5 gpm/sq ft, which resulted in an EBCT of 5 minutes per contactor. Thus, samples could be collected at EBCTs of 5, 10, 15, and 20 minutes. This configuration is referred to as "post-ozone/GAC." Pilot Train #2 at LGTP consisted of a counter-current contactor (10 minute detention time for pre-ozonation of raw water); rapid mix, coagulant addition, and flocculation similar to the full scale plant; and GAC/sand (GAC/S) and A/S filters in parallel. The A/S filter contained 20 inches of 0.9 mm anthracite and 10 inches of 0.45 mm silica sand over a gravel support layer. Calgon Filtrasorb 300 (8 x 20 mesh) was substituted for anthracite in the GAC/S filter. On sampling dates, the mean alum and polymer doses applied in the pilot trains were 8.6 mg/L and 1.0 mg/L respectively, and the mean coagulation pH was 6.6.

The West River Treatment Plant (WRTP) is an in-line direct filtration plant. The treatment process includes pre-oxidation with potassium permanganate, coagulation with ferric chloride and a cationic polymer, and A/S filtration at a hydraulic loading rate of 1.5-3 gpm/sq ft. The WRTP pilot train consisted of a counter current ozone contactor (10 minute contact time), coagulation similar to the full scale plant, and GAC/S filtration at 3 gpm/sq ft. On sampling dates, the mean ferric chloride and polymer doses applied in the pilot train were 7.6 mg/L and 1.3 mg/L respectively, and the mean coagulation pH was 6.6.

The Lake Saltonstall Water Treatment Plant (LSTP) is a conventional treatment plant which includes pre-chlorination, coagulation with alum, rapid mixing, 2-stage flocculation, dual layer sedimentation, and A/S filters which have a design hydraulic loading of 3.5 gpm/sq ft. The LSTP pilot plant was similar to the full scale plant except that ozone was substituted for chlorine as a pre-oxidant, a plate settler was substituted for conventional sedimentation, and GAC media were substituted for anthracite in one of the two parallel filters. On sampling dates, the mean alum dose applied in the pilot train was 35 mg/L and the mean coagulation pH was 7.1. Schematics of the RWA pilot plants are presented in Figure 1. Mean values of raw water quality parameters are presented in Table I. Table II contains mean raw and treated water values for DOC, HAAFP, and THMFP for full scale and pilot scale RWA plants.

Analytical Methods.   Formation potentials (FPs) were used to quantify the concentrations of THM and HAA precursors. Unchlorinated samples were transported to UMass and chlorinated within 24 hours of being collected. The standard chlorine dose for FP samples was 20 mg/L, followed by incubation in the dark at 20 °C for 72 hours. Samples were incubated headspace-free in 300 mL BOD bottles. A phosphate buffer was added to maintain the pH at 7.0. Simulated distribution system (SDS) analysis was used to estimate distribution system THM and HAA concentrations. For SDS tests, a chlorine dose identical to the dose used in the full scale plants at the time of sampling (typically about 3 mg/L) was applied. Samples were then incubated headspace-free in the dark at 20 °C for 48 hours at pH 7.0.

Table I. Raw Water Quality

Parameter	Units	Observed Mean Value at:		
		LGTP	WRTP	LSTP
pH[1]		7.0	6.8	8.2
Turbidity[1]	NTU	0.71	1.12	1.51
Color[2]	Pt-Co Units	21	27	NA[3]
DOC[1]	mg/L	3.0	2.7	2.7
UV absorbance[1]	cm^{-1}	0.095	0.096	0.056
Temperature[1]	°C	1.9-25	NA[3]	4.5-25
Alkalinity[2]	mg/L as $CaCO_3$	12	11	NA[3]
Bromide	mg/L	<0.1[4]	<0.1[4]	<0.1[4]

1. Mean or range of values observed on sampling dates
2. Values for 1993 as reported by Tobiason *(13)*
3. Value not available
4. Accurate value not available. Value is assumed to be ≈ 30-50 µg/L based on typical values for region.

Table II. Mean DOC, HAAFP, and THMFP Levels in Raw and Treated Waters

Process	DOC (mg/L)		HAAFP[1] (µg/L)		CHCl3FP[2] (µg/L)	
	Raw	Treated	Raw	Treated	Raw	Treated
LGTP						
Direct Filtration[3]	3.0	1.8	250	105	176	85
Dir.Filt./Post-O$_3$/GAC[4]	3.0	1.5	257	47	168	38
WRTP						
Direct Filtration	2.7	1.5	228	71	156	59
Pre-O$_3$ with GAC/S Filt.[5]	2.7	1.1	222	48	156	40
LSTP						
Conventional with Pre-O$_3$[5]	2.7	1.3	157	38	101	42

1. HAAFP = sum of DCAAFP and TCAAFP
2. CHCl3FP ≈ THMFP in RWA waters due to low bromide concentration
3. Full scale data
4. Pilot scale data, 10 minute EBCT in GAC contactors
5. Pilot scale data

The concentrations of HAAs and THMs were measured by gas chromatography with electron capture detection after quenching and appropriate sample preparation. Two HAA sample preparation methods were used: a modified version of Standard Method 6322 *(14)*, and an alternate method developed by Xie et al. *(15)*. Both methods involve micro-extractions and methylation. In Standard Method 6322, methylation is accomplished by addition of diazomethane. The method of Xie *(15)* avoids the use of diazomethane and employs acidic methanol instead. A micro-extraction method similar to the method of Koch et al. *(16)* was used for THM

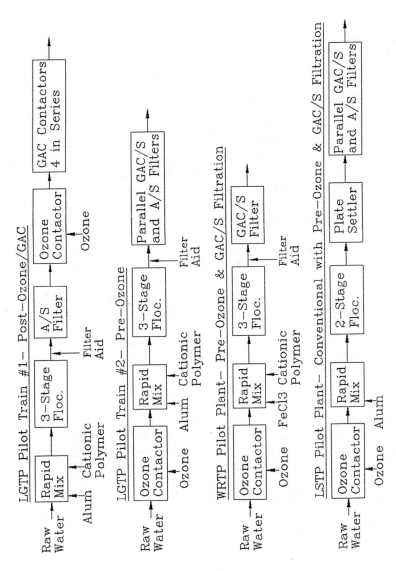

Figure 1. Pilot Plant Schematics

sample preparation. In order to minimize volatilization, extractions were performed immediately after quenching the THM samples. Replicate samples were analyzed for THMs and HAAs; the relative standard deviation was typically less than 5% for DCAA, TCAA, and CHCl3 measurements.

Dissolved organic carbon (DOC) and ultra-violet absorbance at 254 nm (UV) were also routinely measured. Samples were prepared for DOC and UV measurements by filtration through pre-washed 0.45 μm glass fiber filters. DOC concentrations were measured using the UV-persulfate oxidation method after samples were acidified and sparged with nitrogen.

Modeling Chlorination By-products and Their Precursors

In low-bromide systems, three DBP species dominate: trichloracetic acid (TCAA), dichloroacetic acid (DCAA), and chloroform (CHCl3). In RWA waters, these three species typically account for over 90% of the HAAs and THMs which form as a result of chlorination. Although four THM species and five HAA species (six in 1995) were measured, only DCAA, TCAA, and CHCl3 are considered here. Chloroform can be considered approximately equivalent to total THMs and the sum of DCAA and TCAA can be considered approximately equivalent to total HAAs.

Precursor Models. Precursor models can be used to predict the concentrations of DBP precursors, or the extent to which precursors are removed in treatment processes. The dependent variable in these models is typically the ratio of effluent to influent precursor concentration. Initial concentrations, process parameters, other water quality parameters, or combinations of these may be used for predictor variables.

Fully Empirical Model for Precursor Removal Across Coagulation/Filtration. Data collected at the LGTP direct-filtration pilot train (no pre-ozone, A/S filtration) on thirteen sampling dates in 1993, 1994, and 1995 were used to calibrate models for predicting FP in filter effluents. The independent variables for these models are initial concentration, temperature, alum dose, cationic polymer coagulant dose, and anionic polymer filter-aid dose. For each model, the combination of variables which resulted in the highest adjusted correlation coefficient (R^2_{adj}) (17) value was selected. The adjusted correlation coefficient compensates for small sample sizes relative to the number of predictor variables. Maximum values for R^2_{adj} resulted when two of the five predictor variables were used for the DCAA model, and when three of the five were used for the TCAA and CHCl3 models. Log-log transformations resulted in higher values of R^2_{adj} than did untransformed data for all three models. The resulting equations are:

$$\text{DCAAFP:} \quad C_E\big/C_o = 0.53 T^{0.14} F^{0.15}$$
$$N=13, R^2_{adj} = 0.61, P\text{-Value} = 0.007 \tag{1}$$

CHCl3FP: $$\frac{C_E}{C_o} = 162\,A^{-1.08}C_o^{-0.51}F^{0.30}$$ (2)

$N=11$, $R^2_{adj} = 0.70$, P-Value $= 0.022$

TCAAFP: $$\frac{C_E}{C_o} = 6.3C_o^{-0.34}T^{0.082}A^{-0.63}$$ (3)

$N=10$, $R^2_{adj} = 0.83$, P-Value $= 0.002$

where C_E is A/S filter effluent FP in µg/L, C_o is raw water FP in µg/L, A is alum dose in mg/L, F is filter aid dose in mg/L, and T is raw water temperature in degrees C. The P-Value *(17)* is the observed level of statistical significance. The mathematical form is a "power function," which is widely employed for empirical models. The power function form of equations (1), (2), and (3) resulted from taking anti-logs of both sides of the log-transformed regression equations.

In some cases parameters which have a significant effect on the modeled process do not improve the model fit because of a small range of values in the data. Predictor variables which did not improve the model fits were excluded from these models. For example, alum dose and pH clearly have an effect on the removal of DBPs, but alum dose is not included in equation (1) and pH is not included in any of the above equations because of small ranges of values in the data. The mean values of pH, alum dose, and other absent parameters not explicitly included are necessarily incorporated into the constant in each equation. The equations can only be considered valid for pH, temperature, chemical doses, and initial concentrations within the ranges of these parameters in the data, whether or not they appear as predictor variables. Mean values and ranges of operating parameters are given in Table III.

Table III. LGTP Direct Filtration Pilot Plant Operating Parameters

Parameter	Units	Values observed at LGTP Pilot Plant- Feb. 93 - July 95 Mean	Range
Raw Water DCAAFP	µg/L	90	59-137
DCAAFP Removal	%	54	46-62
Raw Water TCAAFP	µg/L	159	101-222
TCAAFP Removal	%	64	54-73
Raw Water CHCl3FP	µg/L	172	128-248
CHCl3FP Removal	%	54	36-68
Raw Water Temperature	°C	9.6	1.9-25.1
Alum dose	mg/L	8.6	7.3-10.7
Filter aid dose	µg/L	52	26-68
Raw Water pH		6.9	6.5-7.4
Filter Effluent pH		6.7	6.4-6.9

Fully empirical models based on process parameters such as those given in equations (1), (2), and (3) may not be very useful for predicting process performance because of the limitations stated above. However, these models can be used for comparing the relative effects of processes parameters. Estimates of the change in the fraction of raw water FP removed per unit change of the indicated parameter at the LGTP direct filtration pilot train are presented in Table IV. These values were computed by evaluating partial derivatives of equations (1), (2) and (3) at the mean values of the parameters.

Table IV. Correlations Between Operating Parameters and DBP Precursor Removal

Parameter	Units	Estimated change in fraction raw water FP removed per unit increase of parameter		
		DCAA	TCAA	CHCl3
Raw Water TCAAFP	μg/L	NI[1]	0.001	NI[1]
Raw Water CHCl3FP	μg/L	NI[1]	NI[1]	0.001
Temperature	°C	-0.02	-0.003	NI[1]
Alum dose	mg/L	NI[1]	0.03	0.06
Filter aid dose	μg/L	-0.004	NI[1]	-0.003

1. Parameter not included because it did not significantly improve model fit

An interesting result shown in Table IV is that higher doses of anionic polymer filter aid correlated with poorer removal of DCAAFP and CHCl3FP. These results suggest that anionic polymer addition may have the undesirable side effect of restabilizing DBP precursor material, although this conclusion cannot be made with certainty based on the results of this study. A controlled experiment is recommended to determine whether filter aid dose and the extent of DBP precursor removal are directly related or whether both are related to some other water quality or operational parameter.

Semi-Empirical Model for Precursor Removal in GAC. Huck and co-workers *(11)* reported a linear relationship between the extent of THMFP removal (and the removal of other organic carbon parameters) and influent concentration in biologically active GAC. To see if a similar relationship exists for THM and HAA precursors in the post-ozone GAC pilot train at the LGTP, the change in FP across each of the GAC contactors was plotted versus influent concentration. The pilot train had been in continuous operation for about 6 months at the LGTP at the time the of the earliest sampling for this data set. In addition, the GAC contactors were operated at the WRTP for about a year before being moved to the LGTP without changing the media. It is therefore assumed that the adsorptive capacity of the GAC was exhausted, and that the removal of DBP precursors in the GAC media was due to biological activity The plots, presented in Figure 2, show significant linear relationships, although correlations are not as good as the correlations observed by Huck et al. *(11)* for THMFP removal ($r^2=0.991$ in a post-filtration deep bed GAC contactor, $r^2=0.748$ in a GAC/sand filter with pre-ozone). The GAC contactors at LGTP are configured in

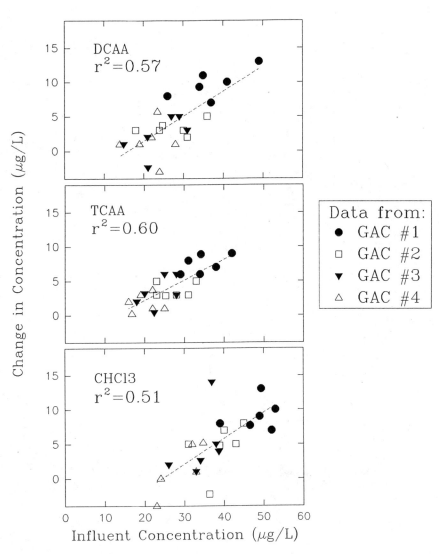

Figure 2. Change in Concentration vs. Influent Concentration in GAC Contactors

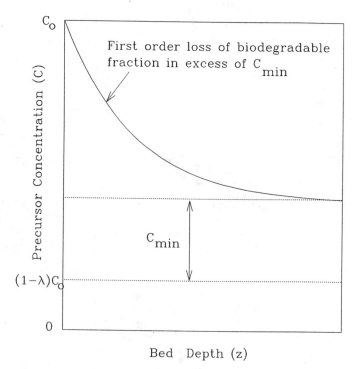

Figure 3. Biologically Acitve GAC Process: 1st Order Model

series. This configuration is analogous to a single deep bed contactor, with "intermediate depth" sampling at the effluents of each contactor. A linear relationship between removal and influent concentration is apparent for all four contactors, suggesting that biological activity occurs throughout the depth of the media. As can be seen in Figure 2, a linear extrapolation of the data shows a positive X intercept, which implies that there are influent concentrations below which no removal occurs. It is also assumed that only a fraction of the THM and HAA precursor material is biodegradable. A first-order model for utilization of THM and HAA precursors in the GAC media is proposed which incorporates the above features. The concentration profile with respect to depth in the proposed model is given by

$$\frac{dC}{dz} = -K\{C - (1-\lambda)C_o - C_{min}\}$$

(4a)

$$C = C_O \text{ when } z = 0$$

(4b)

where C is concentration, z is bed depth penetrated, K is an apparent first order constant, C_0 is influent concentration, λ is the biodegradable fraction of C_0, and C_{min} is the influent concentration of biodegradable material below which no further biodegradation occurs. A graphic representation of the model parameters is presented in Figure 3. The integrated form is:

$$C = C_o(1 - \lambda + \lambda e^{-Kz}) + C_{min}(1 - e^{-Kz})$$

(5)

The parameters K and C_{min} can be determined from plots of removal versus influent concentration, such as those shown in Figure 2, using the following relations:

$$K = \frac{-\ln\left(1 - \frac{m}{\lambda}\right)}{z}$$

(6)

$$C_{min} = \frac{-b}{1 - e^{-Kz}}$$

(7)

where m is the slope of the regression line and b is the y intercept. Bed depth penetrated, z, is used in place of the more common but closely related parameter EBCT in the above equations. The reason is that changes in EBCT (due to flow changes) at the LGTP pilot plant seemed to have little effect on process performance at a given bed depth on the few occasions when the hydraulic loading rate was changed. Bed depth penetrated was therefore considered to be the more meaningful parameter. However, it should be noted that the hydraulic loading rate was 2.5 gpm/sq ft. for most of the calibration data and the model should only be considered valid for flows close to this value.

Values of K and C_{min} were computed using the slopes and intercepts of the regression lines shown in Figure 2. Values of λ were selected to optimize the model fit. These λ values are in reasonable agreement with the values for biodegradability (approximately 65% for CHCl3FP and 80% for DCAAFP) given by Miltner et al. *(18)*. Table V contains values for the model parameters K, C_{min} and λ.

Table V. Constants for First Order Utilization Model

Precursor Type	K (ft^{-1})	C_{min} (μg/L)	λ
CHCl3FP	0.37	19	0.8
DCAAFP	0.21	15	0.9
TCAAFP	0.24	11	0.9

HAA Precursors vs. THM Precursors. Trihalomethanes have been regulated in U.S. drinking waters since 1979, and many U.S. utilities have already conducted studies aimed at adjusting their processes to meet existing THM regulations. A relatively small amount of work to date has focused on HAAs. For this reason and because HAAs are more difficult to measure than THMs, modeling of HAA precursors using THMFP as a predictor parameter was explored. The simplest model of this type is the mean HAAFP:THMFP ratio. In Table VI, HAAFP:THMFP ratios for various raw and treated waters are presented.

Table VI. Mean HAAFP:THMFP Ratios in Raw and Treated Waters

Process	DCAA:CHCl3[1] Raw	Treated	TCAA:CHCl3[2] Raw	Treated	HAA:CHCl3[3] Raw	Treated
LGTP						
Direct Filtration[4]	0.52	0.55	0.95	0.69	1.47	1.24
Dir. Filt./Post-O3/GAC[5]	0.54	0.65	1.00	0.63	1.54	1.28
WRTP						
Direct Filtration[6]	0.57	0.69	0.92	0.56	1.49	1.25
Pre-O3 with GAC/S Filt.[7]	0.57	0.72	0.92	0.50	1.49	1.22
LSTP						
Conventional with Pre-O3[8]	0.69	0.56	0.82	0.39	1.25	0.95

1. None of the raw and treated DCAAFP:CHCl3FP ratios are significantly different at $\alpha = 0.05$.
2. All of the raw and treated TCAAFP:CHCl3FP ratios are significantly different at $\alpha = 0.05$.
3. HAAFP = DCAAFP plus TCAAFP; CHCl3FP $\approx$ total THMFP. The raw and treated HAAFP:CHCl3FP ratios are significantly different at $\alpha = 0.05$ except for the WRTP Direct Filtration and Pre-O3 ratios.
4. Full scale data, N=25
5. Pilot scale data, 10 minute EBCT in GAC contactors, N=18
6. Full scale data, N=12
7. Pilot scale data, N=12
8. Pilot scale data, N=7

These results illustrate the changes in the relative distribution of HAA and THM precursors during treatment processes. A decrease in the TCAAFP:CHCl3FP ratio can be observed between raw and treated waters for all cases. The preferential removal of TCAA precursors compared to CHCl3 and DCAA precursors was observed in bench scale studies *(19)*, and is thought to result from the association of TCAA precursors with humic acids. The humic acids are characterized by large molecular size and a high level of hydrophobicity, and are thus easily removed by coagulation processes *(19)*. The speciation is also different in the different raw waters. The general implication is that empirical models tend to be site specific, and models calibrated using raw waters may not make accurate predictions in treated waters.

It is interesting to note that the lowest treated water TCAAFP:CHCl3FP ratio is seen in the data from the conventional LSTP pilot plant, where a relatively high alum dose, without a cationic polymer, was employed (typically about 13 mg alum/mg DOC). Cationic polymer in addition to a relatively low primary coagulant dose (typically about 2.9 mg alum/mg DOC at the LGTP and about 2.5 mg ferric chloride/mg DOC at the WRTP) were added at the direct filtration plants.

Chlorination By-product Precursors vs. Conventional Water Quality Parameters. Empirical models of chlorination byproduct precursors based on other water quality parameters provide a convenient way of comparing, optimizing, and controlling treatment processes. Turbidity has traditionally been used in drinking water treatment as a surrogate parameter for particle concentrations, but other parameters may be preferable when optimizing and monitoring processes for HAA and THM precursor removal. A comparison was made between UV absorbance, dissolved organic carbon (DOC), and turbidity as predictor parameters for precursor removal models. Fractions of raw water CHCl3FP, DCAAFP, and TCAAFP remaining were plotted against fractions of raw water turbidity, DOC, and UV absorbance remaining in 60 filter and GAC contactor effluent samples from LGTP. The plots are presented in Figure 4 and Figure 5. These results indicate that UV absorbance is superior to DOC or turbidity as a predictor of THM and HAA precursor removal. Regressions (forced through zero) of these data yield the following equations:

CHCl3FP:
$$\frac{C_E}{C_o} = 1.2 \frac{UV_E}{UV_o} \qquad (8)$$

DCAAFP:
$$\frac{C_E}{C_o} = 1.3 \frac{UV_E}{UV_o} \qquad (9)$$

TCAAFP:
$$\frac{C_E}{C_o} = 0.9 \frac{UV_E}{UV_o} \qquad (10)$$

where subscript O and E denote treatment process influent and effluent concentrations respectively. Although most of the data used in developing equations (8), (9), and (10) are from a conventional direct filtration process with no ozone applied, some data

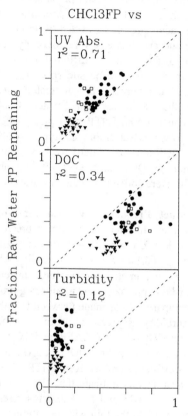

Figure 4.   CHCl3FP Removal vs. Removal of UV Absorbance, DOC, and Turbidity

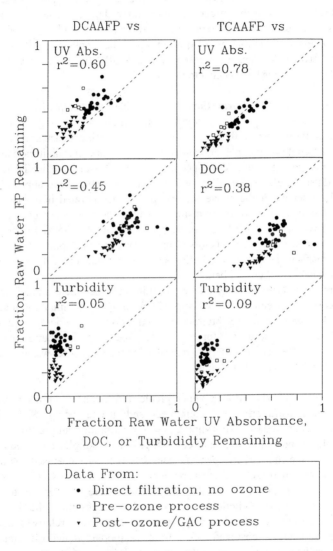

Figure 5. HAAFP Removal vs. Removal of UV Absorbance, DOC, and Turbidity

from the pre- and post-ozone pilot plants at the LGTP are included. In the pre-ozone pilot train, the ozone dose was typically 2 mg/L. In the post-ozone pilot train, a dose of 0.5-0.7 mg/L was applied after filtration but upstream of the GAC contactors. As can be seen in Figure 4, the inclusion of data from the pre- and post-ozone processes did not seem to adversely effect the correlation between the removal of UV absorbance and DBP precursor removal. However, it should be noted that these processes included filtration or adsorption after the application of ozone. The correlation between UV absorbance removal and DBP precursor removal over the ozone contactor alone is be expected to be poor.

Chlorination By-product Formation Models. These models predict concentrations of the DBP species as opposed to precursor concentrations. The important factors that influence the formation of these compounds are chlorine concentration, precursor concentration, contact time, pH, and temperature. Bromide has an important effect on speciation if present in significant concentration. Models that predict THM and HAA formation must therefore take all these factors into account. Two approaches can be envisioned: either all of the above factors can be incorporated into the model, or a predictor parameter which reacts in a similar way to the factors can be employed. As discussed above, models have been published which utilize the first approach, using TOC and UV absorbance as predictors of precursor concentration. An example of the second approach is the use of one DBP species as a predictor of the others. Both of these approaches are discussed below.

Formation potential could also be used in multi-parameter models to represent initial precursor concentration. Models which employ an initial FP measurement, chlorine dose, time, pH, and bromide concentration as predictor variables might ultimately be the most useful for accurately predicting distribution system DBP concentrations. The data necessary to develop this type of model was not collected as part of the RWA/UMass pilot studies.

HAAs vs. THMs. Models which predict HAA concentrations as functions of chloroform concentration were calibrated and tested using SDS data from RWA/UMass pilot studies. All samples used for calibration and testing were treated water samples. The calibration data set consisted of 13 samples collected at LGTP in 1994. To test the models, independent data sets (i.e. data not used for calibration) were employed which included measurements of the parameters being modeled and the predictor parameters. Model predictions were compared to measured concentrations, and the predictions were considered successful if the differences fell within a specified range. Absolute ranges were used instead of relative ranges because the average variance of observed values from the model predictions remained fairly constant over the range of values in the test data. Results using two "success ranges" are reported. The equations were tested on 1995 LGTP data (N=12), and on data collected at LSTP in 1993 and 1994 (N=17). The chloroform model equations and prediction success rates are presented in Table VII.

The HAA vs THM models predicted DCAA and TCAA values from the LGTP (the calibration data was also from LGTP) reasonably well using a success range defined as ± 7 µg/L. Using data from the conventional LSTP, the DCAA predictions were reasonably good, but the HAA vs THM model systematically overpredicted TCAA concentrations. This result illustrates that the extent of TCAAFP removal compared to the removal of CHCl3FP or DCAAFP can be quite different for different coagulation conditions, as discussed above. The TCAA vs. THM model is therefore particularly sensitive to site-specific conditions.

Table VII. Site Calibrated HAA vs THM Models[1]

| | Percent of predictions within: | | | |
| | ± 7 µg/L of measured value at | | ± 3 µg/L of measured value at | |
Equation[2]	LGTP	LSTP	LGTP	LSTP
$DCAA = 0.52*CHCl3$	73	76	45	76
$TCAA = 0.76*CHCl3 - 4$	91	53	36	35

1. These models were calibrated and tested using SDS data. Means and ranges of test data values (µg/L): CHCl3: 29 (14-59), DCAA: 19 (8-37); TCAA: 22 (11-40).
2. Units: µg/L

Multi-parameter THM and HAA Formation Models. The Montgomery Watson models *(9)* for CHCl3, DCAA, and TCAA formation were tested against treated water SDS data from the LGTP and the LSTP. The models were tested on the same data set used to test the HAA/THM models discussed above. Table VIII shows the Montgomery Watson equations *(9)* for CHCl3, DCAA, and TCAA and their prediction success rates.

The Montgomery Watson *(9)* equations were quite successful at predicting HAA concentrations using data from both the LGTP and the LSTP where the success range was defined as ± 7 µg/L, even though most of the UV absorbance values in the test data are outside the "boundary conditions range" (the range of values in the calibration data) given by Montgomery Watson *(9)*. The test data mean UV absorbance value is 0.022 cm[-1] while the lower limit of the boundary conditions range for the DCAA and TCAA equations is 0.05 cm[-1].

The Montgomery Watson *(9)* equation systematically underpredicted CHCl3 concentrations, and the success rate for this equation was consequently very poor. Some of the test data UV absorbance values are outside the boundary conditions range for the CHCl3 equation (the lower limit of the boundary conditions range is 0.029 cm[-1]). In addition, all of the test data in this study are from treated water samples, but a substantial fraction of the database used to develop the CHCl3 model is from raw water samples. The Montgomery Watson equation tended to underpredict CHCl3 concentrations in treated waters during model validation *(9)*.

In this study, the performance of the site-calibrated HAA vs. THM and the general Montgomery Watson *(9)* models for HAA formation was comparable. These models may be useful for making rough estimates of HAA concentrations, but they cannot be considered reliable if a high degree of accuracy is required.

Table VIII. Montgomery Watson Equations[1] for THM and HAA Formation

| | Percent of predictions within: | | | |
| | ±7 µg/L of measured value[3] at | | ±3 µg/L of measured value[3] at | |
Equation[2]	LGTP	LSTP	LGTP	LSTP
$CHCl3=0.064*TOC^{0.33}*pH^{1.16}$ $*T^{1.02}*Cl_2^{0.56}$ $*(Br^-+0.01)^{-0.40}*UV^{0.87}*t^{0.27}$	45	17	18	0
$DCAA=0.605*TOC^{0.29}$ $*T^{0.67}*Cl_2^{0.48}$ $*(Br^-+0.01)^{-0.57}*UV^{0.73}*t^{0.24}$	82	100	55	58
$TCAA=87.2*TOC^{0.36}$ $*pH^{-1.7}*Cl2^{0.88}$ $*(Br^-+0.01)^{-0.68}*UV^{0.90}*t^{0.26}$	82	92	45	50

1. Reference (9)
2. UV = ultraviolet absorbance at 254 nm in cm^{-1}. T = temperature in °C
   t = time in hours. CHCl3, DCAA, and TCAA in µg/L. TOC, Cl$_2$, and Br$^-$
   in mg/L
3. Means and ranges of test data values (µg/L): CHCl3: 29 (14- 59); DCAA:
   19 (8-37); TCAA: 22 (11-40).

Conclusions

Empirical models for DBPs and DBP precursors are useful for summarizing large data sets. They may help elucidate effects of process variables that would otherwise be overlooked, and they may be useful for predicting aspects of water quality, under specific treatment conditions, that have not been directly measured. The use of empirical models, developed from data collected over a four-year period at three RWA plants has led to the following conclusions:

(1) Removal of turbidity in water treatment processes may not indicate good removal of THM and HAA precursors. Ultraviolet absorbance at 254 nm is a superior parameter for process monitoring and optimization for the control of DBPs.

(2) The extent of THM and HAA precursor removal in biologically active GAC media with ozone applied upstream is proportional to influent concentration, and can be predicted using a first order model of the form

$$C = C_o(1-\lambda+\lambda e^{-Kz}) + C_{min}(1-e^{-Kz})$$

where z is the depth of penetration in the GAC bed, K is an apparent first order constant, λ is the biodegradable THMFP or HAAFP fraction, and C_{min} is the concentration of biodegradable material below which no further biodegradation will occur.

(3) Haloacetic acid concentrations can be predicted as linear functions of THM concentrations. Using a small calibration data set (N=13), these models were successful at making rough predictions of haloacetic acid concentrations (± 7 µg/L). The performance of the models for DCAA and TCAA formation developed by Montgomery Watson *(9)* was comparable to the performance of the site-calibrated HAA vs. THM models, although some of the test data values were outside the range of values in the calibration data. The Montgomery Watson model for chloroform was less successful than their HAA models.

(4) The HAAFP:THMFP ratio is lower in treated waters than in raw waters because of the preferential removal of TCAA precursors in coagulation/filtration processes. This effect seems to be more pronounced in conventional treatment than in direct filtration.

(5) A correlation between increased anionic polymer filter aid dose and poorer removal of DBP precursors was observed. It may be that filter aid has a direct inhibiting effect on DBP removal or it may be that filter aid is related to some other water quality or operational parameter that adversely effects removal.

Glossary

Acronyms:

A/S	Anthracite/sand
CHCl3	Chloroform
DBP	Chlorine disinfection by-product
DCAA	Dichloroacetic acid
DOC	Dissolved organic carbon
FP	Formation potential
GAC	Granular activated carbon
GAC/S	GAC/sand
HAA	Haloacetic acid
LGTP	Lake Gaillard Treatment Plant
LSTP	Lake Saltonstall Treatment Plant
RWA	South Central Connecticut Regional Water Authority
SDS	Simulated distribution system
TCAA	Trichloroacetic acid
THM	Trihalomethane
TOC	Total organic carbon
UV absorbance	Ultraviolet absorbance at 254 nm
WRTP	West River Treatment Plant

Symbols:

λ	Biodegradable fraction
A	Alum dose

b	Regression line Y-intercept
C_E	Effluent concentration
C_o	Influent concentration
C_{min}	Minimum biodegradable concentration
EBCT	Empty bed contact time
F	Filter aid dose
K	Apparent first-order constant
m	Regression line slope
P-Value	Observed level of statistical significance
R^2_{adj}	Adjusted correlation coefficient
T	Temperature
UV_E	Effluent UV absorbance
UV_o	Influent UV absorbance
z	Penetration depth in adsorption bed

Acknowledgments

Data collected by UMass graduate students James Ayers, Raashina Humayan, Paul Schmidt, Denise Springborg, Nagaraju Vinod, Jonathan Weiner, and Qing-wen Zhu were used in this work. Operation of the pilot plants was supervised by Howard Dunn and Gary Kaminski, formerly of the South Central Connecticut Regional Water Authority. The funding provided by the South Central Connecticut Regional Water Authority is gratefully acknowledged.

References

1. Pontius, F. W.; *J. AWWA*, **1995**, 87(2), 48.
2. Stevens, A. A. et al.; *J. AWWA*, **1976**, 68, 615.
3. Kavanaugh, M.C.; Trussell, A.R.; Cromer, J.; Trussell, R.R.; *J. AWWA*, **1980**, 72, 578.
4. Engerholm, B. A.; Amy, G. A.; *J. AWWA*, **1983**, 75, 418.
5. Urano, K.; Wada, H.; Takemasa, T.; *Water Res.*, **1983**, 17, 1797.
6. Morrow, C. M.; Minear, R. A.; *Water Res.*, **1987**, 21, 41.
7. Edzwald, J. K.; *Organic Carcinogens in Drinking Water*, Ram, N. M.; Calabrese, E. J.; Christman, R. F.; Eds.; John Wiley and Sons: New York, NY, 1986; pp 199-236.
8. Amy, G. L.; Chadik, P. A.; Chowdhury, Z.; *J. AWWA*, **1987**, 79(7), 89.
9. Montgomery Watson; *Final Report: Mathematical Modeling of the Formation of THMs and HAAs in Chlorinated Waters*; American Water Works Association: Denver, CO, 1993.
10. Edzwald, J. K.; Becker, W. K; Wattier, K. L.; *J. AWWA*, **1985**, 77(4), 22.
11. Huck, P. M.; Zhang, S.; Price, M. L.; *J. AWWA*, **1994**, 86(6), 61.
12. Harrington, G. W.; Chowdhury, Z. K.; Owen, D. M.; *J. AWWA*, **1992**, 84(11), 78.
13. Tobiason, J. E.; *Chemical Water and Wastewater Treatment III*, Klute, R.; Hahn, H. H; Eds.; Springer-Verlag: Berlin, 1994; pp 155-171.
14. *Standard Methods for the Examination of Water and Wastewater*, APHA, AWWA, WPC, Washington, D.C., 1992, 18th Ed.
15. Xie, Y. D.; Springborg, D. C.; Reckhow, D. A.; *Proceedings of AWWA Water Quality Technology Conference*, Miami, FL, 1993.

16.   Koch, B.; Crofts, E. W.; Schimpff, W. K.; Davis, M. K.; *Proceedings of the AWWA Water Quality Technology Conference,* 1988.
17.   Neter, J.; Wasserman, W.; Kutner, Michael H.; *Applied Linear Statistical Models;* Irwin: Burr Ridge, IL, 1990.
18.   Miltner, R. J.; Shukairy, H. M.; Summers, R. S; *J.AWWA,* **1992,** 84(11), 53.
19.   Reckhow, D. A.; Singer, P. C.; *J.AWWA,* **1990,** 82(4), 173.

Chapter 4

Aqueous Chlorination Kinetics and Mechanism of Substituted Dihydroxybenzenes

Alicia C. Gonzalez, Terese M. Olson[1], and Laurence M. Rebenne

Department of Civil and Environmental Engineering,
University of California, Irvine, CA 92697–2175

It is well known that treatment of drinking and wastewater with chlorine results in the formation of chloroform and a variety of organochlorine compounds. Although, it is recognized that 1,3-dihydroxybenzenes moieties in humic matter are responsible for the formation of chloroform and others disinfection by-products, the kinetics of the chlorination of 1,3 dihydroxybenzenes have not been determined. We present here, the chlorination kinetics of several substituted 1,3-dihydroxybenzenes in neutral and alkaline conditions. A series of chlorine-substituted 1,3-dihydroxybenzene substrates were selected to obtain a stepwise evaluation of the successive chlorination of resorcinol through 2,4,6-trichlororesorcinol. The results are useful for predicting the extent of accumulation of chlorinated phenols and resorcinols under varying Cl_2:phenol ratios and for comparing the relative reaction rates of chlorine with phenolic compounds and other substrates. We also present a UV/Vis spectrophotometry study of the chlorination of resorcinol, which in combination with GC\MS analysis, was used to confirm the formation of 2,2,4,6-tetrachloro and pentachlororesorcinol.

The formation mechanisms of chlorination by-products have been extensively investigated since humic substances were first identified as precursors for trihalomethanes in natural water disinfection processes [1]. The majority of these studies have focused on the identification of stable intermediates and products with humic and fulvic acids [2-7] or model compounds [8-13] as substrates. Dihydroxybenzene moieties in humic matter were hypothesized by Rook [9] to be responsible for the formation of chloroform. Later studies by de Leer et al. [5], corroborated this theory. Norwood and Christman [14] have also demonstrated the importance of phenol groups as reactive centers for chlorination by-products.

---

[1]Corresponding author

0097–6156/96/0649–0048$15.00/0
© 1996 American Chemical Society

Early kinetic studies of the chlorination of phenol and chlorine-substituted phenols by Soper and Smith [15] and later by Lee [16] have shown that the overall reaction is second-order and proportional to the concentration of aqueous chlorine and phenol at pH > 6. Both investigators also observed that the rate was highly pH-dependent and has a maximum in the neutral or slightly alkaline pH range. The mechanism over this pH range was found to involve the reaction of the hypochlorous acid species with the phenolate ion:

$$HOCl \xleftrightarrow{K_{Cl}} OCl^- + H^+$$

$$C_6H_5OH \xleftrightarrow{K_{Ph}} C_6H_5O^- \qquad (1)$$

$$HOCl + C_6H_5O^- \xrightarrow{k_2} C_6H_4Cl(O^-) + H_2O$$

$$v = \frac{d[C_6H_5OH]}{dt} = k_{app}[HOCl][C_6H_5O^-]$$

with $pK_a^{Cl} = 7.45$ and $pK_{a, Ph} = 9.89$. Similar findings were reported by Brittain and de la Mare [17].

Few detailed quantitative studies of the chlorination kinetics of dihydroxybenzene compounds are available. Heasley et al. [18,19] determined through the identification of intermediates that successive electrophilic chlorination of resorcinol from monochloro-, to dichloro-, to 2,4,6-trichlororesorcinol occurs before ring-opening. They hypothesized that a pentachloro intermediate was involved in the ring-opening step, but they failed to isolate tetrachloro or pentachlororesorcinol among the reaction products. In these studies, only relative chlorination rates of resorcinol and chlorine-substituted resorcinols were determined at pH 7, however. It is expected that the completely deprotonated resorcinol species (dianion) would be even more reactive than the singly deprotonated species. The contribution of the dianion form to the chlorination reaction at natural water pH conditions, however, can not yet be predicted. Similarly, the reaction mechanism after the formation of 2,4,6- trichlororesorcinol has not been positively established. Moye [20], for example, reported the formation of pentachlororesorcinol upon the chlorination of resorcinol in chloroform. However, Rook [9] suggested a mechanism for the chlorination of resorcinol in aqueous solution that involves the cleavage of trichlororesorcinol (Scheme I):

Scheme I

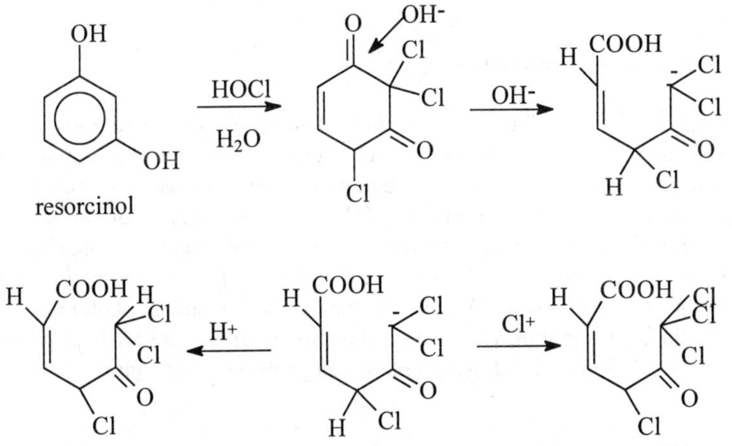

DeLeer and Erkeleus [21] also reported that pentachlororesorcinol was not an intermediate in the chlorination of resorcinol in aqueous solutions. Boyce and Hornig [12] proposed the formation and cleavage of both tetrachloro and penta-chlororesorcinol (scheme II), while Heasley et al. [19] concluded that pentachlororesorcinol is the intermediate that undergoes a ring-opening reaction when resorcinol is chlorinated in CH_3OH.

Scheme II

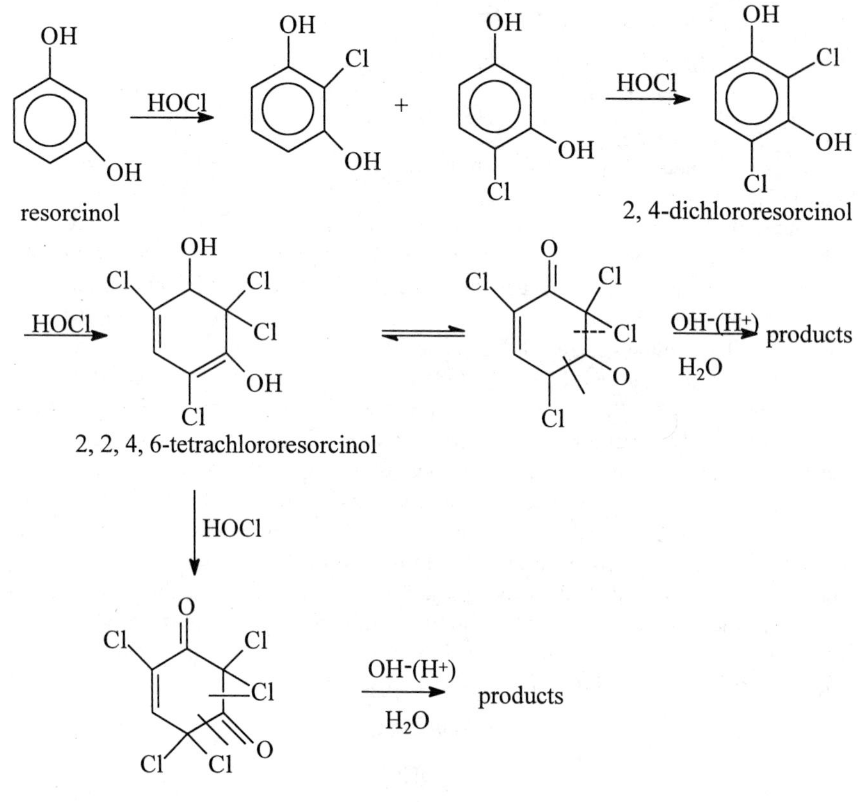

resorcinol 2, 4-dichlororesorcinol

2, 2, 4, 6-tetrachlororesorcinol

2, 2, 4, 4, 6-pentachlororesorcinol

We present here, the chlorination kinetics of several substituted resorcinols in neutral and alkaline conditions. A series of chlorine-substituted resorcinol substrates were selected to obtain a stepwise evaluation of the successive chlorination of resorcinol through 2,4,6-trichlororesorcinol. The results are useful for predicting the extent of accumulation of chlorinated phenols and resorcinols under varying Cl_2:phenol ratios and for comparing the relative reaction rates of chlorine with phenolic compounds and other substrates. We also present a UV/Vis spectrophotometry study of the chlorination of resorcinol, which in combination with GC\MS analysis, was used to confirm the formation of 2,2,4,6-tetrachloro and pentachlororesorcinol.

Experimental Procedures

Materials. Reagents including resorcinol (99% Baker), orcinol or 5-methylresorcinol (ACS reagent grade, Fisher Scientific), 4-chlororesorcinol (98% Aldrich), 4,6-dichlororesorcinol (97% Aldrich), NaOCl (5% solution, Baker), $Na_2S_2O_3$ (Fisher), H_2SO_4 (Fisher), KOH (Fisher), and $NaClO_4$ (99% Aldrich) were used without further purification. Stock solutions (0.2 M) of the dihydroxybenzene compounds were prepared in HPLC grade methanol. All other solutions were prepared with deionized water from a Millipore purification system.

Pentachlororesorcinol was prepared by chlorination of resorcinol in chloroform as described by Moye [20].

Methods. The reactions of resorcinol, monochlororesorcinol, dichlororesorcinol, and orcinol with HOCl were conducted in a batch reactor at room temperature (22°C). Chlorination experiments were conducted under pseudo-first-order conditions where HOCl was in excess. The reaction was quenched with an excess of sodium thiosulfate, at fixed time intervals. Analyses for the resorcinol substrates were performed by HPLC with a Dionex Series 4000i chromatograph equipped with a Supelcosil LC-18 column and a Dionex variable UV/Vis wavelength detector.

UV/Vis spectrophotometry was used to follow the chlorination of resorcinol in order to identify the reaction mechanism after the formation of 2,4,6-trichlororesorcinol. Spectra were taken with a HP 8450A diode array spectrophotometer. In a typical experiment, resorcinol was added to a solution of HOCl at a fixed pH, and the reaction was followed by measuring the absorbance spectra at convenient intervals. The exposure of the reaction mixtures to the light was minimized by using a scan time of 0.1 s., and keeping the shutter closed at other times. 2,4,6-Trichlororesorcinol chlorination rates in the pH range 8 to 12 were followed by monitoring the decay of its absorbance at $\lambda = 384$ nm.

GC\MS analyses of the samples were performed by the UCI mass spectrometry laboratory with a HP5890 gas chromatograph coupled to a Fisons Autospec mass spectrometer. Mass spectra were recorded in EI mode (E=70 eV).

Results and Discussion

The measured acid dissociation constants [22, 23] for each of the resorcinol substrates and several reported pK_a values for resorcinol are tabulated in Table I. A relatively wide range of published pK_{a1} values are available for resorcinol and hence the average of three reported values, 9.43, was used in modeling the proposed mechanism herein. The average literature value is also consistent with the value of $pK_{a1} = 9.33$ obtained from available Hammett correlations for the acidity of phenols, i.e., $pK_a = 9.92 - 2.23S\sigma$ [24]. With the exception of the second pK_a for 4,6-dichlororesorcinol, all of the measured pK_a values in Table I agree closely with published Hammett correlations of phenol acid dissociation constants. The pK_{a2} value for 2,4,6-trichlororesorcinol was estimated by assuming that a similar shift in pK_{a2} exists between 4,6-chlororesorcinol and 2,4,6-trichlororesorcinol as exists between 4-chlororesorcinol ($pK_{a2} = 10.75$) and 4,6-dichlororesorcinol ($pK_{a2} = 10.35$). By this assumption:

$$pK_{a2}^{2,4,6-trichlororesorcinol} \approx pK_{a2}^{4,6-dichlororesorcinol} - \Delta pK_{a2} = 10.35 - (10.75 - 10.35) = 9.9$$

Reaction Rate Orders. The rate of resorcinol disappearance at pH 5, was found to be first-order in the concentrations of chlorine and resorcinol:

$$\frac{d[ArH(OH)_2]_T}{dt} = -k_{app}[ArH(OH)_2]_T[HOCl]_T \tag{1}$$

where $k_{app} = k_{obs}/[HOCl]_T$ is an apparent second-order constant, and k_{obs} is the observed pseudo-first-order rate constant. Similar rate laws were found to describe the disappearance of 4-chlororesorcinol, 4,6-dichlororesorcinol, orcinol, and 2,4,6-trichlororesorcinol.

At all other pH conditions, the pseudo-first-order rate constants for resorcinol, 4-chlororesorcinol, 4,6-dichlororesorcinol, and orcinol were calculated from a single determination of the substrate concentration as follows:

$$k_{obs} = \frac{\ln\left(\dfrac{[ArX(OH)_2]_{T,0}}{[ArX(OH)_2]_{T,t}}\right)}{\Delta t} \tag{2}$$

where Δt was the time period of reaction before quenching. At least three determinations of k_{obs} were performed at each pH.

The reaction of trichlororesorcinol with HOCl was also found to follow a first order kinetic law in HOCl and trichlororesorcinol. Pseudo first order rate constants, k_{obs} over the pH range 8 to 11.3 were obtained from plots of ln $(A_{\infty 384} - A_{t384})/(A_{0,384} - A_{t384})$ vs time.

pH Dependence and Mechanism. The pH dependence of k_{app} is shown in Fig. I for the chlorinated resorcinols up to dichlororesorcinol, and orcinol, and in Fig. II for trichlororesorcinol. In all rate constant vs pH profile, the second order apparent rate constants exhibit maxima between pH 8 and 11. The shape of the rate constant profiles could be partially explained in terms of the reactivity of the ionized and unionized forms of the resorcinol substrates. Upon deprotonation, the -O⁻ substituent is more activating than -OH toward electrophilic substitution. In addition, the maxima in the rate constant profiles might be explained, if HOCl is the only active electrophile and the reactivity of OCl⁻ is negligible.

A reaction scheme for the chlorination of substituted resorcinols at pH 6 -12 is proposed as follows:

$$HOCl \underset{}{\overset{K_{a2}^{Cl}}{\rightleftharpoons}} OCl^- + H^+ \qquad\qquad (fast) \tag{3}$$

$$ArX(OH)_2 \underset{}{\overset{K_{a1}^{R}}{\rightleftharpoons}} ArX(OH)O^- + H^+ \qquad\qquad (fast) \tag{4}$$

$$ArX(OH)O^- \underset{}{\overset{K_{a2}^{R}}{\rightleftharpoons}} ArX(O^-)_2 + H^+ \qquad\qquad (fast) \tag{5}$$

$$ArX(OH)_2 + HOCl \overset{k_1}{\longrightarrow} products \qquad\qquad (slow) \tag{6}$$

TABLE I. Acid dissociation constants for substituted resorcinols[a]

Substrate	pK_{a1} ($\pm\sigma$)	pK_{a2} ($\pm\sigma$)
Resorcinol	9.43 (0.34)[b]	11.21 (0.16)[c]
4-Chlororesorcinol	8.09 (0.08)	10.75 (0.15)
4,6-Dichlororesorcinol	7.53 (0.09)	10.35 (0.11)
2,4,6-Trichlororesorcinol	5.73	9.9[d]
Orcinol	9.35 (0.03)	11.50 (0.02)

[a] Values determined experimentally by Rebenne et al. [22] unless specified otherwise.
[b] Value is the average of three reported values, 9.15 [28], 9.32 [25], and 9.81 [29].
[c] Value is the average of two reported values, 11.1 [25] and 11.32 [28].
[d] Calculated from the pK_{a2} of 2,4 dichlororesorcinol as explained in the text.

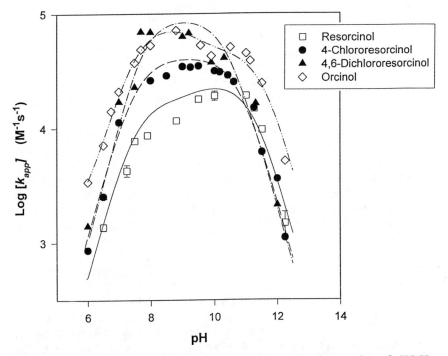

Figure I. Apparent rate constant - pH dependence for the reaction of HOCl and resorcinol, 4-chlororesorcinol, 4,6-dichlororesorcinol, and orcinol at 22°C. Lines are non-linear least-squares regression fit of k_1, k_2, and k_3 to the data.

$$\text{ArX(OH)O}^- + \text{HOCl} \xrightarrow{k_2} \text{products} \qquad \text{(slow)} \qquad (7)$$

$$\text{ArX(O}^-)_2 + \text{HOCl} \xrightarrow{k_3} \text{products} \qquad \text{(slow)} \qquad (8)$$

and the resulting general rate expression for the above reaction mechanism is:

$$v = -\frac{d[\text{ArX(OH)}_2]_T}{dt}$$

$$= k_1[\text{ArX(OH)}_2][\text{HOCl}] + k_2[\text{ArX(OH)O}^-][\text{HOCl}] \qquad (9)$$

$$+ k_3[\text{ArX(O}^-)_2][\text{HOCl}].$$

By introducing the following notation,

$$[\text{ArX(OH)}_2]_T = [\text{ArX(OH)}_2] + [\text{ArX(OH)O}^-] + [\text{ArX(O}^-)_2] \qquad (10)$$

$$[\text{HOCl}]_T = [\text{HOCl}] + [\text{OCl}^-] \qquad (11)$$

α_i^{Cl} = ionization fraction of hypochlorous acid species, where $i = 0$ or 1, for HOCl and OCl$^-$, respectively,

α_i^{R} = ionization fraction of resorcinol substrate species, where $i = 0$, 1, or 2, for ArX(OH)$_2$, ArX(OH)O$^-$, and ArX(O$^-$)$_2$, respectively,

the rate expression can be rewritten as follows:

$$v = (k_1\alpha_0^{R} + k_2\alpha_1^{R} + k_3\alpha_2^{R})\,\alpha_1^{Cl}\,[\text{ArX(OH)}_2]_T\,[\text{HOCl}]_T. \qquad (12)$$

and based on eq 12, the k_{app} term in eq 1 can be rewritten as:

$$k_{app} = \frac{\left(k_1\{H^+\}^2 + k_2 K_{a1}^{R}\{H^+\} + k_3 K_{a1}^{R} K_{a2}^{R}\right)\left(\dfrac{\{H^+\}}{\{H^+\} + K_a^{Cl}}\right)}{\{H^+\}^2 + K_{a1}^{R}\{H^+\} + K_{a1}^{R} K_{a2}^{R}}. \qquad (13)$$

The intrinsic constants k_1, k_2, and k_3 were fit to the data in Figs. I and II with a non-linear least-squares regression and Marquardt-Levenberg algorithm solution method. Acid dissociation constants were held fixed at the values given in Table I for the resorcinol substrates, and pK_a^{Cl} = 7.54 [25]. Fitted values of all of the intrinsic rate constants are presented in Table II. Reasonably close fits were obtained for each of the substrates, as shown by the solid lines in Figs. I and II, and hence the postulated mechanism is consistent with the data. Very poor fits for k_{app}, however, were obtained when the non-linear regresion analysis of the data for 4,6-dichlororesorcinol and 2,4,6-trichlororesorcinol were performed with their predicted pK_{a2} values. The reaction rate-pH dependence, therefore, is a confirmation that the pK_{a2} values for dichlororesorcinol and 2,4,6- trichlororesorcinol are more than 1 pH unit weaker than the pK_{a2} values predicted by Hammett correlations.

The chlorination rate-pH dependence for resorcinol, 4-chlororesorcinol, 2,4-dichlororesorcinol, and orcinol in Fig. I could be fit satisfactorily with or without the k_1 steps. The solid lines in Fig. I for resorcinol and 4-chlororesorcinol correspond to the fit obtained with only the k_2 and k_3 steps. The upper limits estimates of $k_1 \leq 330$ M^{-1}s^{-1} for resorcinol and $k_1 \leq 65$ M^{-1}s^{-1} for 4-chlororesorcinol were obtained as $k_1 + \sigma$, when the k_1 step was included in the regression analysis [23]. The k_1 step of the mechanism,

2,4,6-Trichlororesorcinol

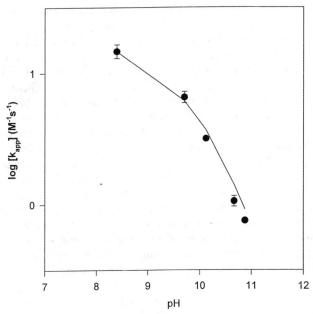

Figure II. Apparent rate constant - pH profile for the reaction of HOCl with 2,4,6-trichlororesorcinol. Solid line is a non-linear least-squares regression fit of k_2, and k_3, to the data.

TABLE II. Rate constants for the proposed reaction mechanism at 22°C.

Substrate	k_1 $(\pm\sigma)$ $(M^{-1} s^{-1})$	k_2 $(\pm\sigma)$ $(M^{-1} s^{-1})$	k_3 $(\pm\sigma)$ $(M^{-1} s^{-1})$
Resorcinol	< 330	1.36 (0.26) × 10⁶	1.15 (0.10) × 10⁸
4-Chlororesorcinol	< 65	1.43 (0.16) × 10⁵	6.73 (0.53) × 10⁷
4,6-Dichlororesorcinol	47 (17)	3.21 (0.76) × 10⁴	5.91 (0.81) × 10⁷
2,4,6-trichlororesorcinol		5.67 (0.61) × 10¹	2.31 (0.14) × 10³
Orcinol	1250 (160)	5.18 (0.34) × 10⁶	4.20 (0.04) × 10⁸

though, should be included for modeling the rate constant-pH profiles for the chlorination of 4,6-dichlororesorcinol and orcinol up to pH 2. The chlorination rate-pH profile for 2,4,6-trichlororesorcinol was fit without the k_1 step.

While the relative order of the intrinsic reactivities of chlorine with resorcinol, 4-chlororesorcinol, 4,6-dichlororesorcinol, and orcinol in Table I is as expected, the order of apparent reactivity, as indicated in Fig. I, is reversed between pH 6 to 10, with the most apparently reactive substrate being 4,6-dichlororesorcinol. This reversal can be explained in terms of the relative acidity of the substrates. Chlorine-substituted resorcinols are more acidic and the deprotonated forms are more activated for electrophilic attack. Trichlororesorcinol chlorination, however, does not follow the same reactivity pattern and was found to be the slowest step in the chlorination of resorcinol at pH 6-12 as is shown in Fig. II. This "abnormally" slow rate of reaction, also observed in the chlorination of trichlorophenol [26] and in the bromination of tribromophenol [27], is an indication of a change in the reaction mechanism. While the chlorination reaction of resorcinol, 4-chloro-, 4,6-dichlororesorcinol, and orcinol can be described as an electrophilic substitution at the o-, p- position to the -OH groups, the reaction of trichlororesorcinol is more likely a phenol - dienone rearrangement which implies a loss of aromaticity.   Consequently, it is expected to be a comparatively slower reaction. This type of rearrangement is shown in scheme III.

Scheme III

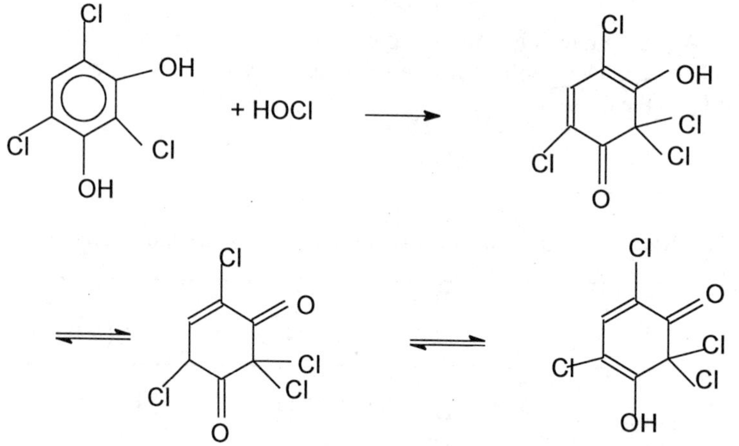

An important implication of these results is that 2,4,6-trichlororesorcinol is the compound most likely to accumulate under mild chlorination conditions, similar to those used in drinking water chlorination.

Reaction Mechanism. The initial products upon electrophilic substitution of chlorine onto resorcinol, 4-chlororesorcinol, and 4,6-dichlororesorcinol are shown in scheme IV.   The rate constants in Table I for resorcinol, 4-chlororesorcinol, and orcinol, therefore, correspond to the sum of rate constants for the "A" and "B" pathways shown in scheme IV. The orientation of resorcinol -OH groups renders the 2, 4, and 6 positions equivalent in terms of reactivity and their reaction-directing effect.

Scheme IV

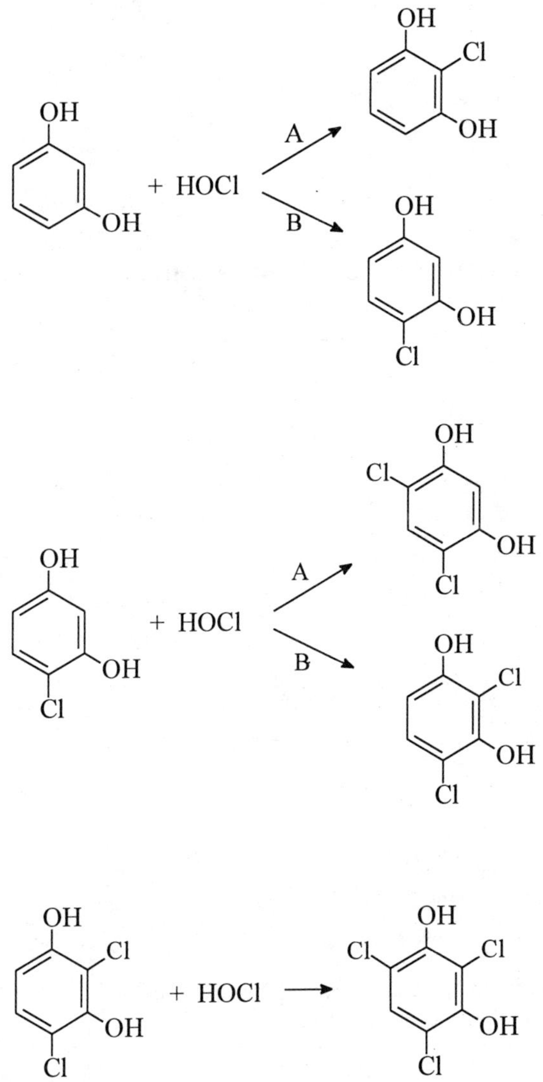

Consequently, the proportion of each of the products can be estimated on the basis of a statistical distribution. In the absence of steric hindrance, the product distribution for the chlorination of resorcinol should be approximately 33% 2-chlororesorcinol and 66% 4-chlororesorcinol. Heasley et al. [18] found a product ratio of 2- and 4-chlororesorcinol of approximately 1:3, respectively, for resorcinol chlorination over the pH range of 2 to 12. The reduced reactivity measured by Heasley at the ortho position

2,4-Dichlororesorcinol + HOCl at pH 12.5

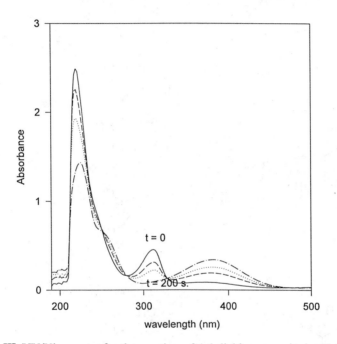

Figure III. UV/Vis spectra for the reaction of 4,6-dichlororesorcinol with HOCl at pH = 12.5. The HOCl solution was used as a blank.

may be due to steric hindrance or to deactivation caused by inductive effects of the neighboring -OH group.In Heasley's study, however, significant fractions of the resorcinol were already converted to dichloro- and trichlororesorcinol when the products were determined. Therefore the actual product formation rate ratio may be closer to the statistically-predicted ratio of 1:2. Nevertheless, Heasley's study is evidence that the three ring positions are similarly reactive towards HOCl. Chlorine substitution rates at either the 2 or 6 ring position of 4-chlororesorcinol should also be similar and hence equal fractions of the products 2,4- and 4,6-dichlororesorcinol are expected. Deviations from this statistical distribution are possible if steric hindrance effects are important. Such effects would likely favor 4,6-dichlororesorcinol.

The introduction of a methyl group to resorcinol at the C_5 position serves to weakly activate meta positions, which are already occupied by OH substitutents. Similar product isomer distributions should therefore be expected for resorcinol and orcinol.

UV/Vis spectrophotometry was used to elucidate the reaction mechanism after the formation of dichlororesorcinol. By working at basic pH conditions, we have been able to follow the chlorination reaction "step-by-step" as evidenced by the changes in the UV/Vis spectra. Fig. III shows the absorption spectra for the reaction of 4,6-dichlororesorcinol with HOCl at pH 12.5. The HOCl equilibrium concentration at this pH is very small, and consequently we were able to adjust the reaction rate to within a time frame compatible with our experimental technique. The first spectra was recorded inmediately after the initiation of the reaction, using the HOCl solution as a blank. The reaction was followed by measuring the spectra at frequent intervals. The spectras show a decay in the absorbance of dichlororesorcinol at $\lambda = 308$ nm, and an increase of the absorbance at 384 nm. The presence of an isosbestic point at $\lambda = 326$ nm is a clear indication that dichlororesorcinol $(ArCl_2(O^-)_2)$ converts stoichiometrically into a product or products absorbing at 384 nm. This product was identified as 2,4,6-trichlororesorcinol by a GC|MS analysis of a reaction mixture (HOCl/dichlororesorcinol = 5/1) that was quenched immediately after the bleaching of the band at 308 nm.

Figure IV shows the absorbance spectras for the reaction of resorcinol with HOCl under pseudo first order condition in HOCl, at pH 11.5 and at several reaction times. The spectras were taken immediately after the initiation of the reaction, using the HOCl solution as a blank. The spectras show that resorcinol and the chlorinated resorcinols, which absorb light at around 300 nm, are immediately consumed after the initiation of the reaction, leading to the formation of trichlororesorcinol with a broad absorption band center at $\lambda_{max} = 384$ nm. The absorption at 384 nm decays rather slowly according to a first order rate law, confirming that trichlororesorcinol chlorination is the slowest step in the chlorination of resorcinol.

In another experiment, the reaction (HOCl:resorcinol = 10) was quenched at a later time, after a decrease in the absorbance at 384 nm was observed. A GC\MS analysis showed the presence of trichlororesorcinol, 2,2,4,6-tetrachlororesorcinol (MS: m/z 246 (4 Cl), 211 (3Cl), 183 (3Cl), 155 (3Cl), 163 (2Cl), 135 (2Cl), 107 (2Cl)), 1,1,3,5,5-pentachloropent-3-en-2-one (m/z 254, (5Cl) and a main peak at 143), tetrachlorinated and pentachlorinated cyclopentenones, and a hexachlorinated pentenone as the main reaction products. We have obtained the 1,1,3,5,5-pentachloropent-3-en-2-one as the

Resorcinol + HOCl at pH 11.5

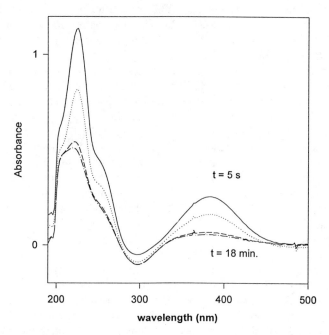

Figure IV. UV/Vis spectra for the reaction of resorcinol with HOCl at pH = 11.5. The HOCl solution was used as a blank.

major product of the hydrolysis of pentachlororesorcinol. This reaction proceeds as shown in scheme V.

Scheme V

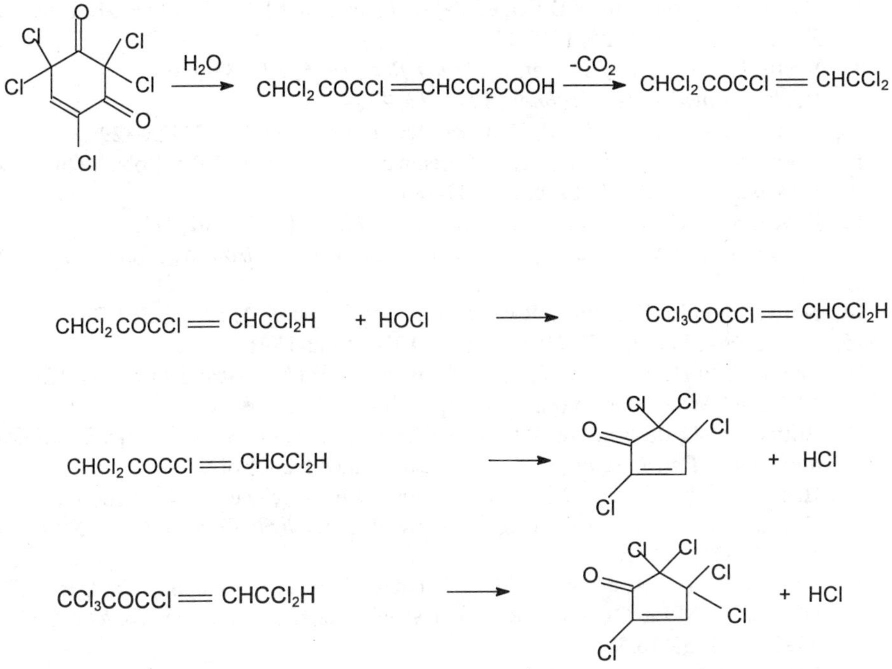

A pentachlorinated pentenone with a similar structure to the one above was reported by Heasley et al. [19] and Boyce and Hornig [12] as the product of the hydrolysis of pentachlororesorcinol. Although we have not measured the kinetics of the reaction after the chlorination of 2,4,6-trichlororesorcinol we can estimate, based on the reaction products reported above, that the chlorination of 2,4,6-trichlororesorcinol is followed by a comparatively much faster set of reaction leading to the formation of the hexachlorinated pentenones and cyclopentenones

Acknowledgements.   The authors gratefully thank the National Science Foundation (Grant #BES-9257896) and the Irvine Ranch Water District (Project #IRWD-15021) for their financial support of this work.

Literature Cited

1.   Rook, J.J. *Water Treat. Exam.* **1974**, *23*, 234-243.
2.   Peters, C.J.; Young, R.J.; Perry, R. *Environ. Sci. Technol.*, **1980**, *14*, 1391-1394.
3.   Christman, R.F.; Norwood, D.L.; Millington, D.S.; Johnson, J.D. *Environ. Sci. Technol.*, **1983**, *17*, 625-628.
4.   Miller, J.W.; Uden, P.C., *Environ. Sci. Technol.*, **1983**, *17*, 150-157.

5. de Leer, E.W.B.; Damsté, J.S.S.; Erkelens, C.; de Galan, L. *Environ. Sci. Technol.*, **1985**, *19*, 512-522.
6. Reckhow, D.A.; Singer, P.C.; Malcolm, R.L. *Environ. Sci. Technol.*, **1990**, *24*, 1655-1664.
7. Hanna, J.V.; Johnson, W.D.; Quezada, R.A.; Wilson, M.A.; Xiao-Qiao, L. *Environ. Sci. Technol.*, **1991**, *25*, 1160-1164.
8. Smith, J.G.; Lee, S-F.; Netzer, A., *Water Res.*, **1976**, *10*, 985-990.
9. Rook, J. *Environ. Sci. Technol.*, **1977**, *11*, 478-482.
10. Larson, R.A.; Rockwell, A.L. *Environ. Sci. Technol.*, **1979**, *13*, 325-329.
11. Norwood, D.L.; Johnson, J.D.; Christman, R.F.; Hass, J.R.; Bobenrieth, M.J. *Environ. Sci. Technol.*, **1980**, *14*, 187-189.
12. Boyce, S.D.; Hornig, J.F. *Environ. Sci. Technol.*, **1983**, *17*, 202-211.
13. Tretyakova, N.Y.; Lebedev, A.T.; Petrosyan, V.S. *Environ. Sci. Technol.*, **1994**, *28*, 606-613.
14. Norwood, D.L.; Christman, R.F. *Environ. Sci. Technol.*, **1987**, *21*, 791-798.
15. Soper, F.G.; Smith, G.F. *J. Chem. Soc.*, **1926**, 1582-1591.
16. Lee, F.C. In *Principles and Applications of Water Chemistry*; Faust, S.D, Hunter, J.V., Eds.; Wiley: New York, 1967; pp 54-74.
17. Brittain, J.M.; de la Mare, P.B.D. In *The Chemistry of Halides, Pseudo-halides, and Azides, Part I, Suppl. D*; Patai, S., Ed.; Wiley: New York, 1983.
18. Heasley, V.L.; Burns, M.D.; Kemalyan, N.A.; McKee, T.C.; Schroeter, H.; Teegarden, B.R.; Whitney, S.E.; Wershaw, R.L. *Environ. Toxicol. Chem.*, **1989**, *8*, 1159-1163.
19. Heasley, V.L.; Anderson, M.E.; Combes, D.S.; Elias, D.S., Gardner, J.T.; Hernandez, M.L., Moreland, R.J.; and Shellhamer, D.F. *Environ. Toxicol. Chem.*, **1993**, *12*, 1653-1659.
20. Moye, C.J. *Chem. Comm.*, **1967**, 196-197.
21. Deleer, E.W.B.; and Erkeleus, C. *Sci. Total Environ.*, **1985**, *47*, 211-216
22. Rebenne, L.M. *Kinetic and Mechanistic Study of the Aqueous Chlorination of Dihydroxybenzenes*, M.S. Thesis, University of California, Irvine, 1994.
23. Rebenne, L.M.; Gonzalez, A.C.; Olson, T.M. *Environ. Sci. Technol.* **1996**, *30*, 2235-2242.
24. Perrin, D.D.; Dempsey, B.; and Serjeant, E.P. *pKa Prediction for Organic Acids and Bases*, Chapman and Hall: New York, 1981.
25. Albert, A.; Serjeant, E.P. *The Determination of Ionization Constants: A Laboratory Manual*, 3ed., Chapman and Hall: New York, 1984.
26. Fischer, A.; Henderson, N.G. *Can. J. Chem.* **1979**, *57*, 552-557.
27. Tee, S.O.; Paventi, M.; Bennett, J.M. *J. Am. Chem. Soc.* **1989**, *111*, 2233-2240.
28. Kortum, G.; Vogel, W.; and Andrussow, K. *Dissociation Constants for Organic Acids in Aqueous Solutions*, International Union of Pure and Applied Chemistry, 1961.
29. Lide, D. ed., *Handbook of Chemistry and Physics*, 72ed., CRC Press, 1991-1992.

Chapter 5

An Economical Experimental Approach to Developing Disinfection By-Product Predictive Models

R. Hofmann and R. C. Andrews

Department of Civil Engineering, University of Toronto, 35 St. George Street, Toronto, Ontario M5S 1A4, Canada

Disinfection by-product (DBP) formation models were developed from bench-scale experiments using a fractional factorial design, with no replicates. This experimental approach minimizes the amount of data required to develop a model while sacrificing some precision. Despite the lower precision, the resultant models still accounted for the majority of the variation in the observed data. This method may be employed to develop site-specific DBP formation models with maximum economy.

Anticipated lower limits on disinfection by-products (DBPs) in drinking water may force certain water treatment utilities to develop strategies which will minimize DBP formation. A model that predicts DBP concentrations, given water quality characteristics and disinfection parameters, would be a useful tool for this purpose. Researchers have identified this need and have attempted to develop such models. For example, two studies led to the development of total trihalomethane (TTHM) and haloacetic acid (HAA_6) predictive models, respectively (1, 2). These studies both involved bench-scale experiments using water samples obtained from many sites in the United States, which represented a cross-section of water types. The DBP formation data obtained from the different water sources were then combined into a single model. While such an approach is useful for illustrating overall trends in DBP formation, the resulting model will not, nor is intended to, always accurately predict DBP concentrations for a specific site, partly due to an inability to characterize organic precursor material. If a specific water treatment facility has identified a need to predict DBP concentrations for alternative treatment scenarios, and if generic DBP models cannot provide adequate levels of accuracy, then the alternative may be to construct a site-specific DBP formation model based on bench-scale experiments.

0097–6156/96/0649–0063$15.00/0

Objective

This research represents a first step towards developing an experimental protocol that can be used to develop site-specific DBP formation models both quickly and economically. The objective of this work was to determine the feasibility of applying a factorial form of experimental design, using no replicates, to bench-scale experiments for developing site-specific DBP formation models.

Factorial Design Background

The discussion provided herein concerning factorial design is intended only to provide the reader with a basic background. For more details, the reader is directed to a statistical reference text, such as *Statistics for Experimenters* (*3*).

A factorial design may be used to quantify the effects of selected variables (factors) on a particular experimental outcome. In this case, the outcome is DBP formation, while the factors to be examined include pH, temperature, TOC, bromide concentration, disinfectant dose, and time. As an example, consider an experiment designed to quantify TTHM formation as a function of pH. The experimenter chlorinates a water sample at pH 6 in a bench-scale batch reactor, and following a selected contact time determines the TTHM concentration to be 50 µg/L. The experimenter simultaneously conducts a similar experiment at pH 8, and observes a TTHM concentration of 150 µg/L. Thus, the experimenter may perform a regression on the two data points, (pH 6, TTHM = 50 µg/L) and (pH 8, TTHM = 150 µg/L), to produce an equation which predicts TTHMs as a function of pH:

$$TTHM \ (\mu g/L) = -210 + 45(pH) \tag{1}$$

The previous example is called a 2-level factorial design, since pH was varied between only 2 levels; a low level (pH 6) and a high level (pH 8). It should be noted that Equation (1) assumes that TTHM increases linearly with respect to pH. If the relationship is non-linear, then the equation is inaccurate. However, in order to investigate non-linearity, more experiments would be required. When several factors are being investigated (such as pH, temperature, TOC, etc.), the increase in the amount of experimental work in order to investigate non-linearity is geometric. For example, to investigate 3 factors, each at 2 levels, requires $2^3=8$ experiments, while investigation of 3 factors, each at 3 levels, requires $3^3=27$ experiments. Since the primary goal of this research was to develop an easy, economical means of developing site-specific DBP formation models, it was decided that investigations would be directed towards the use of 2-level designs rather than those of a higher level (i.e. 3- or 4- level designs). The argument for adopting this approach is that the extra accuracy associated with the higher level designs would not justify the much greater amount of work and cost involved.

When investigating the effect of pH and temperature on TTHM formation, the general form of the resulting equation would be:

$$\text{TTHM} = \text{constant} + \text{pH effect} + \text{temp effect} + (\text{pH} \times \text{temp}) \text{ interaction} \qquad (2)$$

Note that a factorial design also allows the identification and quantification of interactions between factors, as shown in an example experimental matrix for investigating 2 factors at 2 levels each ($2^2 = 4$ experiments total) (Figure 1). It is observed that at 10°C, an increase in pH from the low to the high level only results in 10 µg/L more TTHM, while at 20°C, the increase in pH results in an associated increase in TTHMs of 18 µg/L. Thus, the pH effect may be viewed as being temperature dependent (it can also be stated that the temperature effect is pH dependent). Such a dependence implies that there is a pH×temperature interaction. The mathematics involved in quantifying interactions are not described herein; the reader is referred to a statistics text for a detailed explanation.

In order to investigate the effects of n factors on a particular outcome (such as DBP formation), with each factor considered at 2-levels, 2^n experiments would be required. For example, in Figure 1 pH and temperature are each varied between 2 levels, and therefore there are $2^2 = 4$ possible combinations. Consider the case where the influence of 6 factors on TTHM formation are to be investigated: $2^6 = 64$ experiments would be required. Each of the 64 experiments contributes to the estimation of 1 statistic which describes TTHM formation. Thus, with 64 experiments, an equation which describes TTHM formation as a function of 6 factors has 64 terms, as shown in Table I.

A hierarchy exists with respect to the influence of the factor effects. Single-factor effects are typically more significant than two-factor interaction effects, which in turn are more significant than three-factor interaction effects, and so on. Based on this information the experimenter can make certain assumptions which will lead to a reduction in the experimental workload, by making an informed assumption that the information which is lost is not significant. For example, in an experiment designed to investigate the effects of 6 factors on TTHM formation, with each factor evaluated at two-levels (2^6), the experimenter may choose to assume that all three-, four-, five-, and six-factor interactions are negligible. A model describing TTHM formation as a function of single-factors and two-factor interactions can then be constructed using only 16 experiments. This is called a *fractional* factorial design, and is denoted as 2^{6-2}, implying that a two-level investigation of six factors is accomplished in only $2^4 = 16$ experiments.

The use of fractional factorial designs is a powerful way of reducing experimental work and cost, with limited loss of precision. The theory behind the construction of a fractional factorial design cannot be described in detail here. In the example previously described, where instead of conducting all 64 experiments only 16 were selected, this selection process is not random. It is based on a specific algorithm which can be found in an appropriate statistics text.

Determination of Experimental Error in Factorial Design. When investigating the effects of selected variables (i.e. pH, temperature) on a particular outcome (i.e. DBP formation), it is possible that some of the variables will not influence the outcome at all. It is also possible that the effects of certain variables are so small

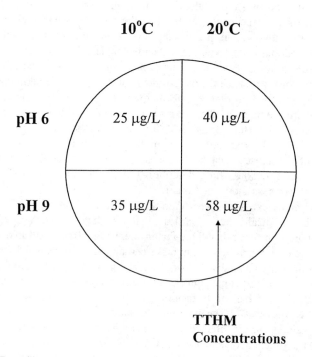

Predictive Equation:

TTHM $(\mu g/L) = 6 + 0.67(pH) - 0.1(Temperature) + 0.267(pH)(Temperature)$

Figure 1. A 2^2 factorial design experiment to determine the effects of pH and temperature on TTHM formation.

Table I. Terms in an Equation Derived Using a 2^6 Factorial Design.

			Interactions			
constant	1-factor effects	2-factor	3-factor	4-factor	5-factor	6-factor
# of terms: ($\Sigma = 64$)						
1	6	15	20	15	6	1
examples:	1. pH 2. Time 3. TOC 4. Temp 5. Bromide 6. Dose etc.	1. pH×Time 2. Dose×Temp 3. TOC×pH . . etc.	1. pH×Time×TOC 2. Temp×TOC×pH 3. Time×Dose×pH . . etc.	1. pH×Time× TOC×Temp 2. Dose×pH× Time×TOC etc.	1. pH×Time×TOC× Temp×Dose 2. Temp×Bromide× pH×TOC×Time etc.	1. pH×Time×TOC× Temp×Dose×Bromide

that they are not distinguishable from experimental error. An important part in developing a DBP formation model is determining which factors are important enough to include in the model. For example, if the change in HAA concentration when varying pH from a low to high level was 2 μg/L, it must be determined whether pH was truly responsible for the observed change in HAAs, or whether the change was due to random error in the experiment. For this case the traditional way to distinguish real effects from random effects would be to determine the standard deviation of a series of HAA sample replicates, and then convert this information into a specific confidence interval. For example, replicate measurements of the same concentration of HAAs may reveal that 95% of the replicates will vary by ±3 μg/L. If this is the case, then the observed pH effect of 2 μg/L may simply be due to experimental error, and it may be inappropriate to include pH in an HAA formation equation. In order to make such judgments, the standard deviation of the DBP measurements must first be defined, which typically requires that replicate analyses be conducted. However an alternative method exists which can be used to estimate the experimental error in a factorial design without conducting replicates. A benefit of this approach is the minimization of the amount of experimental work required to develop DBP formation models.

The estimation of experimental error when conducting a factorial (or fractional factorial) design is not presented in detail here, however a conceptual example is provided. Consider a 2-level factorially designed experiment to investigate the effects of 4 factors (pH, temperature, chlorine dose, and time) on TTHM formation. Thus $2^4=16$ combinations of the 4 factors are prepared as individual experiments, and the TTHM formation corresponding to each of the 16 combinations is measured. An initial hypothesis may be proposed that none of the 4 factors has an effect on TTHM formation. Therefore, it may be assumed that any variation in the 16 TTHM measurements is due to experimental error alone. Since experimental error is random, it should follow a Gaussian (Normal) distribution, and results from the 16 experiments can be examined to determine if they follow such a distribution. Suppose that 3 of the 16 measurements display a significant departure from the Gaussian distribution; this suggests that the particular combination of the 4 factors in those 3 samples had a real effect on TTHM formation. It can then be assumed that the remaining 13 experiments were unaffected by the 4 factors. The standard deviation and 95% confidence interval of the experimental error can be calculated from these 13 experiments which may now be viewed as replicates. Using the calculated 95% confidence interval, it can be determined if the 3 samples are, in fact, significantly affected by the 4 factors, or whether their observed TTHM concentrations were due to random error. Using this method, true factor effects can be distinguished from experimental error without conducting replicate experiments. The disadvantage of this approach is that the estimate of the experimental error is the sum of the scatter due to true experimental error plus the scatter due to real factor effects. Hence, this method overestimates experimental error, which may mask the smaller and mid-size effects. However, since the models being developed in this research are intended to provide information that can be used to develop DBP minimization strategies, it is the factors with the most significant effects on DBP formation that are of practical interest, and it is these factors that are identified in a non-replicated factorial design.

Experimental Design

Models which describe DBP formation were developed as a function of the following six factors: pH, disinfectant dose, contact time, total organic carbon (TOC), temperature, and bromide concentration.   The selection of these variables included consideration of those which were known to affect DBP formation, as well as variables that can be routinely measured by a treatment facility.   In particular, pH, disinfectant dose, TOC, and contact time, are parameters which may influence DBP formation, and can be controlled in a treatment train.   Bromide concentration and temperature cannot be controlled, however these factors were included in the experiment to determine the sensitivity of the DBP predictive models to these two parameters.   A series of models were developed using both raw and clarified water (see Methods section) from a surface water source.   Models were developed which address both chlorine and chloramine application to clarified water, while only chlorine was added to the raw water.   It was considered unnecessary to add chloramines to raw water, since chloramination is typically a secondary disinfectant and therefore normally not applied to raw water.

An experiment based on a 2^{6-2} fractional factorial design, using no replicates, was conducted in order to examine the effects of the chosen six factors on the formation of AOX (adsorbable organic halides), four THMs, five haloacetic acids (monochloro-, monobromo-, dichloro-, dibromo-, and trichloro-), trichloronitromethane (TCNM), and dibromoacetonitrile (DBAN).   A 2^{6-2} design requires DBP formation to be measured in 16 different combinations of the variables.   The combinations of the variables are shown in Table II.   All experiments were conducted in bench-scale, batch reactors.

Methods

The water used in this experiment was obtained from the Ottawa River, a large river located on the southern border of the provinces of Ontario and Quebec, Canada. Water quality characteristics of this river are summarized in Table III.   Experiments were conducted using both raw and clarified water from this site.   Clarification was performed in the laboratory, using an optimum alum dose as determined by turbidity reduction in a jar test, followed by flocculation using a custom-built paddle at a standardized velocity gradient ($G = 50$ s^{-1}).   The water was allowed to settle for 30 minutes prior to supernatant being decanted for use in an experiment.

Sixteen 2.5L amber glass reactors were filled with either raw or clarified water and adjusted to the water quality characteristics dictated by the factorial design (Table II).   Samples were adjusted to pH 6.5 by the addition of a phosphate buffer, while samples at pH 8.5 involved the addition of a borate buffer.   Buffers were prepared using Milli-Q water which contained the respective salts at near-saturation levels to minimize any dilution effects resulting from the subsequent addition of buffers to the samples. High TOC levels corresponded to the initial TOC of the water samples; low TOC levels were achieved by diluting the samples with Milli-Q water using a 2:3 Milli-Q:sample volume ratio. Bromide concentrations were adjusted to the required levels by spiking with potassium bromide (low bromide levels corresponded to the ambient concentrations in the initial water samples). Temperature control was maintained by placing the reactors in temperature-controlled chambers.

Chlorine stock solutions were prepared from commercially available Javex bleach, with the chlorine concentration measured prior to every experiment. Pre-formed chloramines were prepared between 15 to 90 minutes prior to addition to water samples by slowly adding a chlorine solution (pH 8.5) to an ammonium chloride solution at a $Cl_2:NH_3$-N mass ratio of 3:1. Thus, monochloramine was the only significant chloramine species present in the stock solution. The concentration of the resulting stock chloramine solution was always measured prior to addition to the water samples. All glassware was made chlorine demand free by exposure to a chlorine solution of at least 40 mg/L overnight in the dark.

Standards for THMs, HAAs, HANs, and TCNM were obtained from Supelco, and used for calibration purposes within 1 week of receipt. The AOX standard stock solution was pentachlorophenol. All standards were stored at 4°C prior to use. Trihalomethanes, chloropicrin, and the haloacetonitriles were analyzed using a Hewlett Packard 5890 Series II Plus gas chromatograph, according to EPA Method 551 (4). Haloacetic acids were analyzed according to EPA Method 552 (5).

Chlorine and chloramine residuals were measured using Standard Method 4500-Cl D (6). TOC was measured using Standard Method 5310 C (6). Bromide concentrations were determined using EPA Method 300.0 (7), with a Dionex DX500 ion chromatograph.

Results and Discussion

Equations relating DBP formation to pH, TOC, time, disinfectant dose, bromide concentration, and temperature, as developed using a 2^{6-2} fractional factorial design with no replicates, are shown in Tables IV, V, and VI. Tables IV and V contain results which relate to the chlorination of raw and clarified water, respectively. Table VI shows results for chloramination of clarified water.

The focus of this discussion is not the equations themselves, but rather whether or not a fractional factorial design approach using no replicates is capable of providing DBP formation models that may be readily applied. Therefore discussion of the experimental data focuses on the statistical parameters shown in Tables IV to VI, rather than the resulting equations. In these tables, R-squared values are shown, as well as the 95% confidence intervals of the estimate of the experimental error, as expressed by a percentage of the average DBP concentration.

R-squared values provide measures of the percentage of the variation observed in the data that can be accounted for by the model. Hence, an R-squared value of 0.8 means that 80% of the overall variance of the data is described by the model. The remaining 20% is due to unidentified experimental error. Recall

Table II.   Randomised 2^{6-2} Factorial Design Protocol.

RUN	DISINFECTANT DOSAGE (mg/L)	CONTACT TIME (hours)	FACTOR pH	TOC (mg/L)	BROMIDE (μg/L)	TEMPERATURE (°C)
1	5	24	6.5	5	300	1
2	8	24	8.5	5	300	20
3	8	24	6.5	5	15	20
4	5	24	6.5	3	15	20
5	8	1	8.5	5	300	1
6	8	1	6.5	3	300	20
7	5	1	6.5	3	15	1
8	8	1	6.5	5	15	1
9	8	1	8.5	3	15	20
10	5	1	8.5	3	300	1
11	8	24	8.5	3	15	1
12	5	24	8.5	3	300	20
13	8	24	6.5	3	300	1
14	5	1	6.5	5	300	20
15	5	24	8.5	5	15	1
16	5	1	8.5	5	15	20

Table III.   Water Quality Characteristics of the Ottawa River Water.

PARAMETER	TYPICAL VALUE
Turbidity	3.4 NTU
TOC	5.9 mg/L
Bromide	<10 μg/L
pH	7.3
Color	29 TCU

Table IV. DBP Formation Models for Chlorination of Raw Water.

COMPOUND	EQUATION[a]	R^2	95% CI[b]
Trihalomethanes			
$CHCl_3$	-214 + 1.14*Temp + 17.6*TOC - 0.10*Br + 24.3*pH + 1.86*Time + 0.11*Temp*Time	0.93	13%
$CHBrCl_2$	-4.48 + 0.15*Br + 0.84*Time	0.67	33%
$CHBr_2Cl$	-10.7 + 0.0029*Br - 0.056*Time - 0.38*Br*Time	0.75	47%
CHBr3	0.60 + 0.010*Br	0.69	29%
TTHM	-31.5 + 3.5*Temp + 25.1*TOC + 30.5*pH + 3.6*Time	0.83	18%
Haloacetic Acids			
MCAA	0.21 + 0.045*Temp + 0.10*Time	0.51	50%
DCAA	-19.8 + 1.0*Temp + 5.5*TOC + 1.1*Time	0.82	17%
TCAA	67 + 8.1*TOC - 11.2*pH + 1.3*Time	0.70	23%
MBAA	0.14 + 0.0070*Br	0.57	42%
DBAA	-0.26 + 0.11*Temp + 0.018*Br	0.73	28%
HAA_5	-45.0 + 2.0*temp + 14.2*TOC + 2.5*time	0.83	14%
Miscellaneous			
AOX	-84.7 + 84.5*TOC + 8.12*Time	0.73	12%
DBAN	11.4 + 0.011*Br - 1.30*pH	0.75	21%
TCNM	0.31 - 0.0015*Br + 0.021*Time	0.53	59%

[a] Units are as follows: All DBPs = µg/L, Time = hours, Disinfectant dose = TOC = mg/L, Bromide concentration = µg/L, Temperature = °C

[b] 95% CI = 95% confidence interval for the estimate of the experimental error, expressed as a percentage of the average DBP concentration.

Table V. DBP Formation Models for Chlorination of Clarified Water.

COMPOUND	EQUATION[a]	R^2	95% CI[b]
Trihalomethanes			
$CHCl_3$	-79.5 + 1.7*Temp - 0.096*Br + 13.7*pH + 1.7*Time	0.80	23%
$CHBrCl_2$	-10.0 + 0.60*Temp + 0.096*Br + 0.69*Time	0.90	20%
$CHBr_2Cl$	No Observed Factor Effects[c]	N/A	85%
CHBr3	0.96 + 0.017*Br	0.62	29%
TTHM	-156 + 1.0*Temp + 12.0*TOC + 18.8*pH + 1.2*Time + 0.11*Temp*Time	0.90	14%
Haloacetic Acids			
MCAA	No Observed Factor Effects	N/A	95%
DCAA	4.83 + 0.68*Temp + 0.68*Time	0.89	22%
TCAA	-11.5 + 0.50*Temp + 6.9*TOC - 0.034*Br + 0.89*Time	0.89	16%
MBAA	0.0043*Br + 0.000292*Temp*Br	0.75	27%
DBAA	0.23 + 0.00066*Temp + 0.0012*Br + 0.0010*Temp*Br	0.84	33%
HAA_5	-23.3 + 1.4*Temp + 10.6*TOC + 1.7*Time	0.84	13%
Miscellaneous			
AOX	-111 + 3.3*Temp + 62*TOC + 5.8*Time	0.93	8%
TCNM	-0.33 - 0.0019*Br + 0.019*Time	0.57	51%
DBAN	7.6 + 0.017*Temp + 0.014*Br - 0.85*pH + 0.00099*Temp*Br	0.93	16%

[a] Units as follows: All DBPs = µg/L, Time = hours, Disinfectant dose = TOC = mg/L, Bromide concentration = µg/L, Temperature = °C
[b] 95% CI = 95% confidence interval for the estimate of the experimental error, expressed as a percentage of the mean DBP concentration.
[c] None of the factors influenced the average DBP concentration by more than the 95% confidence level.

Table VI. DBP Formation Models for Chloramination of Clarified Water.

COMPOUND	EQUATION[a]	R^2	95% CI[b]
Trihalomethanes			
$CHCl_3$	1.05 - 0.015*pH*Time + 0.013*pH*Dose	0.51	10%
$CHBrCl_2$	6.5 - 0.77*pH	0.74	33%
$CHBr_2Cl$	Below Detection Limits (<0.1 µg/L)	N/A	N/A
CHBr3	22.8 - 2.38*pH - 0.046*Time + 0.030*pH*Time	0.96	11%
TTHM	32.5 - 3.6*pH	0.94	12%
Haloacetic Acids			
MCAA	Below Detection Limits (< 1 µg/L)	N/A	N/A
DCAA	No Observed Factor Effects[c]	N/A	100%
TCAA	0.39 + 0.012*Time	0.26	19%
MBAA	Below Detection Limits (< 0.5 µg/L)	N/A	N/A
DBAA	No Observed Factor Effects	N/A	90%
HAA_5	1.25 + 0.11*Time	0.38	48%
Miscellaneous			
AOX	98.5 + 2.8*TOC - 13.0*pH + 0.79*Time	0.87	14%
TCNM	Below Detection Limits (< 0.1 µg/L)	N/A	N/A
DBAN	24.4 - 2.69*pH - 0.038*Time + 0.025*pH*Time	0.94	14%

[a] Units as follows: All DBPs = µg/L, Time = hours, Disinfectant dose = TOC = mg/L, Bromide concentration = µg/L, Temperature = °C

[b] 95% CI = 95% confidence interval for the estimate of the experimental error, expressed as a percentage of the mean DBP concentration.

[c] None of the factors influenced the average DBP concentration by more than the 95% confidence level

however, that in a factorial design with no replicates, the "experimental error" is actually the sum of the true experimental error plus the "scatter" due to the effects of factors which are not included in the model. For example, the equation shown in Table IV for AOX describes its formation as a function of TOC concentration and contact time. The R-squared for this equation is 0.73. This means that TOC concentration and contact time together contributed 73% of the variance observed in the data. The remaining 27% of the variance is due to "scatter" arising from random error, as well as the effects of the other four factors (pH, chlorine dose, bromide concentration, and temperature), plus all of the 2-, 3-, 4-, 5-, and 6-factor interactions. These other factors likely each contributed to the variance in AOX measurements, however the individual effects were not large enough to be distinguishable from the estimated experimental error, using the 95% confidence level. The percentage shown to the right of the R-squared value for AOX in Table IV indicates that for an individual factor to appear in the equation, it would have to influence the average AOX concentration by more than 12%, otherwise it would not be distinguishable from the experimental error. Therefore it may be stated that any of the other factors that were examined (eg. pH) may have an effect on AOX concentration, but that the effect is relatively small, influencing the AOX concentration by less than 12%.

The R-squared values shown in Tables IV, V, and VI typically range from 0.7 to 0.9. This means that the models developed using a fractional factorial design with no replicates were capable of accounting for 70 to 90% of the variance observed in the data. This implies that most of the major factors which influence DBP formation are included in the models, and also that some minor factors have been omitted. However, since it is the major factors that would likely be examined when developing a DBP minimization strategy, it can be argued that the models as described provide useful information. The 95% confidence intervals of the estimated experimental errors typically ranged from ±10% to 30%. This implies that when using the fractional factorial design approach with no replicates, individual factors must influence the average DBP concentration by more than 10 to 30% before they can be identified above the experimental noise, and included in the model. Again, if it is assumed that minor factor influences are not as important as the major influences, then the lack of precision may be acceptable, however this must be determined on an application specific basis.

It is the intention that models such as the ones described in this research may be useful in the development of DBP minimization strategies for individual water treatment facilities. In a treatment facility, parameters that can typically be controlled include disinfectant dose (or the dose:TOC ratio), pH, and, if options exist concerning the location of chlorine and ammonia feed points, then free-chlorine contact time as well. While bromide concentration and temperature were included as variables in this study, in practice these would likely remain constant during extended periods of the year. Therefore, the number of factors involved in a bench-scale investigation to develop site-specific DBP formation equations may be easily limited to 3 or 4, providing further experimental economy.

Conclusions

In this study, the feasibility of using a 2-level fractional factorial design with no replicates to develop site-specific DBP formation models was investigated. It was found that despite the inherently large estimates of experimental error associated with this approach, models which are capable of describing data with R-squared values typically ranging from 0.7 to 0.9 could be developed. In addition, factors which influence average DBP concentrations by more than 10 to 30% could be consistently identified. It may therefore be generalized that this approach is capable of easily identifying and quantifying the major factors which influence DBP formation. DBP minimization strategies would likely be based on a consideration of major factors, and therefore the great economy in terms of experimental time, effort, and cost associated with non-replicated fractional factorial designs may justify its use.

Future Work

The work which has been described represents the first step in a larger program to develop a protocol for creating site-specific DBP formation models. This first step has demonstrated the feasibility of applying non-replicated fractional factorial designs. The approach will be further refined. Specific goals include the determination of which factors should generally be used in a factorial design for prediction of a given DBP, the number of replicates (if any) that should be included, whether 3-level designs would be appropriate in specific cases, and the extent to which follow-up experiments need to be considered. Field validation studies will also be conducted.

Acknowledgments

This work was funded in part by Health Canada, and the Natural Sciences and Engineering Research Council of Canada (NSERC). The authors also wish to acknowledge Dr. David T. Williams of Health Canada, and Ian Douglas and Jim Guthman of the Region Municipality of Ottawa-Carleton for their valued participation in the project.

Literature Cited

1. Amy G.L., Chadik P.A. and Chowdhury Z.K.. *Jour. AWWA*, **1987**, 79,7,89-97.
2. Amy G.L., Siddiqui M., Ozekin K., Wang X., Zhu H.W. and Westrick J.. *Proc. 1993 AWWA WQTC*, Miami, FL, **1993**, pp. 113-127.
3. Box G., Hunter W. and Hunter J.. *Statistics for Experimenters*. John Wiley & Sons, Inc.. Toronto, ON, **1978**, 653 pp.

4. *Methods for the Determination of Organic Compounds in Drinking Water,* *Supplement I.* EPA/600/4-90/020, NTIS Publ. PB91-146027. **1990**.

5. *Methods for the Determination of Organic Compounds in Drinking Water,* *Supplement II.* EPA/600/R-92/129, NTIS Publ. PB92-207703. **1992**.

6. *Standard Methods for the Examination of Water and Wastewater,* 18th ed. American Public Health Association, Washington D.C., **1992**.

7. *Methods for the Determination of Inorganic Substances in Environmental* *Samples.* EPA/600/R/93/100. NTIS Publ. PB94-120821. **1993**.

Chapter 6

The Influence of Operational Variables on the Formation of Dissolved Organic Halogen During Chloramination

J. M. Symons[1], R. Xia[1], A. C. Diehl[2], G. E. Speitel, Jr.[2],
Cordelia J. Hwang[3], Stuart W. Krasner[3], and S. E. Barrett[3]

[1]Department of Civil and Environmental Engineering,
University of Houston, Houston, TX 77204–4791
[2]Department of Civil Engineering, University of Texas,
Austin, TX 78712
[3]Water Quality Division, Metropolitan Water District of Southern
California, 700 Moreno Avenue, La Verne, CA 91750–3399

The objective of this study was to extend previous work on disinfection by-product (DBP) formation during free chlorination to situations in which chloramines are used as the disinfectant. Although four trihalomethanes (THMs), two cyanogen halides (CNX), and the sum of six haloacetic acids (HAA6) were measured, the focus of this phase of the work was on the production of dissolved organic halogen (DOX), a group parameter that measures "all" of the halogen-substituted DBPs. Chloramination was carried out in three waters, at three pH levels, at three chlorine-to-nitrogen (Cl_2:N) ratios, at three total chlorine residual concentrations, and under a variety of mixing conditions. The results of all the DBP and DOX analyses on these samples showed that (1) in the range of 1-4 mg/L total residual (as used herein, total residual is the sum of the concentrations of monochloramine, dichloramine, and free chlorine), 2-d DOX production was not significantly influenced by disinfectant concentration; (2) usually, a small percentage of the measured DOX (typically 5-20 percent) could be identified when compared to the molar sum of the total THMs (TTHMs), CNX, and HAA6; (3) mixing techniques had little influence on DOX production; and (4) chloraminating at the highest possible pH and lowest possible Cl_2:N ratio typically minimized DOX formation.

0097–6156/96/0649–0078$15.00/0
© 1996 American Chemical Society

Producing microbiologically and chemically safe drinking water is the most important challenge facing water utilities today. Historically, chlorine has been the most popular disinfectant in the United States. In 1974 trihalomethanes (THMs), which are suspected human carcinogens, were found to form during disinfection of water with chlorine (*1*). In addition to THMs, subsequent research has identified a variety of other undesirable and potentially dangerous compounds that form during water treatment (*2*). Collectively these compounds are referred to as disinfection by-products (DBPs). The current U.S. Environmental Protection Agency (USEPA) primary drinking water maximum contaminant level (MCL) for total THMs (TTHMs) is 0.10 mg/L. Promulgation of the Disinfectants/DBP (D/DBP) Rule will set Stage 1 MCLs for TTHMs at 0.080 mg/L and 0.060 mg/L for the sum of five haloacetic acids (HAA5). In addition, the D/DBP Rule includes a proposal to lower the MCLs in Stage 2 to 0.040 mg/L for TTHMs and 0.030 mg/L for HAA5 (*3*).

In response to current and anticipated DBP regulations, three primary areas of research have been conducted to solve this problem: the use of alternative disinfectants, the removal of DBP precursors and the removal of DBPs after their formation. The removal of DBP precursors prior to disinfection provides the best assurance that DBP formation will not occur. Despite some successes in this area, however, removing, or destroying organic precursors in source water can be difficult for completely controlling THMs and other DBPs (*4*). Allowing THMs and other DBPs to form and then removing them prior to their entry into the distribution system may not be practical. These limitations in precursor and DBP removal techniques highlight the desirability of utilizing a disinfectant that does not significantly contribute to DBP levels when added to source or treated water containing precursor materials. Chloramines have received considerable attention because much lower levels of DBPs are produced during chloramination than during chlorination (*5*).

However, some DBPs associated with chlorination—*e.g.*, THMs, HAAs, and haloacetonitriles (HANs)—have been observed during chloramination as well (*6*). Dissolved organic halogen (DOX) formation also is observed during chloramination (*2, 6*). Formation of DBPs during chloramination may result from unusual chemical characteristics of the water or the disinfection process. These might include a high total organic carbon (TOC) concentration, a high bromide concentration, a low pH, the relative ratio of mono- and dichloramines, and the chloramine dosage. The importance of these parameters in the formation of DBPs during chloramination is not well defined and needs detailed investigation. Formation of DBPs also may result from inadequate mixing at the point of application of free chlorine and ammonia; therefore, the impact of mixing intensity on DBP formation also must be evaluated.

To address the above issues, the project sponsored by the American Water Works Association Research Foundation (AWWARF)—"Factors Affecting Disinfection By-Products Formation During Chloramination," RFP 803—covered three primary aspects of work: (1) what factors influence DBP formation during chloramination, (2) what DBPs are formedduring chloramination, and (3) what treatment steps can be implemented to lower DBP concentrations during chloramination?

Current Knowledge

DOX as a Surrogate for DBPs. Surrogate parameters often provide a relatively easy approach to the measurement of overall DBP formation in water systems. The most commonly used surrogate measurement of DBPs is DOX. This parameter represents the "total" amount of chlorine- and bromine-substituted organic matter present in a water.

Krasner et al. (*6*) correlated the measured molar concentration of DOX with the arithmetic molar sum of 19 individual DBPs (i.e., four THMs, five HAAs, four HANs, and six miscellaneous chlorine-substituted DBPs) measured in a 1-year quarterly survey of 35 utilities, nationwide, that used chlorine, chloramines, or both. The data indicated that 25 percent of the measured DOX (on the average) was accounted for by summing the molar concentration of the 19 measured DBPs.

Singer et al. (*7*) reported on a study where eight water treatment plants in North Carolina, all using free chlorine, were sampled during three seasons of the year. Plotting the molar sum of the 12 individually measured DBPs (four THMs, four HAAs, two HANs, and two other chlorine-substituted DBPs) versus the molar DOX concentration produced a best-fit line with an r^2 value of 0.83 and a slope of 0.36, indicating that 36 percent of the molar DOX concentration (on average) was accounted for by the molar sum of the 12 DBPs measured.

Chloramines Versus Free Chlorine for the Control of DOX. In general, as cited in the examples below, most reports indicate that the use of chloramines in place of free chlorine, within the range of conditions most commonly seen in potable water treatment, results in lower DOX concentrations in treated water.

For example, a pilot-plant study at the Louisville (Ky.) Water Company compared the effect of different disinfectants on DOX concentrations (8). The study found that the use of chloramines resulted in a mean DOX concentration 45 percent lower than the mean DOX value when chlorine was the disinfectant. Stevens et al. (5) found a DOX concentration 85 percent lower when chloramines were used in place of chlorine in formation-potential tests with a humic acid solution. A USEPA-sponsored field study at Jefferson Parish (La.) examined the effects of different disinfectant treatment schemes on DOX formation (9). The average DOX concentration after a 30-min contact period was 117 µg Cl⁻/L when chloramines were used and 263 µg Cl⁻/L when chlorine was used—i.e., about 55 percent less with chloramines. In further work at Jefferson Parish, the samples were stored for 5 days with a disinfectant residual to assess the DOX formation potential (DOXFP) (10). The results of this testing indicate that the use of chloramines instead of chlorine can lessen DOXFP by as much as 90 percent. Jensen et al. (11) compared the reactions of monochloramine and chlorine with aquatic fulvic acid precursors. Monochloramine produced about one-fifth the DOX concentration that chlorine did. They did note, however, that monochloramine-produced DOX is more hydrophilic than chlorine-produced DOX.

Chloramination Treatment Variables and DBP Formation. The formation of DBPs resulting from chloramination is influenced by several treatment variables, as discussed in the following paragraphs.

Chloramine Dosage. Formation of DBPs relative to chloramine dosage is not a linear relationship. Although nonpurgeable organic chlorine (NPOCl) will form rapidly at low concentrations of monochloramine, Fleischacker and Randtke (12) observed that the amount of NPOCl increases at a much slower rate at higher concentrations of chloramines. These researchers also found that chloramine dosages of more than 20 mg/L in several source waters with different precursor types did not result in additional increases in NPOCl formation (12). Within the range of chloramine residuals commonly used in the water industry (1 to 5 mg/L), chloramine dosage did not appear to be significant factor in DBP formation.

Chlorine:Ammonia-Nitrogen Ratio. The ratio of chlorine to ammonia has been shown to directly influence the amount of TTHMs formed in studies done by the Metropolitan Water District of Southern California (MWDSC) (*13*). TTHMs remained quite low at chlorine-to-nitrogen (Cl_2:N) ratios less than 5:1 (on a mass basis, where ammonia was the primary reduced nitrogen species) and then increased dramatically above the 5:1 Cl_2:N ratio. Also, above the 5:1 Cl_2:N ratio, the relative concentration of monochloramine decreased rapidly and the concentration of dichloramine started to increase significantly. The presence of a larger amount of dichloramine above the 5:1 Cl_2:N ratio may be a reason for the greater level of TTHM formation observed at higher ratios.

pH. Stevens et al. (*2*) reported that the concentrations of NPOX formed at pH 11.5 and pH 7.5 were found to be about one-fourth and two-thirds, respectively, of the NPOX concentration formed at pH 5.9 during the chloramination of humic acid solutions, regardless of the incubation time, whereas THM levels remained low and unchanged with increasing pH. A pilot-plant study at the Louisville Water Company found that during chloramination, average DOX concentrations tended to decrease when the pH was increased (*8*).

Objectives

The specific objectives of performing DOX measurements in the current AWWARF study were (1) to determine the influence of water quality and operational parameters (total residual—as used herein, total residual is the sum of the concentrations of monochloramine, dichloramine, and free chlorine, pH,Cl_2:N ratio, bromide concentration) on two-day (2-d) DOX formation in three different waters using preformed chloramines; (2) to determine the influence of mixing intensity on 2-d DOX formation and (3) to determine what fraction of the 2-d DOX during chloramination could be accounted for by summing the molar concentration of the 12 individually measured DBPs (chloroform, bromodichloromethane, dibromochloromethane, bromoform, monochloroacetic acid, dichloroacetic acid, trichloroacetic acid, monobromoacetic acid, dibromoacetic acid, bromochloroacetic acid, cyanogen chloride, and cyanogen bromide.

Experimental and Analytical Methods

Overview. Batch experiments were conducted on three water sources, Lake Austin water (LAW), Lake Houston water (LHW), and California State Project water (CSPW). One large sample of water was collected from each source, except for Lake Houston where two samples were collected. Collection of water occurred during periods of typical source water quality. The water was shipped as rapidly as possible and was stored at 4°C prior to testing. The experiments were conducted directly on the raw water, so that the maximum DBP precursor concentrations were present to simulate worst case conditions.

Two types of batch experiments were conducted on each water source. The first focused on how chemical conditions (i.e., total residual concentration, residual speciation,Cl_2:N ratio, pH, and bromide concentration) affected the production of 12 measured DBPs and DOX. The second set of experiments investigated the effect on DBP formation of mixing at the point of chlorine and ammonia addition. In particular, the relative importance of system chemistry versus mixing conditions in DBP formation was of interest.

Chemistry Experiments. The major variables were TOC concentration, bromide concentration, chloramine dose, pH, and Cl_2:N ratio. For any given water, the TOC concentration was constant, leaving four variables to study. Each variable was studied at different levels to establish its importance in DBP formation. To avoid confounding effects from imperfect mixing, the experiments were conducted with preformed chloramines.

Two levels of bromide concentration were studied, the ambient concentration and the ambient concentration plus 0.5 mg/L. The added bromide provided some indication of the importance of the bromide ion in DBP formation during chloramination. Three chloramine doses were selected to provide target residual concentrations after 48 h (i.e., 2 d) of incubation at ~20°C. The target concentrations were 1, 2, and 4 mg/L, which spans the range of current practice.

Three levels of pH (6, 8, and 10) were studied, also covering the range of current practice. For systems utilizing enhanced coagulation, the settled water pH levels may be close to 6; whereas softened water systems can produce water at pH 10. Chloramination of such waters can yield different by-products than for waters near a neutral pH. Three $Cl_2:N$ ratios (3:1, 5:1, and 7:1) were studied to span the broadest possible range of operation in practice. These experiments represent a 2x3x3x3 matrix, resulting in 54 experimental conditions for each water.

The batch experiments were conducted in 1-L amber glass bottles. The bottles were filled with an appropriate volume of raw water, bromide and pH levels were adjusted to the desired level, and the bottles were then dosed with a concentrated stock solution of pre-formed chloramines. Pre-formed chloramines were created by mixing aqueous ammonium sulfate and sodium hypochlorite solutions. These solutions were formulated so that approximately equal volumes of the two, when combined, would produce the desired $Cl_2:N$ ratio. Both solutions were adjusted to pH 9 with nitric acid and/or sodium hydroxide prior to mixing. The concentration of the chlorine solution was approximately 5 mg/mL and was measured immediately before the ammonium solution was added. Small adjustments were made, as needed, to the volume of ammonium solution to ensure the correct $Cl_2:N$ ratio. The chlorine solution was added slowly to the ammonium solution with constant mixing in an ice bath at 1°C. After 15 minutes of mixing, the concentration of the chloramine solution was measured, and the sample water was dosed. Before performing the 2-d simulated distribution system (SDS) tests, an initial test was run to determine the chlorine demand of the sample water under the specific SDS conditions and the effect on sample pH of addition of pre-formed chloramines. When the SDS test was set up, these values were used to determine dose and initial pH.

At the completion of the 2-d period, the THM, total residual disinfectant and DOX concentrations were measured on all samples. Broamine species are highly reactive, and appear as free chlorine in the standard analytical techniques used to measure free and combined chlorine. (14) (15). THMs were measured to quantify DBPs of current regulatory concern, whereas DOX was measured as an indicator of a broad spectrum of DBPs. In addition, HAA, cyanogen bromide (CNBr), and cyanogen chloride (CNCl) concentrations were measured on selected samples from the element matrix for each water. Analyses for HAA and cyanide halide (CNX) (made up of CNCl + CNBr) were performed at all three pH values, nominal 2 mg/L disinfectant residual, ambient bromide levels, and both the 3:1 and 7:1Cl_2:N ratios to survey a broad range of conditions. The DBP concentrations after incubation were taken as 2-d simulated distribution system (SDS) values, indicative of concentrations that would be present in a consumer's tap water after 2-d of detention time. Details of the analytic procedures are contained in Reference *16*.

Mixing Experiments. The experiments were conducted in a jar-test apparatus at various known values of G (mean velocity gradient, a measure of mixing intensity). The main experimental variables were the mixing intensity and the relative timing in dosing the chlorine and ammonia solutions. Results from the chemistry experiments were used to select conditions in which mixing intensity may play an important role. Therefore, a smaller number of conditions were evaluated for each water than in the chemistry experiments.

A standard jar-test apparatus allows six samples to be run simultaneously, and the mixing intensity is the same in each sample. One sample was used as a control and was dosed with preformed chloramines; therefore, five chemical conditions could be conveniently investigated in each experiment. Five different combinations of mixing intensities and timing of the dosing solutions were studied to span the spectrum of practical applications. Thus, for each water a matrix of five chemical conditions by five mixing and dosing conditions resulted, plus five controls.

The five mixing and dosing conditions were the same for each of the three waters: (1) low G (60 sec^{-1}) with simultaneous addition of chlorine and ammonia and 1 min of mixing after chemical addition; (2) intermediate G (500 sec^{-1}) with simultaneous addition of chlorine and ammonia and 1 min of mixing after chemical addition; (3) high G (1000 sec^{-1}) with simultaneous addition of chlorine and ammonia and 1 min of mixing after chemical addition; (4) low G (60 sec^{-1}), chlorine addition with a 30-sec delay before ammonia addition and 1 min of mixing after ammonia addition; and (5) intermediate G (500 sec^{-1}), chlorine addition with a 30-sec delay before ammonia addition and 1 min of mixing after ammonia addition.

The experiments were conducted in reactors that were calibrated for mixing intensity (i.e., G) as a function of the stirrer paddle speed. The mixing pattern was established in each beaker before the dosing solutions were applied. Mixing continued for 1 min after the completion of dosing. Samples were then transferred rapidly to amber bottles and held for 2 d at 22°C. Because the time was short prior to placing samples in head-space free bottles, little if any loss in DBPs was expected to occur. As with the chemistry experiments, the THM, residual disinfectant and DOX concentrations were measured for all samples at the end of the 2-d incubation period. HAA and CNX concentrations were measured on selected samples.

The five chemical conditions were selected for each water based on the results of the batch chemistry experiments and are presented in Table 1. The first three conditions for LAW were selected because significant DBP formation was expected. Minimal DBP formation was expected for the fourth and fifth conditions, unless inadequate mixing or delayed dosing of ammonia promoted DBP formation. For LHW, the first condition was selected to obtain the maximum DBP production with bromide addition. The second condition was selected as a control on bromide addition and as a typical operating condition for this type of water. The third condition was selected because it seemed to produce the most DBPs in the batch experiments. The fourth condition was selected to simulate a typical operating condition for this type of water, whereas the fifth condition was selected to produce minimal DBP formation based on chemical conditions alone. For CSPW, the first condition was selected to obtain the maximum DBP production with bromide addition at a realistic Cl_2:N ratio. The second condition was selected as a control on bromide addition. The third and fourth conditions were selected to simulate the full range of conditions for pH 8. The fifth condition was selected to provide an indication of performance at pH 10.

Table 1. Chemical Conditions for the Mixing Experiments*

Parameter	LAW Test No.					LHW Test No.					CSPW Test No.				
	1	2	3	4	5	1	2	3	4	5	1	2	3	4	5
pH	6	8	10	8	10	6	6	8	8	10	6	6	8	8	10
Cl_2:N ratio	7:1	5:1	5:1	3:1	3:1	3:1	3:1	7:1	3:1	5:1	3:1	3:1	7:1	3:1	3:1
Bromide	A**	A	A	A	A	+***	A	A	A	A	+	A	A	A	A

*2-d residual = 2 mg/L.
**A = ambient bromide concentration.
***+ = +0.5 mg/L bromide.

Source Water Quality. As noted above, one "batch" of LAW and CSPW was used for all of these experiments. Because dosing with the proper amount of chlorine to achieve the target total residual proved to be quite difficult for LHW, the first batch was exhausted before this task was completed, so a second batch of LHW was collected to complete the study. The selected water quality parameters for the three water sources are listed in Table 2. LAW has an average TOC concentration, a high alkalinity, and a high bromide concentration. LHW has high TOC and moderate bromide concentrations and a low alkalinity. CSPW had low TOC and moderate bromide concentrations and a moderate alkalinity.

Table 2. Selected Water Quality Parameters for the Three Water Sources

Parameter	LAW	LHW	LHW	CSPW
Date of collection	9/17/93	10/28/93	2/22/94	12/9/93
TOC (mg/L)	3.1	9.2	6.7	2.4
Bromide (mg/L)	0.24	0.08	0.075	0.103
pH	8.1	7.4	NA*	7.6
Turbidity (NTU)	0.53	52	56	0.5
Alkalinity (mg CaCO3/L)	156	25	47	73
Background DOX (μg Cl⁻/L)	6.8	17.5	35.7	2.5
Cl_2 demand$_4$ (mg/L)**	4.1	14.9	12.3	5.3
DOXFP$_4$, pH 6 (μg Cl⁻/L)**	840	Invalid data	2513	367
DOXFP$_4$, pH 8 (μg Cl⁻/L)**	648	Invalid data	2125	376
DOXFP$_4$, pH 10 (μg Cl⁻/L)**	524	Invalid data	1158	399

*NA = Not available.

**Subscripts indicate elapsed time in days until measurement was made; 4-d tests all performed with free chlorine.

DOX Analysis.

Methodology. DOX is a group parameter that measures "all" of the halogen-substituted organic compounds in a sample. DOX was analyzed using a Mitsubishi Chemical Industries Total Organic Halogen Analyzer Model TOX-10 (currently distributed by Cosa Instruments). The concentration of DOX was determined by Standard Method No. 5320B (*17*).

The general procedure involves adsorption of the water sample acidified to pH 2 onto activated carbon, washing with a nitrate solution to desorb any inorganic halide ions adsorbed, pyrolysis of the adsorbed sample, and analysis for DOX content using a

microcoulometric titration. The detection limit was 10 µg Cl⁻/L. In this study, each
DOX sample was analyzed twice. If they deviated from one another by 10 µg Cl⁻/L, a
third DOX determination was performed on a duplicate sample bottle. The DOX
concentrations reported are the average of replicate samples.

Before conducting the DOX experiments, control studies were performed to
determine the recovery of the 12 measured DBPs in the DOX analysis using standard
procedures. The results are presented in Table 3. Quantitative recoveries of most of
the DBPs were found, thus justifying the use of DOX analysis as an indicator of a
broad spectrum of DBPs. Low recoveries could be caused by poor adsorption on the
activated carbon or loss during the nitrate-wash step or both.

Table 3. Recovery of 12 DBPs by DOX Procedure

DBP	Recovery (%)	Std. Dev. (%)	(n)
Chloroform	86.3	1.1	2
Bromodichloromethane	92.5	0.2	2
Dibromochloromethane	94.0	3.9	2
Bromoform	97.9	4.0	2
Monochloroacetic acid	9.2	4.2	4
Dichloroacetic acid	33.6	3.2	2
Trichloroacetic acid	97.1	2.6	3
Monobromoacetic acid	73.2	9.6	3
Dibromoacetic acid	90.8	10.7	4
Bromochloroacetic acid	87.0	4.8	3
Cyanogen bromide	102.0	18.2	2
Cyanogen chloride	9.1	0.5	2
2,4,6-Trichlorophenol*	87.4	**	2

*Calibration standard; 10 mg Cl⁻/L concentration evaluated to estimate the
method detection limit.
**The two replicates yielded 83.4 and 91.3 percent recoveries.

Quality Control/Quality Assurance. Three different sets of tests were conducted
to evaluate the performance of the DOX determination. One consisted of eight
replicates of a City of Houston tap water sample, which gave a measure of the precision
of the test. The average DOX value was 156.1 mg Cl⁻/L, with a standard deviation of
9.9 mg Cl⁻/L and a relative standard deviation of 6.3 percent. Secondly, to ensure that
the equipment was performing properly each day, a standard of 49.9 µg Cl⁻/L of 2,4,6
trichlorophenol in deionized water was analyzed. This indicated whether or not the
analysis was "in control" that day. The average recovery was high, 95.6 percent,
with a standard deviation of 5.2 percent. Thirdly, sample activated carbon blanks were

tested frequently. The blanks—in which the same amount of carbon as used in the test was pyrolyzed—contained 0.41 mg Cl⁻ on average, with a standard deviation of 0.08 mg Cl⁻.

Sample Preservation. Concentrated nitric acid (0.50 mL) was used as a preservative and 0.50 mL of a 50-g/L solution of sodium sulfite was used as a dechlorination agent. Samples were stored at 4°C, typically for up to 21 days.

Results and Discussion

Chemistry Experiments.

Effects of Total Residual on DOX Concentration. The concentration of DOX after 2 d of incubation at various pH levels, Cl_2:N ratios, and bromide concentrations shows that, within the range of total chloramine residuals used in these batch studies (1 to 4 mg/L), chloramine dosage had a minor influence on the DOX concentration (Figure 1 shows an example in LHW). Thus, only data collected at a nominal total residual concentration of 2 mg/L will be discussed in this paper.

Effects of pH, Cl_2:N Ratio, and Bromide Concentration on Chloramine Residual Species. Table 4 presents the dichloramine residual (as a percentage of total residual) as a function of Cl_2:N ratio and bromide concentration at pH 6 after 48 h for the three source waters. The production of dichloramine was favored as the Cl_2:N ratio increased. At pH 6, dichloramine represented a significant fraction of the total chloramine residual (21-94 percent), whereas at pH 8 and 10 very little dichloramine was formed (≤4 percent of the total residual) or it dissipated before the 48-h measurement. The alkaline pH favored monochloramine formation. The effects of the bromide ion on residual species varied case by case. For LAW and CSPW, when the water was spiked with bromide, the percentage of dichloramine decreased at all Cl_2:N ratios. This may be caused by chlorine consumption during bromide oxidation in preference to dichloramine production from monochloramine. This also implies that more bromine substitution of organic matter took place under these conditions. In addition, bromamines may have formed, which are not as stable as chloramines. LHW, however, showed comparable dichloramine percentages in response to bromide

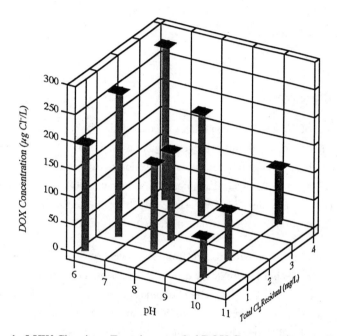

Figure 1. LHW Chemistry Experiments: 2-d DOX Concentration as a Function of Total Cl$_2$ Residual and pH at a Cl$_2$:N Ratio of 5:1 at Ambient Bromide Concentration.

addition. This suggests that less formation of free and combined bromine may have occurred in this water.

Table 4. Dichloramine Residual (as a Percentage of Total Residual) as a Function of Cl_2:N Ratio and Bromide Concentration at pH 6*

Source Water		Cl_2:N Ratio	
	3:1	5:1	7:1
LAW, ambient Br⁻	32%	51%	68%
LAW, +0.5 mg/L Br⁻	21%	25%	44%
LHW, ambient Br⁻	51%	58%	94%
LHW, +0.5 mg/L Br⁻	40%	64%	92%
CSPW, ambient Br⁻	26%	57%	73%
CSPW, +0.5 mg/L Br⁻	23%	31%	52%

*Incubation for 2 d with a total chlorine residual of 2 mg/L.

Effects of pH, Cl_2:N Ratio, and Bromide Concentration on DOX Concentration. Figures 2-4 are plots of the 2-d DOX concentration data collected at a nominal total residual concentration of 2 mg/L in three source waters under varying conditions of pH, Cl_2:N ratio, and bromide concentration. Generally, significant concentrations of DOX were observed with chloramines under many different exposure conditions.

One of the most important treatment variables affecting DOX concentrations during chloramination is pH. The chemistry experiment results have demonstrated that variations of pH between 6 and 10 can significantly alter the formation of DOX concentrations resulting from chloramination. Most of the DOX data follow the general trend of decreasing DOX formation with increasing pH (Figures 2-4). In part, this may result from the effect of pH on chloramine speciation between dichloramine and monochloramine. Limited data show that dichloramine gives a much greater production of DOX than monochloramine (*18*). Exceptions to the trend were noted in some instances at pH 8, for example in LAW at a Cl_2:N ratio of 5:1 (Figure 2). This may have been caused by analytical error or because of some complex bromide-chloramine chemistry. Usually, DOX formation was highest during chloramination at pH 6.

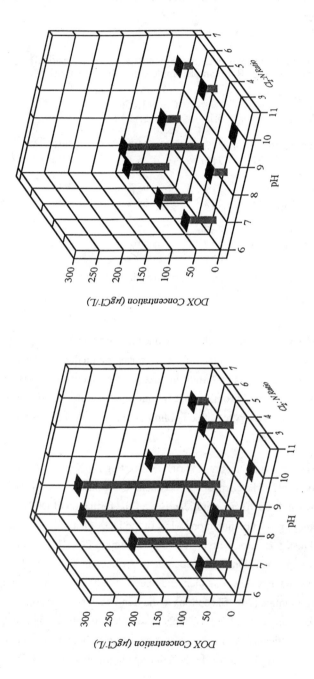

0.74 mg/L Bromide   0.24 mg/L Bromide

Figure 2. LAW Chemistry Experiments: 2-d DOX Concentration as a Function
of Cl$_2$:N Ratio and pH at a Nominal Total Chlorine Residual of 2 mg/L.

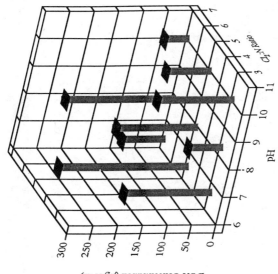

0.58 mg/L Bromide

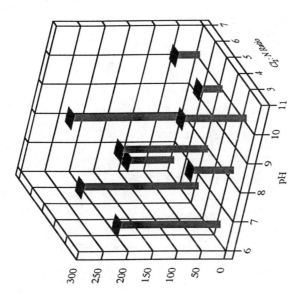

0.08 mg/L Bromide

Figure 3. LHW Chemistry Experiments: 2-d DOX Concentration as a Function of Cl_2:N Ratio and pH at a Nominal Total Chlorine Residual of 2 mg/L.

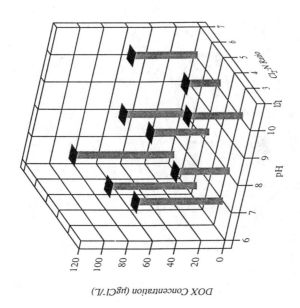

0.10 mg/L Bromide

0.60 mg/L Bromide

Figure 4. CSPW Chemistry Experiments: 2-d DOX Concentration as a Function of Cl$_2$:N Ratio and pH at a Nominal Total Chlorine Residual of 2 mg/L.

However, a significant concentration of DOX was formed at pH 8 and 10, showing that monochloramine can also react with humic substances and other natural organic matter to form halogen-substituted DBPs (measured by DOX).

The Cl_2:N ratio also influenced the concentration of DOX formed in the three waters (Figures 2-4). The DOX concentration typically increased as the Cl_2:N ratio increased toward the breakpoint ratio of 7.6:1 (theoretical). Chloramination at different Cl_2:N ratios placed the water on different parts of the breakpoint curve and involved different chloramine species. At pH 6, the dichloramine concentration (as a percentage of total residual) increased significantly as the Cl_2:N ratio increased (Table 4). At other pH values (pH 8 and 10) where the disinfectant was almost all monochloramine, the DOX concentration also varied. Although not always, generally higher concentrations of DOX were formed at the higher Cl_2:N ratios. At some time during the 2-day incubation period, some dichloramine may have been present. Testing of different Cl_2:N ratios on a water source undergoing chloramination may be warranted to find a ratio that meets disinfection needs while minimizing DBP formation, although the lower ratios may be better from a DOX perspective.

In addition to pH and Cl_2:N ratio, the bromide concentration has been observed to affect the formation of DOX (Figures 2-4). DOX measurement underestimates the effects of the bromide ion on a mass basis because it measures all halogens as chloride, whereas bromide has a higher molecular weight than chloride. Although the DOX analysis cannot differentiate between chlorine- and bromine-substituted organics, the production of DOX increased with the addition of bromide for LAW and CSPW. This is because when the bromide ion is present in the water, hypobromous acid/hypobromite ion ($HOBr/OBr^-$) or bromamines or both are likely to form. $HOBr/OBr^-$ should lead to a greater production of DBPs than that produced by bromamines or chloramines because of the high reactivity of $HOBr/OBr^-$; thus, an increase in bromide concentration should increase the production of DBPs.

Moreover, dibromamine is the least stable bromamine and HOBr is a likely decomposition product. The half-life of dibromamine is 30 min. at pH 8 and the decomposition rate increases as the pH decreases (*19*). Therefore, conditions that select for dibromamine formation may accentuate DOX production because of the reaction with the HOBr generated during decomposition. This is most obvious at pH 6 and possibly to some extent at pH 8. Because bromamine species are short-lived, their contribution should be minimal after the 48 hour incubation time used to simulate DBP formation in water distribution systems. No attempt was made to differentiate between

free chlorine and the various bromamine species that would be measured as free
chlorine. The "free chlorine" residual measured was generally insignificant compared to
the total residual after 48 hours. The DOX concentration for LHW showed little
dependence on the added bromide, which is consistent with the observations of the
effects of the bromide ion on the percentage of dichloramine in the total chlorine
residual.

DOX Accounted for by the 12 Measured DBPs. Among the samples collected
during the batch-scale chemistry experiments, certain samples were selected for
additional DBP analyses beyond the DOX and THM determinations. These selected
samples were also tested for six HAAs (HAA6), CNCl, and CNBr. Thus, in these
samples, after the 2 d of incubation, 12 DBPs were measured as well as DOX.

In an effort to determine the percentage of the DOX that was accounted for by
the measured DBPs that were being formed in these samples, each of the 12 DBPs was
converted to $\mu mol/L$ of DOX (i.e., DBPOX) that it would have contributed to the DOX
measurement. The recoveries noted in Table 3 were used in making this calculation.
These 12 molar "DOX equivalencies" were then summed and compared to the
measured molar DOX concentration in the same sample.

As an example, if a sample contained 119.5 $\mu g/L$ of chloroform and 253 $\mu g/L$
of bromoform each would contain three $\mu mol/L$ of halogen, for a total of six μmol
halogen/L in the sample. Further, if their recoveries in the DOX determination were 100
percent (actual values in Table 3), each THM would produce a response in the DOX
determination of 106.5 μg Cl^-/L for a total DOX of 213 $\mu g/L$ Cl^-/L or six $\mu mol Cl^-/L$. In
this example, the "percentage of DOX accounted for by the 2 measured DBPs" would
be 100. For this study, this type of comparison was reported as [(Σ 12 measured
DBPOX/DOX) X 100], see Table 5.

Table 5. Percentage of DOX Accounted for by the 12 Measured DBPs*

| Source | Cl_2:N = 3:1 | | | Cl_2:N = 7:1 | | |
Water	pH 6	pH 8	pH 10	pH 6	pH 8	pH 10
LAW	19%	14%	NA**	17%	34%	12%
LHW	4.0%	13%	3.2%	18%	10%	33%
CSPW	11%	4.3%	2.9%	11%	13%	20%

*Chemistry studies, 2 mg/L nominal total residual, ambient bromide
  concentration.
**NA = Not available; no DOX detected in this sample.

The most striking finding shown in this table is that the percentage of DOX accounted for was 34 percent or less, with several results less than 5 percent. This shows that during chloramination, a large concentration of "unidentifiable" halogen-substituted DBPs are formed. These data tend to be quite variable, perhaps in part because of analytic error in each of the 12 DBP measurements and in the DOX determination.

Mixing Experiments.

Effects of Mixing on Residual Species. The effects of mixing on disinfectant residual species at pH 6, where significant dichloramine formation occurred, is presented in Table 6. As noted above, a control with preformed chloramines was run at one of the five mixing conditions in each experiment, and these data are presented as well. Taken as a whole, the experiments on the three water sources indicate that mixing conditions do not significantly affect the dichloramine fraction. Rather, the system chemistry is the controlling factor.

Table 6. Dichloramine Residual (as a Percentage of Total Residual)
as a Function of Mixing and Dosing Conditions at pH 6

Source Water	Cl_2:N Ratio	Mixing and Dosing Conditions					
		High	Med.	Low	Med. w/ Delay	Low w/ Delay	Pre-formed
LAW	7:1	59%	78%	78%	67%	69%	67%
LHW	3:1	57%	49%	78%	69%	76%	87%
LHW + 0.5 mg/L Br⁻	3:1	60%	51%	71%	63%	68%	84%
CSPW	3:1	31%	34%	44%	38%	27%	56%
CSPW + 0.5 mg/L Br⁻	3:1	0%	28%	32%	23%	18%	19%

The percentage of dichloramine in LAW at all mixing conditions was higher than or comparable to the percentages found in LHW and CSPW because of the higher Cl_2:N ratio (i.e., 7:1) in the LAW studies. The chemical conditions for LHW and CSPW were similar (pH 6.0, Cl_2:N ratio 3:1), but the percentage of dichloramine in the total residual was higher in LHW than in CSPW. Possibly the very different TOC concentration in these two waters accounted for the difference. For CSPW, the experiment at ambient bromide concentration had a larger percentage of dichloramine at all mixing conditions than the experiment with bromide added, whereas comparable percentages were observed for LHW with or without added bromide. The differences in CSPW are consistent with consumption of chlorine for the production of bromine and bromamines at the expense of dichloramine production.

Effects of Mixing on DOX Concentration. The results of the mixing experiments for 2-d DOX data are presented graphically in Figures 5-7. Each figure presents DOX concentration in a given source water for the five different mixing conditions, including results from the corresponding control experiment with preformed chloramines and the corresponding batch experiment conducted in the chemistry study reported above. Ideally, the data from the preformed chloramine mixing control and batch experiments should be identical. Note that for the graphical representation of these data, the individual datum points are connected with straight lines to aid in assessing trends, even though each datum point was generated from an individual sample. Although the influence of the chemical conditions chosen on the DOX concentration formed is important as previously noted in the batch studies, the evaluation of the influence of mixing on the DOX concentration formed can be assessed by determining whether or not the DOX concentrations change as the mixing scheme changes (i.e., whether the lines in Figures 5-7 are generally horizontal or not).

For LAW (Figure 5), the highest DOX concentrations were typically observed with medium-intensity mixing and delayed ammonia addition. In LAW, except for the unexpected spike for medium mixing with delay, the mixing scheme typically had no significant effect on the formation of DOX. These DOX concentrations, however, were unexpectedly insensitive to system chemistry, which is different from the data shown in Figure 2. This arouses a certain degree of suspicion about the data for this mixing condition. In LHW, the mixing conditions had little effect on the DOX formation (Figure 6.)

Several interesting results are apparent from the CSPW mixing experiments (Figure 7). In CSPW, the high- and medium-intensity mixing conditions produced somewhat lower DOX concentrations, whereas low-intensity mixing (with simultaneous addition of chlorine and ammonia) and both mixing-intensity conditions that were evaluated with delayed ammonia addition produced higher concentrations of DOX. In general, the data for the mixing study showed the same trends observed in

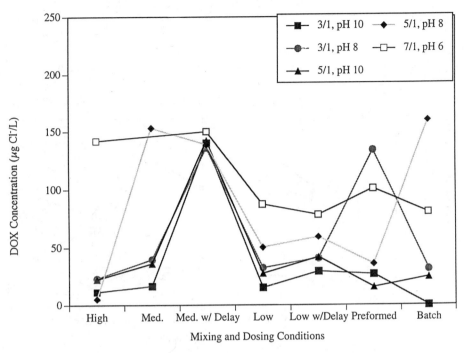

Figure 5. LAW Mixing Experiments: Effects of Mixing on 2-d DOX Formation.

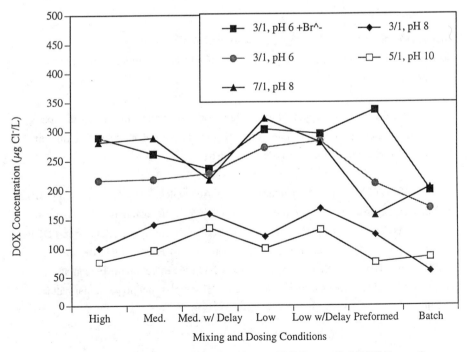

Figure 6. LHW Mixing Experiments: Effects of Mixing on 2-d DOX Formation.

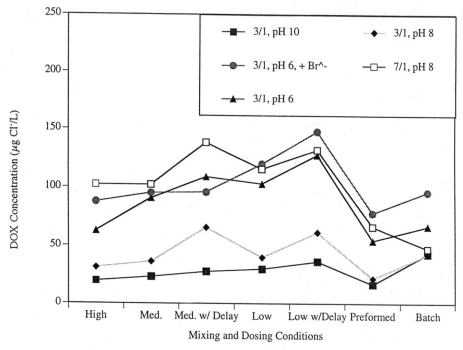

Figure 7. CSPW Mixing Experiments: Effects of Mixing on 2-d DOX Formation.

chemistry experiments of decreasing DOX formation with increasing pH and decreasing Cl_2:N ratios. The expected increase in DOX concentrations when bromide was added is evident (pH 6, Cl_2:N ratio 3:1).

No clear trends are apparent for all three source waters on the effects of mixing conditions on DOX production. This points again to the secondary role of mixing on DOX formation in comparison to system chemistry.

Effects of Mixing on Percentage of DOX Accounted for by 12 Measured DBPs. The mixing experiments data are presented in Table 7. Here again, as in the chemistry experiments (Table 5), the percentage of DOX accounted for by the 12 measured DBPs is relatively low: ≤22 percent for LAW, ≤16 percent for LHW, and ≤31 percent for CSPW. For LHW and CSPW, mixing intensity had little effect on the percentage accounted for. Differences were noted in the LAW samples, but no particular pattern developed.

Table 7. Percentage of DOX Accounted for by the 12 Measured DBPs*

| | | | | | | Mixing and Dosing Conditions | | |
| | Cl$_2$:N | | | | | Med w/ | Low w/ | |
Source Water	Ratio	pH	High	Med.	Low	Delay	Delay	Batch
LAW	7:1	6	NA**	12%	22%	6%	17%	17%
LAW	3:1	8	11%	13%	2%	2%	4%	13%
LAW	3:1	10	14%	21%	0%	2%	3%	0%
LHW + Br⁻ ***	3:1	6	11%	13%	14%	11%	16%	NT #
LHW	3:1	8	6%	5%	10%	NT	NT	13%
LHW	5:1	10	16%	13%	14%	8%	14%	NT
CSPW + Br⁻ ***	3:1	6	21%	30%	20%	31%	26%	NT
CSPW	3:1	10	7%	10%	9%	15%	9%	3%

*Mixing studies.
**NA = Not available; CNX data not available.
***+ Br⁻ = + 0.5 mg/L bromide.
# NT = Not tested.

Conclusions

(1) For the ranges of chemical characteristics studied (pH 6,8,10, total disinfectant residual 1, 2,4 mg/L, Cl$_2$:N mass ratios 3:1, 5:1, 7:1) substantial concentrations of DOX were formed in all three waters after 48 hours of contact with preformed chloramines.

(2) In the range 1 to 4 mg/L of total chlorine residual, 2-d DOX production was not significantly influenced by disinfectant concentration.

(3) Dichloramine was present at significant concentrations only at pH 6 and increased as a percentage of the total residual as the Cl$_2$:N ratio increased from 3:1 to 7:1.

(4) In general, the highest DOX concentrations occurred when higher percentages of dichloramines were present (low pH, high Cl$_2$:N ratio).

(5) In LAW and LHW, where high DOX concentrations were formed after 48 hours of contact with preformed chloramines, less DOX was formed at higher pH (i.e., pH 10).

(6) In general, chloraminating at the highest possible pH and lowest possible Cl_2:N ratio commensurate with the other constraints of treatment will minimize DOX formation .

(7) Mixing techniques had little influence on DOX production, whereas chemical conditions had a significant effect.

(8) Low percentages of DOX (35 percent maximum, mostly <20 percent) could be accounted for by summing the molar concentration of the 12 measured DBPs. Thus, large quantities of "unknown" DBPs are formed during chloramination.

Acknowledgments

The authors thank Mr. Louis A. Simms, staff chemist at the University of Houston, for his assistance with the analytic portion of this study; AWWARF (Mr. Joel Catlin, project manager) for their financial support of the project "Factors Affecting Disinfection By-Products Formation During Chloramination," RFP 803; and Ms. Peggy Kimball of MWDSC for her technical editing of the manuscript. The contributions of the Project Advisory Committee—Mr. Bill Lauer, Dr. Marco Aieta, Ms. Susan Teefy, Mr. Richard Miltner, and Dr. Paul Heffernan—are gratefully acknowledged. Finally, the authors wish to thank the anonymous reviewers whose comments greatly improved the paper.

References

1.   Bellar, T.A.; Lichtenberg, J.J.; Kroner, R.C. Jour. AWWA 1974, 66 (12), 703-706.

2.   Stevens, A.A.; Moore, L.A.; Miltner, R.J. Jour. AWWA 1989, 81 (8), 54-60.

3.   USEPA. Fed. Reg. 1994, 59 (145), 36668-38829.

4.   Kreft, P.; Umphres, M.; Hand, J.; Tate, C.; McGuire, M.J.; Trussell, R.R. Jour. AWWA 1985, 77 (1), 38-45.

5. Stevens, A.A.; Dressman, R.C.; Sorrell, R.K.; Brass, H.J. Jour. AWWA 1985, 77 (4), 146-154.

6. Krasner, S.W.; McGuire, M.J.; Jacangelo, J.G.; Patania, N.L.; Reagan, K.M.; Aieta, E.M. Jour. AWWA 1989, 81 (8), 41-53.

7. Singer, P.C.; Obolensky, A.; Greiner, A. Jour. AWWA 1995, 87 (10) 83-92.

8. AWWARF. Chloro-Organic Water Quality Changes Resulting From Modification of Water Treatment Practices; AWWARF: Denver, Colo., 1986.

9. Lykins, B.W.; Koffskey, W.E.; Miller, R.G. Jour. AWWA 1986, 78 (11), 66-75.

10. Lykins, B.W. In AWWA Seminar Proceedings: Current Research Activities in Support of USEPA's Regulatory Agenda; AWWA: Denver, Colo., 1990; pp 95-111.

11. Jenson, J.N.; St. Aubin, J.J.; Christman, R.F.; Johnson, J.D. In Water Chlorination: Chemistry, Environmental Impact and Health Effects; Jolley, R.L.; Bull, R.J.; Davis, W.P.; Katz, S.; Roberts, M.H.; Jacobs V.A., Eds.; Lewis Publishers: Chelsea, Mich, 1985; Vol. 5, pp 939-949.

12. Fleischacker, S.J.; Randtke, S.J. Jour. AWWA 1983, 75 (3), 132-138.

13. Barrett, S.E. Trihalomethane Concentration at Various Locations on the Breakpoint Curve; Water Quality Laboratory Memorandum; MWDSC: La Verne, Calif., 1985.

14. Palin, A.T., Jour AWWA 1975, 67 (1), 32-33.

15. Gordon, G.; Cooper, W.J.; Rice, R.G.; Pacey, G.E. Disinfectant Residual Measurements, AWWARF:Denver, Colo. , 1987.

16. Symons, J.M.; Speitel, G.E. Jr.; Hwang, C.J.; Krasner, S.W.; Barrett, S.E.; Diehl, A. C.; Xia, R. Factors Affecting Disinfection By-Product Formation During Chloramination; AWWARF, In Press.

17. American Public Health Association (APHA). Standard Methods for the Examination of Water and Wastewater; 18th Edition; APHA, AWWA, and Water Environment Federation: Washington, DC, 1992.

18. Fujioka, R.S.; Tenno, K.M.; Loh, P.C. In Water Chlorination: Chemistry, Environmental Impact and Health Effects; Jolley, R.L.; Brungs, W.A.;

Cotruvo, J.A.; Cumming, R.B.; Mattice, J.S.; Jacobs, V.A., Eds.; Ann Arbor Science Publishers: Ann Arbor, MI, 1983; Vol. 4, Book 2, pp 1067-1076.

19.   Jolley, R.L.; Carpenter, J.H. In Water Chlorination: Chemistry, Environmental Impact and Health Effects; Jolley, R.L.; Brungs, W.A.; Comming, J.S.; Mattice; Jcaobs, V.A., Eds. Ann Arbon Science Publishers, Ann Arbor, Mich., 1983, Vol. 4, Book 1, pp 1-47.

Chapter 7

Application of Product Studies in the Elucidation of Chloramine Reaction Pathways

Peter J. Vikesland, Richard L. Valentine, and Kenan Ozekin[1]

Department of Civil and Environmental Engineering, 122 Engineering Research Facility, University of Iowa, Iowa City, IA 52242

Chloramines are commonly used for drinking water disinfection in systems where it is difficult to maintain a free chlorine residual or where concern over DBP formation exists. They are however, intrinsically unstable and also are lost via an autodecomposition reaction resulting in nitrogen oxidation. This work discusses the measurement of chlorine and nitrogen containing species as a tool in understanding chloramine decay mechanisms and reaction pathways. In order to account for the formation of nitrogen gas by monochloramine decay, a $^{15}N_2$ isotope technique was developed and is discussed fully.

Chloramines have long been used to provide a disinfecting residual in distribution systems in which it is difficult to maintain a free chlorine residual. This is because chloramines are generally less reactive than free chlorine with constituents such as dissolved organic matter (DOM). The EPA has suggested the use of chloramines to replace free chlorine as a disinfectant because they are believed to produce fewer trihalomethane (THM) disinfection by-products (DBPs) (1). A number of studies, however, have shown that a variety of halogenated and non-halogenated DBPs are produced in chloraminated water containing DOM (2-7).

While chloramines are generally believed to be less reactive than free chlorine, they are, however, inherently unstable even in the absence of organic matter. This is because a complex set of reactions occur, ultimately resulting in the oxidation of ammonia and reduction of the active chlorine. The rate of the reactions depends on the ratio of chlorine to ammonia nitrogen (Cl:N ratio) as well as on pH. In general, the greater the ratio of chlorine to nitrogen, the faster the oxidation of ammonia occurs. However, at dose ratios where monochloramine is the dominant chloramine in solution, the redox reactions take hours to days. Recent work indicates that ammonia nitrogen may be oxidized to N_2, NO_3^- , and at least one unidentified product (8). During these reactions, chlorine goes from the +1 valence state in monochloramine to -1 in chloride. The net reaction is also expected to increase the free ammonia concentration by an amount which depends on the specific products formed, as is shown for the formation of nitrate and nitrogen gas:

[1]Current address: 5739 South Andes Street, Aurora, CO 80015

$$4 \text{ NH}_2\text{Cl} + 3 \text{ H}_2\text{O} \rightarrow 4 \text{ Cl}^- + 3 \text{ NH}_3 + \text{NO}_3^- + 5 \text{ H}^+ \qquad (1)$$

$$3 \text{ NH}_2\text{Cl} \rightarrow \text{N}_2 + \text{NH}_3 + 3 \text{ Cl}^- + 3 \text{ H}^+ \qquad (2)$$

Loss of chloramines in a distribution system is therefore attributable to both the normally occurring redox reactions involving ammonia production (auto-oxidation which is presumably the *good* pathway), as well as to reactions which involve organic and inorganic species, some of which might lead to DBP formation (possibly *bad* pathways). The relative importance of the two general decomposition pathways, i.e ammonia oxidation vs. oxidation/substitution reactions involving organic and other matter, is important because it relates chloramine loss to the potential formation of DBPs. It is the two concerns, DBP formation and disinfection, which require that chloramine dosages be small enough to minimize DBP formation and yet still provide an adequate disinfecting residual throughout the system.

Product studies incorporating all known inorganic products may be a very useful tool in understanding the fate of monochloramine in distribution systems, especially in differentiating between auto-decomposition and DOM reaction pathways. This provides an understanding of the limits on the formation of halogenated and non-halogenated DBP formation and monochloramine loss mechanisms which govern the stability of disinfectant residuals. To date, work has not been done to simultaneously measure all known inorganic species which include chloride, nitrate, nitrogen gas, and ammonia. This paper discusses our approach to making detailed analyses of all major products and presents results at two different reaction conditions. This was accomplished by utilizing nitrogen isotope techniques for the measurement of nitrogen gas, by using ion chromatography for the measurement of chloride, nitrate and total ammonia, and by quantifying the monochloramine concentration with the DPD-FAS titrimetric method.

Experimental

Materials and Methods. The water used for all experiments and for the cleaning of glassware was at a minimum deionized using a Barnstead ULTRO pure water system. For the mass balance experiments which required highly purified water, a Barnstead OrganicPure system was employed. All chemicals used in these experiments were analytical laboratory grade or better.

The DPD-FAS method was employed for the measurement of free chlorine and chloramine residuals (9). A Dionex 2000I IC, equipped with an AS4A anion separatory column and an AS4G guard column was used for the chloride and nitrate measurements. Ammonia concentrations were measured using a Dionex 4000I IC equipped with an AS10 cation separatory column and an AG10 guard column. To eliminate any possible interactions which might occur between the cation resin and monochloramine, all of the samples were reduced with sodium sulfite prior to injection. The pH was measured with a Fisher Model 420 meter after appropriate calibration.

In order to obtain accurate chloride measurements it is essential that a low background chloride concentration be obtained in the free chlorine stock solution. To this end, the sodium hypochlorite solution used for the preparation of the monochloramine samples was produced using a modification of a method developed by Reinhard and Stumm (10). This method involves the dissolution of chlorine gas in an aqueous suspension of mercuric oxide (HgO). After approximately one hour, the solution was distilled using a Büchler rotary evaporator. The resultant solution had a HOCl concentration of approximately 8000 mg/L, and the chloride concentration in the sample was subsequently determined to be 500 mg/L. This corresponds to a 7.89:1 Cl_2/Cl^- molar ratio. For comparison the Cl_2/Cl^- molar ratio in a commercially available HOCl solution from Fisher is approximately 1:2.

To facilitate the mass balance experiments detailed herein it was also necessary to eliminate any chlorine demand which might exist between compounds adsorbed onto the glassware and the free and/or combined chlorine found in solution. This was done by placing all glassware into a concentrated chlorine bath (~5,000 mg/L Cl_2) for a period of at least 24 hours. Adsorbed compounds react with the chlorine and exert any chlorine demand which they may have. After the glassware was allowed to soak, it was thoroughly rinsed with chlorine free OrganicPure water and then allowed to dry.

The experimental chloramine solutions were prepared with the following conditions: pH 6.5 and 7.5, 4 mM $NaHCO_3$, incubation at 25 °C in darkness, and $[NH_2Cl]_o$ = 0.25 mM (17.75 mg/L). These solutions were prepared using a preformed monochloramine stock solution, using $(^{15}NH_4)_2SO_4$ for the production of the ammonium stock instead of unlabeled ammonium sulfate (*11*). Aliquots of 100 mL were then placed into 120 mL septa capped vials (Supelco #3-3111), leaving a 20 mL headspace. These vials were capped using Teflon coated silicone septa (Supelco #2-7236) and were then placed upside down in a 25 °C incubator. In order to ensure that mass transfer effects were minimized, the vials were periodically shaken.

At the beginning and at each point during the experimental trial where the system was analyzed, the following measurements were taken: Cl^-, NO_3^-, total ammonia, the total oxidant residual, and the N_2 gas concentration in the headspace.

GC/MS Analysis for $^{15}N_2$. The GC/MS system employed for these experiments was composed of a Hewlett Packard 5890 gas chromatograph containing a J&W Scientific DB-1 column (0.32 mm x 30 m). This GC was used as the inlet to the mass spectrometer and was operated isothermally at 50 °C throughout the experiments. A VG TRIO-1 single quadrapole mass spectrometer with electron impact (EI) ionization was used for the mass spectra measurements. The standard operating parameters for the GC/MS, are given in Table I.

The system was calibrated using $^{15}N_2$ standards which were produced by removing aliquots of 98% $^{15}N_2$ (Cambridge Isotope Labs, Lot ED-314) from a two liter gas cylinder and then transferring them to Mininert capped vials of known total volume using gas tight syringes. The prepared nitrogen gas standards were then allowed to equilibrate for one hour prior to analyzing them by injecting a 100 µL aliquot of gas sample into the GC/MS. Each sample was measured in at least quadruplicate in order to minimize the effect of sampling and analysis errors. By monitoring the peak area of the mass spectral peak at 30 m/z for a number of different concentrations it was possible to construct a calibration curve for the instrument (Figure 1). This calibration curve is given in terms of nmol $^{15}N_2$ injected in order to make it applicable to other injection volumes. Using the calibration curve it is then possible to calculate the mass (nmol) of $^{15}N_2$ injected for each of the monochloramine samples. This value was then converted into an equivalent water concentration by taking into account the injection volume and the partitioning between the solution and the headspace.

$$^{15}N_2 \text{ Concentration (mM)} = \frac{M(\mu mol)}{V_{inj}(\mu L)} \times \frac{V_{hs}}{V_w} \times \frac{1}{f_{hs}} \qquad (3)$$

Where M = mass $^{15}N_2$ injected (µmol), V_{inj} = injection volume, V_{hs} = volume of reactor headspace, V_w = monochloramine solution volume, and f_{hs} = the fraction of nitrogen which partitions into the headpace. In order to calculate f_{hs} it is necessary to

Table I.   Operation parameters for the VG TRIO-1 GC/MS.

Parameter	Set Point	Parameter	Set Point
Ion Repeller	4.0 V	Low Mass Resolution	20.0
Electron Current	150 mA	High Mass Resolution	12.5
Electron Energy	70 eV	Multiplier	700 V
Focus 1	15.0 V	Focus 2	8.0 V
Focus 3	120.0 V	Focus 4	25.0 V
Ion Energy	4.0 V	Ion Energy Ramp	3.0 mV/amu

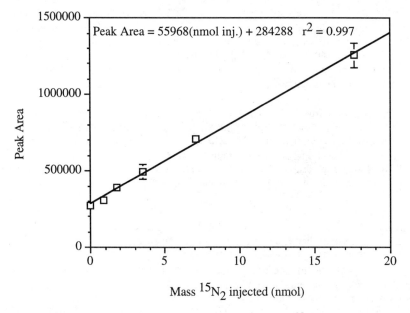

Figure 1 Example calibration curve for the measurement of $^{15}N_2$ via GC/MS. 100 mL injection volume.

utilize the headspace volume, the Henry's constant for $^{15}N_2$ at the temperature of interest, and the ideal gas law (11).

$$f_{hs} = \left(1 + \frac{V_w}{H V_{hs}}\right)^{-1}$$ (4)

Where V_{hs} = volume of headspace, V_w = volume of aqueous solution, and H= dimensionless Henry's constant (60 @ 25 °C).

Reactor Integrity. Because the sample protocol used for the measurement of nitrogen gas developed in this paper involves the use of septa capped serum vials it is vital that the reliability of the seals and the GC/MS system be determined. Without this information it is impossible to say with any validity that the measured $^{15}N_2$ concentrations are correct. In order to make sure that the sample vials did not have any systematic leaks, the following procedure was used: Using the same 120 mL sample vials used for the monochloramine decay experiments, a set of standards containing 100 mL of water with a headspace of 20 mL was produced. These vials were then sealed using the same silicone-Teflon septa and crimptop seals as were used for the decay experiments. A sample aliquot (varying from 0.5 to 6.0 mL) of $^{15}N_2$ gas was taken from the $^{15}N_2$ gas cylinder and then transferred to each vial using a gas-tight syringe. Once the gas had been added to each vial, the hole which had been punctured in the septa was sealed by applying a thin coat of silicone sealant to each septa. The silicone was used to cover the hole because it was desired that the control experiments be conducted in vials that approximated as closely as possible the experimental reactor vials. Since it was not possible to add $^{15}N_2$ gas to the control vials without puncturing the septa, it was necessary to add the external sealant in order to approximate the virgin septa used in each of the decay experiments.

Once the control experiment vials had been sealed, they were treated in the same manner as the monochloramine decay experiments. The controls were stored in this manner for thirteen days after which they were removed from the incubator and analyzed for their $^{15}N_2$ content using GC/MS.

On the same day that the stored controls were analyzed, a second set of controls was produced. These controls were produced using the exact same procedure and concentrations as had been used to produce the original set of controls. This second set of controls was then analyzed at the same time as the original set. By comparing the measured $^{15}N_2$ concentrations for the two sets of controls it was possible to ascertain if any leakage of $^{15}N_2$ gas occurred. A 'two-sample' t-test was utilized to determine if the measured $^{15}N_2$ concentrations in the two sets of vials were statistically indistinguishable.

For a situation like this one, where the absolute variance (σ) of the concentration measurements is unknown it is appropriate to assume that the variances of the two data sets are equal (12). By making this assumption, it can be shown that if the means (μ) of the two data sets are equal (i.e. null hypothesis (H_o): $\mu_1 = \mu_2$) then the following statistic has a $t_{n1+n2-2}$ distribution:

$$\frac{\overline{Y}_1 - \overline{Y}_2}{s_p \sqrt{1/n + 1/n_2}} \tag{5}$$

Where $\overline{Y}_1$ and $\overline{Y}_2$ are the sample means determined from two sets of samples, s_p is the pooled variance of the two samples, and n_1 and n_2 correspond to the number of individual trials in each experiment. The t-statistic calculated in this manner can then be compared to the t-statistics tabulated in any statistics handbook.

The calculated average and standard deviations for ten measurements of each experimental vial are tabulated in Table II along with the corresponding statistical information. Because the t-statistic obtained for the two data sets, -0.583, is less than the critical 2-sided t-statistic, -2.1, the two data sets are statistically indistinguishable at the 95% confidence level. This suggests that it can be said with a 95% confidence level that the experimental reactor vials do not leak.

Table II.   Two sample t-test statistics for ten measurements of each $^{15}N_2$ standard. (100 μL injection volume)

	Thirteen Day Old Vials	Fresh Vials
Mean	5.89 (nmol)	6.08 (nmol)
Variance	0.26	0.83
Observations	10	10
Pooled Variance	0.54	
Hypothesized Mean Difference	0	
df	18	
t Stat	-0.58	
P(T<=t) two-tail	0.57	
t Critical two-tail	2.10	

Results and Discussion

Chloride production. It is commonly accepted that the major chlorine atom containing decay product of monochloramine decomposition is chloride ion (13-15). The measurement of the chloride produced by the decay reaction is important not only for the formation of a mass balance relationship for chlorine, but also because chloride is assumed to be the only reduced product of monochloramine decay. The production of chloride due to monochloramine decay for experiments conducted at pH 6.5 and 7.5 was observed (Figure 2). In these experiments, chloride was measured in two different fashions. First, unreduced monochloramine samples were directly injected into the ion chromatograph. The chloride measured in these injections is labeled Cl^-_{unred}. These chloride values correspond to any background chloride initially present in the monochloramine solution plus any chloride produced by monochloramine decay. Second, the monochloramine samples were reduced by adding a stoichiometric amount of sodium sulfite (Na_2SO_3) to the solution and then injecting it into the ion chromatograph. The chloride concentration measured in the reduced monochloramine samples is labeled Cl^-_{red}. The chloride measured in the reduced samples corresponds to the chloride measured by the unreduced injection (Cl^-_{unred}), in addition to the chloride produced by the reduction of monochloramine and any other chlorine containing species which react with the sulfite reagent.

From Figure 2 it is apparent that the measured unreduced chloride concentration (Cl^-_{unred}) in the reactor vials increases with time and that the total chloride concentration (Cl^-_{red}) in the vials is relatively constant over the course of the experiment. This indicates that when the samples are reduced, all of the chloride in solution is recovered. This is an important observation because it implies that by reducing the samples, it is possible to obtain closure on the chloride mass balance (i.e. $Cl^-_{initial} = Cl^-_{final}$).

It is possible to measure the amount of chloride produced by the decay reaction by subtracting the unreduced chloride measured on the initial day of each experiment $(Cl^-_{unred})_0$ from the chloride value measured on any subsequent day $(Cl^-_{unred})_t$. This relationship may be described by:

$$(Cl^-_{unred})_t - (Cl^-_{unred})_0 = \Delta Cl^-_{unred} \tag{6}$$

The ΔCl^-_{unred} value obtained may then be compared with the measured monochloramine decay to ascertain how well the observed production of chloride correlates to the decay of monochloramine. This calculation was made for the experiments described above and is tabulated in Table III. Based on the calculated percent differences it is apparent that the chloride production corresponds quite well to monochloramine decay, with less than a 10 % difference for each sample.

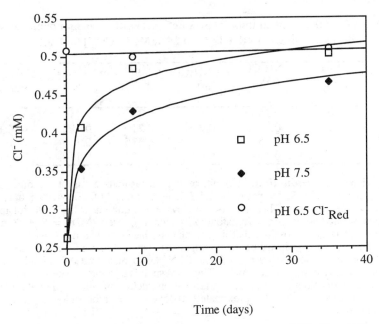

Time (days)

Figure 2 Chloride production due to monochloramine decay. $[NH_2Cl]_o = 0.25$ mM, $[HCO_3^-] = 4$ mM, pH 6.5 and 7.5

Table III. A comparison between observed chloride production and monochloramine decay. $[NH_2Cl]_o = 0.25$ mM, $[HCO_3^-] = 4$ mM.

pH	Time (day)	ΔNH_2Cl (mM)	Cl^-_{unred} (mM)	ΔCl^-_{unred} (mM)	$\Delta NH_2Cl - \Delta Cl^-_{unred}$ (mM)	% Difference
6.5	0	0	0.263	0	0	---
	9	0.236	0.484	0.221	0.015	6.36 %
	35	0.250	0.502	0.239	0.011	4.40 %
7.5	0	0	0.263	0	0	---
	9	0.092	0.353	0.090	0.002	2.17 %
	35	0.222	0.465	0.202	0.020	9.01 %

Nitrogen Products

Ammonia production. Because the samples were reduced, it was necessary to subtract the monochloramine concentration from the measured N(III) concentration in order to account for the reduced monochloramine.

$$NH_{3,free} = \text{Measured } N(III) - NH_2Cl \qquad (7)$$

The $NH_{3,free}$ values which result from this manipulation are tabulated in Table IV along with the corresponding monochloramine decay values. It is apparent that the ammonia concentration increases with monochloramine decay. Unfortunately, due to a malfunction with the ion chromatograph no ammonia measurements were obtained at the 35 day sample time.

Table IV. Observed ammonia production relative to monochloramine decay values. $[NH_2Cl]_0 = 0.25$ mM, $[HCO_3^-] = 4$ mM.

pH	Time (day)	NH_2Cl (mM)	ΔNH_2Cl (mM)	$N(III)$ (mM)	$NH_{3,free}$ (mM)	$\Delta NH_{3,free}$ (mM)
6.5	0	0.256	0	0.375	0.119	0
	9	0.020	0.236	0.210	0.190	0.071
7.5	0	0.256	0	0.375	0.119	0
	9	0.164	0.092	0.314	0.150	0.031

Nitrate Production. No nitrate was measured at any time during the experiment for pH 6.5. However, at pH 7.5, approximately 0.001 mM nitrate was measured after 35 days. This value is quite low and is quite near the MDL for nitrate analysis on the Dionex ion chromatograph which was used (MDL = 0.0008 mM). Therefore it is quite possible that any nitrate produced at pH 6.5 was below the detection limit for the instrument. Leung (16), using an initial monochloramine concentration of 1.51 mM (at pH 7.5 and Cl/N = 0.1) measured 0.0222 mM nitrate for a 98% decay in monochloramine. This corresponds to approximately 1.5% of the nitrogen in monochloramine going toward the production of nitrate. For the pH 7.5 experiment described here, approximately 0.45% of the monochloramine nitrogen ended up as nitrate.

Nitrogen Gas Production. Nitrogen gas has been theorized to be a product of monochloramine decomposition since the early decay studies of Chapin (15), but it has never actually been measured. The results presented in Figure 3 give conclusive evidence that nitrogen gas is produced. It is apparent that nitrogen production occurs more rapidly in the pH 6.5 sample than in the pH 7.5 sample. This finding is not surprising due to the increased rate of monochloramine decay at lower pH. What is surprising is the fact that for roughly the same level of monochloramine decay in the two samples on day 35, the observed nitrogen gas concentration is much lower in the pH 7.5 sample. This indicates that the monochloramine decay product distribution is affected by the pH of the solution, and that the difference in the observed decay rate at each pH may be due to the corresponding change in reaction mechanism.

Nitrogen Mass Balance. The measurements for all of the nitrogen containing species were tabulated in order to derive a nitrogen mass balance for monochloramine decay (Table V). This mass balance is symbolized in the following equation where each species is given in mM (as N):

$$N \text{ Balance} = \Delta NH_2Cl - (\Delta NH_{3,free} + \Delta NO_3^- + 2\Delta N_{2,gas}) \qquad (8)$$

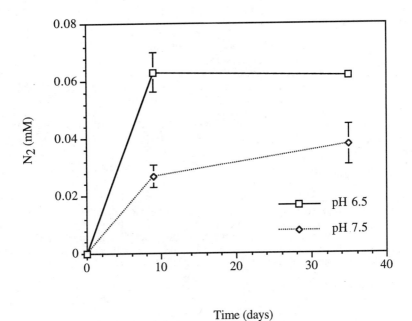

Time (days)

Figure 3 $^{15}N_2$ production due to monochloramine decay. Error bars indicate the pooled standard deviation of five injections from three separate experimental vials. $[NH_2Cl]_o = 0.25$ mM, $[HCO_3^-] = 4$ mM, pH 6.5 and 7.5

By subtracting the sum of the ammonia production, the nitrate production, and the nitrogen gas production (as N) from the measured monochloramine decay it is possible to get a mass balance for nitrogen.

The mass balance information tabulated in Table V indicates that the sum of the measured nitrogen containing product concentrations correlate reasonably well with the measured monochloramine decay. For the pH 6.5 experiment the nitrogen balance is within 16% and for the pH 7.5 experiment the balance is within 7.6% of closure. Based upon these measurements it is apparent that most of the nitrogen released by monochloramine decay may be accounted for.

Table V. Nitrogen mass balance derivation. $[NH_2Cl]_o = 0.25$ mM, $[HCO_3^-] = 4$ mM, pH 6.5 and 7.5. N Balance $= \Delta$ NH$_2$Cl $- (\Delta NH_{3,free} + \Delta NO_3^- + 2\Delta N_{2,gas})$

pH	Time (day)	Δ NH$_2$Cl (mM)	ΔNO_3^- (mM)	$\Delta NH_{3,free}$ (mM)	$\Delta N_{2,gas}$ (mM)	N Balance	% N Unaccounted For
6.5	0	---	---	---	---	---	---
	9	0.236	0	0.071	0.0632	0.039	16.5%
7.5	0	---	---	---	---	---	---
	9	0.092	0	0.031	0.0269	0.007	7.61%

Conclusions

The decay of monochloramine was studied at two different pHs, 6.5 and 7.5 and all known inorganic products were measured. The use of GC/MS to measure isotopic nitrogen gas using headspace analysis in batch reactors consisting of septa closed vials was successfully applied. Chloride, ammonia, and nitrogen gas are all significant products of monochloramine decay. At pH 7.5, nitrate was also shown to be a product of the decay process. Based upon the high percent recoveries (> 80%) of chlorine and nitrogen containing decay products it is apparent that under the conditions studied that most of the initial chlorine and nitrogen in monochloramine may be accounted for.

Acknowledgments

The authors would like to express their appreciation to the AWWA Research Foundation (AWWARF) for financial support. AWWARF assumes no responsibility for the content of the research documented in this article or for the opinions expressed herein. The use of trade names for commercial products does not imply endorsement by AWWARF but is included solely for reference. The authors express their thanks to AWWARF project officer Joel Catlin for his assistance.

Literature Cited

(1)   Cotruro, J.A. *ES&T*, 1981, *14*, 268.
(2)   Fleischacker, S.J. and Randtke, S.J. *J. AWWA*, 1983, *75*, 1323-144.
(3)   Amy, G.L., et al. In *Water Chlorination: Chemistry, Environmental Impact, and Health Effects*; R.L. Jolley, et al. Eds.; Lewis Publishers: Chelsea, MI, 1990, Vol. 6; pp 605-622.
(4)   Jenson, J.N., et al. In *Water Chlorination: Chemistry, Environmental Impact, and Health Effects*; R.L. Jolley, et al. Eds, Lewis Publishers: Chelsea, MI, 1985, Vol. 5; pp. 939-950.
(5)   Arber, A., et al. In *Water Chlorination: Chemistry, Environmental Impact, and Health Effects*; R.L. Jolley, et al. Eds, Lewis Publishers: Chelsea, MI, 1985, Vol. 5; pp. 951-964.
(6)   Kanniganti, R., et al. *ES&T*, 1992, *26*, 1988-2004.
(7)   Stevens, A.A., et al. In *Water Chlorination: Chemistry, Environmental Impact, and Health Effects*; R.L. Jolley, et al. Eds, Lewis Publishers: Chelsea, MI, 1990, Vol. 6; pp. 579-604.
(8)   Valentine, R.L. and Wilber, G.G., In *Water Chlorination: Chemistry, Environmental Impact, and Health Effects*; R.L. Jolley, et al. Eds, Lewis Publishers: Chelsea, MI, 1990, Vol. 6; pp. 819-832.
(9)   APHA, ANWA, and WPCF, *Standard Methods for the Examination of Water and Wastewater*; American Public Health Association, American Water Works Association, and Water Pollution Control Association: Washington, D.C.; 16th Edition
(10) Reinhard, M. and Stumm, W. In *Water Chlorination: Chemistry, Environmental Impact, and Health Effects*; R.L. Jolley, et al. Eds, Lewis Publishers: Chelsea, MI, 1980; Vol. 3; pp. 209-218.
(11) Vikesland, P. *MS Thesis,* The University of Iowa: Iowa City, 1995.
(12) Fisher, L. D. and Van Belle, G. *Biostatistics: A Methodology for the Health Sciences*; John Wiley and Sons, Inc.: New York, 1993.
(13) Diyamandoglu, V. Ph.D. Dissertation; University of California-Berkeley,1989.
(14) Leao, S. F. Ph.D. Dissertation,: University of California, Berkeley, 1981.
(15) Chapin, R. *J. Am. Chem. Soc.* 1931, *53*, 912-921.
(16) Leung, S. Ph.D. Dissertation; The University of Iowa: Iowa City, 1989.

Chapter 8

Modeling the Decomposition of Disinfecting Residuals of Chloramine

Kenan Ozekin[1], Richard L. Valentine, and Peter J. Vikesland

Department of Civil and Environmental Engineering,
122 Engineering Research Facility, University of Iowa,
Iowa City, IA 52242

Chloramines have long been used to provide a disinfecting residual in distribution systems when it is difficult to maintain a free chlorine residual. In spite of a long history of chloramine use, however, the fate of chloramines in distribution systems as well as characteristics and processes which influence its stability are largely unknown. The notion of chloramine stability remains vaguely defined although dosage requirements are frequently determined by the rate and extent of chloramine loss in a distribution system. In this study, chloramine decay was studied using chloramine concentrations from 0.01 mM to 0.1 mM as Cl_2 (0.71 to 7.1 mg/L as Cl_2) , a range important in drinking water treatment and where no previous kinetic investigations have been conducted. Experimental results compared well with the output from a comprehensive reaction model that incorporates 10 reactions. Also a simple coefficient of use in defining the limit of chloramine stability is presented.

Chloramines, produced by a reaction between free chlorine and ammonia in a process called chloramination, have long been used to provide a disinfecting residual in distribution systems when it is difficult to maintain a free chlorine residual. Strictly speaking, chloramines include monochloramine, dichloramine, and trichloramine. However, for disinfection purposes, monochloramine is the desired dominant form preferentially created by careful control of the chlorine to ammonia ratio to a value generally less than 5:1 on a weight basis (about 1:1 on a molar basis). Typical chloramine residuals desired for good protection range from approximately 1.0 to 5.0 mg/L as chlorine. The EPA has promulgated an MCL of 4.0 mg/L as chlorine.

The EPA has suggested the use of chloramines to replace free chlorine as a disinfectant because they are believed to produce fewer trihalomethane (THM) disinfection by-products (DBPs) (1). A number of studies, however, have shown that a variety of DBPs are produced and significant formation of non THM organic halogen containing compounds has been observed (2-6). In comparison to free chlorine, chloramines are generally believed to produce fewer oxidation and more substitution products. However it is expected that, as found for the application of

[1]Current address: 5739 South Andes Street, Aurora, CO 80015

0097–6156/96/0649–0115$15.00/0

free chlorine, chloramination also produces a significant quantity of non halogenated oxidation products, which may be considered a health concern (7-9).

The disappearance of chloramines in water distribution systems depends on many factors, including the presence of constituents that react with chloramines. While chloramines are generally believed to be less reactive than free chlorine, they are however, inherently unstable even in the absence of organic matter. This is because of a complex set of reactions that occur ultimately resulting in the oxidation of ammonia and reduction of the active chlorine. The rate of chloramine loss by this route depends primarily on the ratio of chlorine to ammonia nitrogen as well as on pH. In general, the greater the ratio of chlorine to nitrogen, the faster the oxidation of ammonia. As the weight ratio of chlorine to ammonia approaches approximately 7:1 (about 1.6 on a molar basis) the rates greatly increase by several orders of magnitude. At or above this dose ratio, measurable amounts of free chlorine begin to exist and the reactions become extremely fast with the redox reactions complete in a few minutes. This process has been used to remove ammonia, and is called "breakpoint" chlorination. However, at dose ratios where monochloramine predominates, the redox reactions take hours to days and the net effect of decomposition is to *increase* the free ammonia concentration. This can be shown by writing a net monochloramine decomposition reaction leading to nitrogen gas formation, which is generally believed to be the primary oxidation product

$$3\,NH_2Cl \rightarrow N_2 + NH_3 + 3\,Cl^- + 3\,H^+$$

In spite of a long history of chloramine use, the fate of chloramines in distribution systems as well as factors which influence their stability are largely unknown. Little is known about water quality and treatment practices, and their relationship to chloramine loss or fate. The notion of chloramine stability is vaguely defined. While recent work done in clean systems by Jafvert and Valentine (10) has helped to identify important reactions involving ammonia and has provided a kinetic model capable of predicting decomposition rates in clean laboratory systems, the model has been verified only at unrealistically high chloramine concentrations in phosphate buffered water. The model must be evaluated at conditions that are representative at drinking water treatment.

In this paper we report on 1) modeling the decay of chloramines at initial concentrations important from the utility perspective (less than 4.0 mg/L as Cl_2), and 2) characterization of the effect of Cl/N ratio, pH, and ionic strength on monochloramine decay. Results are also analyzed to show that a simple coefficient can be used to describe chloramine decomposition with time.

Experimental Section

Reagents. All chemicals were reagent grade. All the water used in the decomposition studies was purified using a Barnstead OrganicPure UV system that was attached to a Barnstead NanoPure cartridge system.

Procedures: To eliminate any chlorine demand that might be due to adsorbed components, the glassware was prepared by soaking it in a concentrated chlorine bath for at least 24 hr. Prior to use, the glassware was rinsed with copious amounts of deionized water and then allowed to air dry.

Monochloramine was preformed in all studies. Monochloramine stock solution was prepared by dropwise addition of free chlorine stock solution (Fisher) into a well stirred, buffered solution containing ammonium chloride (5.64 mM).

Bicarbonate was used to buffer the system and the pH of the solution was then adjusted to 8.5 or above to stabilize monochloramine decomposition. A designated amount of free chlorine (200 and 280 mg/L) was added to the solution, and a 30 minute aging period was allowed. The pH was monitored for an initial 5 minutes to assure stability at pH 8.5 or above. The baseline experiments used a Cl/N molar ratios of 0.5 and 0.7. Chloramine decay experiments were performed by spiking aliquots of stock chloramine solution into a batch of the bicarbonate buffered source water. Ionic strength, when adjusted, was controlled by addition of sodium perchlorate. Except for the experiments to measure ionic strength effect on chloramine decomposition, the ionic strength was set at 0.1 M. In those experiments, ionic strength was adjusted from 0.005 M to 0.1 M. Batch reactors were 128 mL amber bottles that were sealed and stored in an incubator set at 25°C in the dark until they were analyzed. The concentration of chloramine was measured using DPD-FAS titrimetric method (*12*).

Model Formulation

Speciation and redox reactions and their corresponding rate expressions were adopted from the chloramine decomposition model proposed by   Jafvert and Valentine (*10*). Only the reactions that apply under chloramination conditions where measurable free chlorine is absent were adopted. Table I shows the chloramine decomposition reactions and appropriate rate constants that was used in this paper to model the monochloramine decomposition. The model includes three principle reaction schemes as well as their pertinent equilibrium reactions: 1) speciation reactions of HOCl with ammonia and the chlorinated derivatives of ammonia, 2) disproportionation reactions of chloramine species, and the corresponding back reactions, and 3) redox reactions that occur in the absence of measurable free chlorine.

The overall rate of chloramine oxidant loss near neutral pH values is primarily limited by the rate of formation of dichloramine. Dichloramine formation occurs by monochloramine hydrolysis pathway (reaction 2 and 3) and by a carbonate catalyzed monochloramine disproportionation reaction (reaction 5). Once dichloramine forms, it rapidly decomposes. In general, the importance of each pathway will depend on several factors such as pH, ionic strength, temperature, and alkalinity.

A computer subroutine (DDRIV2) developed by National Institute of Standards and Technology (NIST) and Los Alamos National Laboratory (LANL) was obtained from SLATEC Public Domain Library. The program uses the method of Gear to solve N ordinary differential equations. It was used to solve the set of five simultaneous differential equations to obtain molar concentrations of NH_3 and NH_4, HOCl and OCl^-, NH_2Cl, $NHCl_2$, and pH. The model considers ionic strength effects on all equilibrium species using the extended DeBye-Huckel law to calculate activity coefficients. The pH may either be fixed or allowed to vary according to the presumed reaction stoichiometry using appropriate buffer intensity relationships. The model utilized constants previously evaluated at 25 °C (Table I).

Results and Discussion

The behavior of monochloramine at concentrations which may actually be encountered in drinking water is important from the utility perspective. Studies previously conducted aimed at understanding the fundamental chemistry of chloramines utilized relatively high concentration of chloramines (> 0.1 mM as Cl_2).

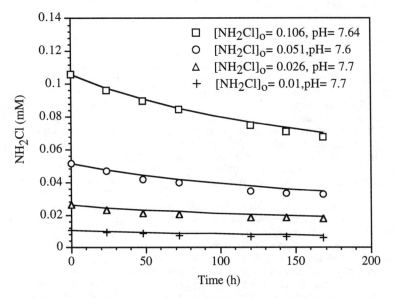

Figure 1. Monochloramine decomposition as a function of initial monochloramine concentration. Symbols are data points and lines are model calculations. (Cl/N= 0.7, C_{T,CO_3} =4 mM, $[NH_3]_T$ = 0.0203 mM, I= 0.1M, T= 25°C)

There is a need to study the kinetics at much lower concentrations (≤ 0.05 mM) since the MCL for chloramines may be set at 4 mg/L (0.056 mM as Cl_2). Not only may rate determining reactions differ at high and low concentrations, but the chloramine "demand" of the water may not be noticeable at high chloramine concentrations. To address these issues, experiments were conducted using initial monochloramine concentrations of : 0.1, 0.05, 0.025, and 0.01 mM (7.1, 3.55, 1.8, and 0.71 mg/L). Figure 1 presents the experimental results obtained over this range in monochloramine at Cl/N = 0.7 and pH 7.7 with lines showing the exact model predictions. A good correspondence between measured and predicted values was obtained over the entire range of chlorine concentration utilized.

Table I Chloramine Decomposition Kinetics and Associated Rate Constants

	Reaction	Rate Constant (25°C)	Ref
(1)	$HOCl + NH_3 \longrightarrow NH_2Cl + H_2O$	$k_1 = 1.5 \times 10^{10}\ M^{-1}h^{-1}$	12
(2)	$NH_2Cl + H_2O \longrightarrow HOCl + NH_3$	$k_2 = 0.1\ h^{-1}$	12
(3)	$HOCl + NH_2Cl \longrightarrow NHCl_2 + H_2O$	$k_3 = 1.26 \times 10^6\ M^{-1}h^{-1}$	13
(4)	$NHCl_2 + H_2O \longrightarrow HOCl + NH_2Cl$	$k_4 = 2.3 \times 10^{-3}\ h^{-1}$	13
(5)	$NH_2Cl + NH_2Cl \longrightarrow NHCl_2 + NH_3$	$k_d{}^*$	14
(6)	$NHCl_2 + NH_3 \longrightarrow NH_2Cl + NH_2Cl$	$k_6 = 2.2 \times 10^8\ M^{-2}h^{-1}$	15
(7)	$NH_2Cl + NHCl_2 \longrightarrow N_2 + 3H^+ + 3Cl^-$	$k_7 = 55.0\ M^{-1}h^{-1}$	16
(8)	$NHCl_2 + H_2O \longrightarrow NOH + 2HCl$	$k_8 = 6.0 \times 10^5\ M^{-1}h^{-1}$	14
(9)	$NOH + NHCl_2 \longrightarrow N_2 + HOCl + HCl$	$k_9 = 1.0 \times 10^8\ M^{-1}h^{-1}$	16
(10)	$NOH + NH_2Cl \longrightarrow N_2 + H_2O + HCl$	$k_{10} = 3.0 \times 10^7\ M^{-1}h^{-1}$	16
(11)	$NH_4^+ \longleftrightarrow NH_3 + H^+$	$pK_a = 9.3$	17
(12)	$H_2CO_3 \longleftrightarrow HCO_3^- + H^+$	$pK_a = 6.3$	17
(13)	$HCO_3 \longleftrightarrow CO_3^{-2} + H^+$	$pK_a = 10.3$	17

$^*k_d = k_H\,[H^+] + k_{H_2CO_3}\,[H_2CO_3] + k_{HCO_3}\,[HCO_3^-]$

$k_H = 2.5 \times 10^7\ M^{-2}h^{-1}$ (18)

$k_{HCO3} = 800\ M^{-2}h^{-1}$ (19)

$k_{H_2CO_3} = 40000\ M^{-2}h^{-1}$ (19)

Ionic strength could affect chloramine decomposition by 1) affecting equilibria involving reacting species and 2) directly affecting the kinetics of elementary reactions involving charged transition states. Experimental and predicted chloramine concentrations for three ionic strength values and pH of approximately 6.5 and 7.5 are shown in Figure 2. Model results compare very well with measured values. More importantly, decay rates are not a strong function of ionic strength as is more clearly shown in Figure 3 in which all the ionic strength results are plotted together.

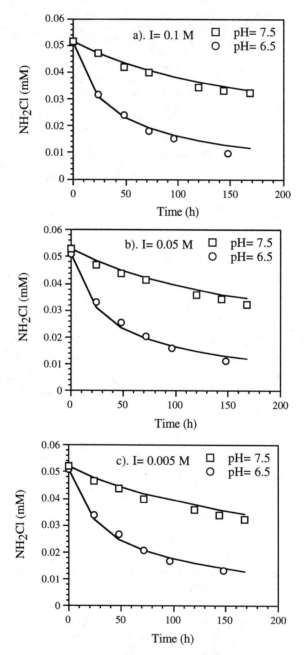

Figure 2. Effect of ionic strength on monochloramine decomposition at pH 6.5 and 7.5. Symbols are data points and lines are model calculations. (Cl/N= 0.7, C_{T,CO_3} =4 mM, $[NH_3]_T$ = 0.0203 mM, T= 25°C)

The Cl/N ratio is another important variable since it affects the free ammonia concentration initially introduced into the system. At a fixed initial monochloramine concentration, the concentration of free ammonia increases as the Cl/N ratios decreases. Figure 4 shows measured and predicted monochloramine concentrations at Cl/N ratios of 0.5 and 0.7 The correspondence of model to measured monochloramine concentrations is good over all reaction conditions including lower pH values where the rate is fastest.

A Simple Coefficient to Define Limit of Chloramine Stability

Chloramine decay can be adequately predicted by the detailed mechanistic model. Its use, however, is complicated and beyond the capabilities of many that would like to know if observed chloramine decay rates are reasonable and should therefore be expected. Excessive rates of decay could also be used to point to problems in the system such as the existence of oxidizable iron or organics. Evaluating the effects of changes in operational parameters such as pH would also be facilitated if a more simple and user friendly approach to describing chloramine loss over time was available.

We explored the use of several simplifying assumptions and found that chloramine decay may, under most reaction conditions, be adequately described by a simple second order relationship (Figure 5) in terms of a single coefficient k_{VSC} (here coined the "Valentine Stability Coefficient") which defines the limit of chloramine in the absence of reactions at the pipe-water interface or with oxidizable substances other than ammonia.

$$\frac{d[NH_2Cl]}{dt} = k_{VSC} [NH_2Cl]^2$$

where $[NH_2Cl]$ = monochloramine concentration at time t
           $[NH_2Cl]_o$ = initial monochloramine concentration
           k_{VSC} = Valentine Stability Coefficient

This relationship can be derived by considering the simplified mechanism involving only five reactions shown in Table I (reactions 1, 2, 3, 5, and 7). In this mechanism, the rate of monochloramine loss is governed by the rate of dichloramine formation occurring by a second order general acid catalyzed pathway (reaction 5) involving two monochloramine molecules, and from reaction of HOCl with monochloramine (reaction 3). Further assumptions are that HOCl is in equilibrium with monochloramine (reactions 1 and 2), and that once formed, dichloramine reacts rapidly with excess monochloramine to produce nitrogen gas (reaction 7). Dichloramine is therefore always at a very low concentration in comparison to monochloramine, and can be considered at pseudo-steady state. This is consistent with observations of monochloramine decay at pH values greater than about 7.0 and initial monochloramine concentrations less than approximately 5 mg/L as Cl_2.

Based upon this analysis we are able rationalize the dependency of k_{VSC} on the known rate and equilibrium constants, and a number of water quality parameters such as pH, alkalinity, ammonia concentration, and temperature.

$$k_{VSC} = 3 \left\{ k_{H^+} [H^+] + \alpha_o\, k_{H2CO3}\, C_{T,CO3} + \alpha_1\, k_{HCO3}\, C_{T,CO3} \right\} + \frac{2\, k_3\, K_e}{\alpha_{o,N}[NH_3]_T}$$

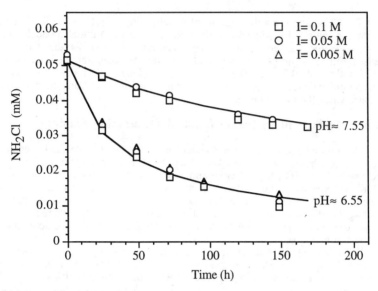

Figure 3. Effect of ionic strength on monochloramine decomposition at three different ionic strength values. Symbols are data points and lines are model calculations at I= 0.1 M. (Cl/N= 0.7, pH= 7.5, C_{T,CO_3}=4 mM, $[NH_3]_T$ = 0.0203 mM, T= 25°C)

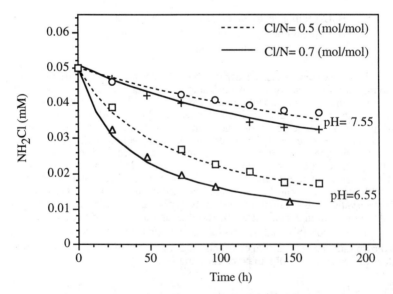

Figure 4. Monochloramine decomposition as a function of pH and Cl/N ratio. Symbols are data points and lines are model calculations (C_{T,CO_3}=4 mM, I= 0.1M, T= 25°C)

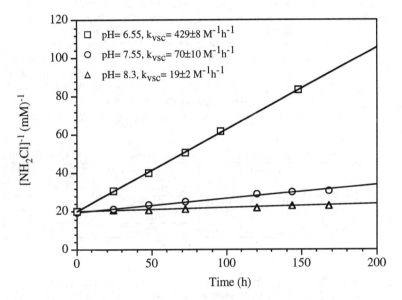

Figure 5. Modeling chloramine decay as second order relaitonship. Errors indicate 95% confidence intervals.(Cl/N= 0.7, C_{T,CO_3} =4 mM, $[NH_3]_T$ = 0.0203 mM, I= 0.1M, T= 25°C)

where, $[NH_3]_T$ = total ammonia concentration in excess of monochloramine , M

$C_{T,CO3}$= total carbonate concentration, M

$$\alpha_o = \frac{[H_2CO_3]}{C_{T,CO_3}} = \text{ionization fraction for } H_2CO_3$$

$$\alpha_1 = \frac{[HCO_3]}{C_{T,CO_3}} = \text{ionization fraction for } HCO_3$$

$$\alpha_{o,N} = \frac{[NH_3]_{free}}{[NH_3]_T} = \text{ionization constant for ammonia}$$

k_{VSC} = Valentine Stability Coefficient

The value of k_{VSC} increases with decreasing pH and initial excess free ammonia in the system. It also increases with increasing total inorganic carbon and temperature. It can be rationalized in terms of measurable parameters and known rate constants and therefore does not need to be measured for each water. Table II compares the measured and predicted k_{vsc} values for the results presented in Figure 5. In general, predicted results are very close to the measured values especially at higher pH values. This suggests that simple k_{vsc} equation can be used to calculate the limit of chloramine stability in the water. Rates that are excessive of k_{vsc} could be used to point to problems in the water source such as the existence of oxidizable iron or organics.

Table II. Summary of calculations of Valentine Stability Coefficients

pH	NH₂Cl (M)	$[NH_3]_T$ (M)	$C_{T,CO3}$ (M)	k_{vsc} - Measured ($M^{-1}h^{-1}$)	k_{vsc} - Predicted ($M^{-1}h^{-1}$)
6.55	5.0E-5	2.03E-5	4.0E-03	429±8	506
7.55	5.0E-5	2.03E-5	4.0E-03	70±10	62
8.3	5.0E-5	2.03E-5	4.0E-03	19±2	19

Conclusions

A model describing chloramine decomposition at concentrations typically found in drinking water was developed. The model incorporates 10 chemical reactions and accounts for a number of water quality parameters. A simple index (Valentine Stability Coefficient or k_{VSC}) has been proposed as a measure of the limit of chloramine stability in drinking water in the absence of other reactive constituents. Our work contributes to establishing a rationale approach to chloramination practices and an understanding of processes leading to chloramine loss in distribution systems.

Acknowledgments

The authors would like to express their appreciation to the AWWA Research Foundation (AWWARF) for financial support. AWWARF assumes no responsibility for the content of the research documented in this article or for the opinions expressed herein. The use of trade names for commercial products does not imply endorsement by AWWARF but is included solely for the readers reference.

The authors express their thanks to AWWARF project officer Joel Catlin for his assistance.

Literature Cited

(1) Cotruvo, J.A. *ES&T*. 1981, 14, 268.
(2) Fleischacker, S. J.; Randtke, S. J. *J. AWWA*. 1983, 75, 132.
(3) Amy, G.L. et al. In *Water Chlorination: Chemistry, Environmental Impact and Health Effects* . Jolley, R.L. et al.,Ed.; Lewis Publishers: Chelsea, Michigan, 1990, Vol 6; pp 605-621.
(4) Jensen, J.N., et al. In *Water Chlorination: Chemistry, Environmental Impact and Health Effects* . Jolley, R.L. et al.,Ed.; Lewis Publishers: Chelsea, Michigan, 1985, Vol 5; pp 939-949.
(5) Arber, R., et al. In *Water Chlorination: Chemistry, Environmental Impact and Health Effects* . Jolley, R.L. et al.,Ed.; Lewis Publishers: Chelsea, Michigan, 1985, Vol 5; pp 951-963.
(6) Kanniganti, R., et al. *ES&T*. 1992, 26, 1988.
(7) Stevens, A.A., et al. In *Water Chlorination: Chemistry, Environmental Impact and Health Effects* . Jolley, R.L. et al.,Ed.; Lewis Publishers: Chelsea, Michigan, 1990, Vol 6; pp 579-604.
(8) Thompson, G.P., et al. In *Water Chlorination: Chemistry, Environmental Impact and Health Effects* . Jolley, R.L. et al.,Ed.; Lewis Publishers: Chelsea, Michigan, 1990, Vol 6; pp 171-178.
(9) Christman, R.F., et al. *Chemical Reactions of Aquatic Humic Materials with Selected Oxidants.* EPA 600/D-83-117. Washington, D.C.: USEPA, 1983.
(10) Jafvert, C. and Valentine, R.L.*ES&T*. 1992, 26, 577.
(11) APHA, Standard Methods for the Examination of Water and Wastewater, 16th Edition., American Public Health Association, 1980.
(12) Morris, J.C., Isaac, R.A. In *Water Chlorination: Chemistry, Environmental Impact and Health Effects* . Jolley, R.L. et al.,Ed.; Lewis Publishers: Chelsea, Michigan, 1983, Vol 4; pp 49-63.
(13) Margerum, D.W., et al. In *Organometals and Organometalloids; Occurrence and Fate in The Environment.* Brinckman, F.E. and Bellama, J.M., Ed., American. Chemical. Society: Wahington, DC, 1978; pp 278-291.
(14) Jafvert, C., Ph.D. Dissertation., The University of Iowa, Iowa City, IA, 1985.
(15) Hand, V.C., and Margerum, D.W. *Inorganic Chemistry.* 1983, 12, 1449.
(16) Leao, S.L., Ph.D. Dissertation., University of California, Berkeley, California, 1981.
(17) Snoeyink, V.L., Jenkins, D. *Water Chemistry.* New York: John Wiley and Sons, 1980.
(18) Granstrom, M.L., Ph.D. Dissertation., Harvard University, Cambridge, MA, 1954..
(19) Valentine et al., *Chloramine Decomposition Kinetics And Degradation Products In Distribution System And Model Waters,* Final Report Submitted to AWWARF, 1996.

Chapter 9

A Comparison of Analytical Techniques for Determining Cyanogen Chloride in Chloraminated Drinking Water

Michael J. Sclimenti, Cordelia J. Hwang, and Stuart W. Krasner

Water Quality Division, Metropolitan Water District of Southern California, 700 Moreno Avenue, La Verne, CA 91750–3399

This study was undertaken to evaluate various analytical techniques for the determination of cyanogen chloride (CNCl) in chloraminated drinking water. CNCl will be included in the Federal Information Collection Rule and is a possible candidate for regulation in Stage 2 of the Disinfectants/Disinfection By-Products Rule. Analytical techniques for the measurement of CNCl include purge-and-trap (P&T)/gas chromatograph (GC)-mass spectrometer (MS) analysis, headspace/GC-electron capture detector (ECD) analysis, and micro-liquid/liquid extraction (micro-LLE) with GC-ECD analysis. Currently, the official U.S. Environmental Protection Agency method for CNCl is P&T/GC-MS analysis, although it is recognized that this method has its limitations. This research has demonstrated that the micro-LLE/GC-ECD, P&T/GC-MS, and headspace/GC-ECD methods were comparable analytical techniques for the determination of CNCl in chloraminated drinking water. Moreover, the micro-LLE/GC-ECD method is applicable for CNCl analyses in various matrix waters and should be usable in more laboratories in which GC-ECD equipment is more common. A cost comparison showed that the micro-LLE/GC-ECD method was the least expensive analytical technique compared to the P&T/GC-MS and headspace/GC-ECD methods.

Many utilities today use chloramines as an alternative to free chlorine as a secondary disinfectant in their distribution systems to minimize further formation of chlorination by-products. However, chloramines also form disinfection by-products (DBPs) of a different chemical nature. One chloramine DBP of interest is cyanogen chloride (CNCl), a highly volatile organic compound. The formation of CNCl has been shown to result from the chlorination of aliphatic amino acids in the presence of the ammonium ion (1). F. E. Sculley, Jr. ("Reaction Chemistry of Inorganic Monochloramine:

0097–6156/96/0649–0126$15.00/0

Products and Implications for Drinking Water Disinfection," presented at the 200th ACS National Meeting, Washington, DC, 1990), and E. J. Pedersen et al. ("Formation of Cyanogen Chloride from the Reaction of Monochloramine and Formaldehyde," presented at the 210th ACS National Meeting, Chicago, IL, 1995) have shown CNCl formation to result from the reaction of formaldehyde--known to be an ozonation DBP--and monochloramine. CNCl was first reported as a chloramine DBP by Krasner and co-workers in a nationwide survey of 35 water utilities (*2*). This study found CNCl levels ranging from approximately 0.4 to 12 µg/L. CNCl is to be included in the Federal Information Collection Rule for systems using chloramines (*3*) and is a possible candidate for regulation in Stage 2 of the Disinfectants/DBP Rule.

This study was undertaken to evaluate various analytical techniques for the determination of CNCl in chloraminated drinking water. Analytical techniques for the measurement of CNCl include purge-and-trap (P&T) gas chromatograph (GC)/mass spectrometer (MS) analysis, headspace/GC-electron capture detector (ECD) analysis, and micro-liquid/liquid extraction (micro-LLE) with GC-ECD analysis. Currently, the official U.S. Environmental Protection Agency (USEPA) method for CNCl is P&T/GC-MS analysis, although it is recognized that this method has its limitations. Thus, the purpose of this research was to evaluate alternate analytical techniques for CNCl.

The P&T/GC-MS method utilizes a modification to USEPA Method 524.2, as described by Flesch and Fair (*4*). Method 524.2 is ideally suited for the analysis of many volatile organic compounds (VOCs), but it is difficult to obtain precise and accurate results for chemicals such as vinyl chloride and CNCl, which are gases at room temperature. In addition, because CNCl is a highly reactive compound, an all-glass system with inert or deactivated internal surfaces must be used. Xie and Reckhow developed a headspace/GC-ECD method that is more reliable (*5*). However, only 1 percent of the CNCl is recovered on an absolute basis by this headspace analysis. Xie and Reckhow estimated the Henry's Law constant of 0.9 atm-L/mol for CNCl, thus explaining the low absolute recovery. Also, this method is not easily automated without investment in a special headspace autosampler. A micro-LLE/GC-ECD technique was developed at the Metropolitan Water District of Southern California, using salted, methyl *tert*-butyl ether extraction with GC-ECD analysis (*6*). The micro-LLE/GC-ECD method recovers 14 percent of the CNCl on an absolute basis. Procedural calibration standards were used for all three methods to facilitate accurate quantitation in lieu of 100 percent absolute recoveries.

The analytical techniques available for CNCl determination are varied. The purpose of this study was to compare the three methods and evaluate each method simultaneously to determine whether or not there is equivalency between them. The need for an accurate and precise method for CNCl determination is important for regulatory purposes. In addition, ease of use and availability of equipment are important considerations.

Experimental Section

Reagents and Chemicals. Information on the analytical standards used in this research is provided in Table I. CNCl is a gas at room temperature and is highly toxic. Standards can be prepared by dissolving the pure gas into methanol. However, a prepared standard can be purchased at concentrations up to 2000 μg/mL. The internal standard for the micro-LLE/GC-ECD and headspace techniques was 1,2-dibromopropane, and for the P&T/GC-MS technique the internal standard was fluorobenzene. The surrogate used for the P&T/GC-MS technique was 4-bromofluorobenzene.

Table I. Analytical Standards

Compound [CAS No.]	Source	Purity (percent)	Molecular Weight (g/mole)	Boiling Point (°C)	Density (mg/mL)
Cyanogen chloride[1] [506-77-4]	Island[2]	99.5	61.47	13.9	1.186
1,2-Dibromopropane[3] [78-75-1]	Aldrich[4]	99	201.90	140	1.9324
Fluorobenzene[5] [462-06-6]	Aldrich[4]	99	96.11	85.11	1.0225
4-Bromofluorobenzene[6] [460-00-4]	Aldrich[4]	99	175.01	152	1.4946

[1]CNCl is also available as a solution (concentration = 2000 μg/mL) from Protocol Analytical Supplies, Inc. (Middlesex, NJ).
[2]Island Pyrochemical Industries (Great Neck, NY).
[3]Internal standard used for micro-LLE and headspace/GC analyses.
[4]Aldrich Chemical Company, Inc. (Milwaukee, WI).
[5]Internal standard for P&T/GC-MS analysis.
[6]Surrogate used for P&T/GC-MS analysis.

The extraction solvent used for the micro-LLE/GC-ECD technique was "Omnisolv"-grade methyl *tert*-butyl ether from EM Science (Gibbstown, NJ). The salt used in the micro-LLE/GC-ECD technique was sodium sulfate (Na_2SO_4) from J. T. Baker, Inc. (Jackson, TN).

The sulfuric acid used for sample preservation was Fisher Scientific Co. (Pittsburgh, PA) A.C.S. reagent grade. The dechlorinating/dechloraminating agent, L-ascorbic acid, was obtained from Sigma Chemical Co. (St. Louis, MO). Samples were preserved in the same manner for all three analytical techniques. Preservation included the addition of 2.5 mg of ascorbic acid to a 40-mL sample plus 0.2 mL of a $1M$ sulfuric acid solution. Sample pH was ~3-3.5, which minimized base-catalyzed hydrolysis. With these preservatives, samples could be held for up to 14 days (6).

Analytical Methods.

Sample Preservation. Samples were collected in nominal 40-mL vials with Teflon-faced polypropylene septa and screw caps (I-Chem Research, Inc., Hayward, CA). The sample vials were filled so as to ensure that no air bubbles passed through the sample, thus minimizing aeration. Approximately 0.1 mL of a freshly prepared $0.142M$ ascorbic acid solution and 0.2 mL of $1M$ sulfuric acid were added to each vial prior to sampling. It was important that the ascorbic acid solution be fresh, as it had a very short shelf life. The volume of acid was adjusted as needed to achieve the desired pH of ~3-3.5. The bottles were not rinsed before filling and were not allowed to over-fill, since the bottles contained preservatives. The sample vials were sealed headspace-free. Ascorbic acid and sulfuric acid were employed as dechlorinating/dechloraminating and preservation agents, respectively. The ascorbic acid reduced any free chlorine or monochloramine residual present, thus preventing further production of CNCl. The $1M$ sulfuric acid reduced the pH of the sample to ~3-3.5. Samples were stored in a refrigerator at 4°C. All samples were brought up to room temperature prior to analysis.

The GC conditions used in this comparison are outlined in Table II. A brief summary of each method used is given below.

Micro-LLE/GC-ECD Analysis. Samples were first brought up to room temperature prior to extraction. A 30-mL aliquot of sample was extracted by addition of 10 g of Na_2SO_4 and 4 mL of methyl *tert*-butyl ether containing 100 μg/L internal standard, 1,2-dibromopropane. The purpose of the salt was to increase the extraction efficiency by increasing the ionic strength of the sample matrix, thus reducing the solubility of the CNCl in water, and the purpose of the internal standard was simply to monitor the autosampler injections for constancy. The sample was then shaken in a mechanical shaker for 10 min. The methyl *tert*-butyl ether layer was transferred between two 1.5-mL autosampler vials. The second vial was used as a backup extract in the event that reanalysis became necessary. The analysis was conducted on a GC (model 3600; Varian Instrument Group, Sunnyvale, CA) with a ^{63}Ni ECD and a fused-silica capillary column (DB-624, 30-m length, 1.8-μm film thickness, 0.32-mm internal diameter; J&W Scientific, Folsom, CA) to obtain baseline resolution of CNCl from vinyl chloride.

P&T/GC-MS Analysis. Samples were first brought up to room temperature prior to analysis. A 25-mL gas-tight syringe was filled with sample from a 40-mL vial. Internal standard and surrogate were then added to this aliquot. The aliquot was then transferred to a fritted sparger attached to a P&T concentrator (Tekmar LSC2000; Rosemount Analytical, Inc., Tekmar Co., Cincinnati, OH). Connections were made with all glass-lined tubing because CNCl is highly reactive and can be degraded very easily on hot metal surfaces. The sample was then sparged for 4 min with helium onto a Tenax #1 cartridge trap (Enka Research Institute, Arnhem, Netherlands) where CNCl was retained. The analyte was then desorbed from the trap by heating, and the desorbed gas was cryofocused (Tekmar Cryofocusing Module; Rosemount Analytical) prior to injection onto the GC and analysis on the MS. The GC was a model HP 5890 (Hewlett-Packard Co., Avondale, PA)--with a fused-silica capillary column (DB-5,

Table II. GC Conditions and Parameters

	Micro-LLE/ GC-ECD	*P&T/GC-MS*	*Headspace/ GC-ECD*
Injector/ sample introduction	Septum-equipped, programmable temperature injector (Varian model 1093)	P&T concentrator with cryofocusing unit	Split/splitless injector (Varian model 1077); split ratio set at 50:1
Injector/ sample introduction programming	Hold at 35°C for 0 min, ramp 180°C/min to 200°C, hold for 12.59 min	Purge 4.0 min Desorb preheat = 175°C Desorb 1.5 min at 180°C Capillary cooldown = -150°C Inject 0.85 min at 220°C Bake 2.0 min at 220°C	Isothermal, set at 150°C
Column temperature program	25°C, 1 min; 10°C/min; 120°C, 0 min; 35°C/min; 190°C, 1 min	10°C, 4 min; 20°C/min; 184°C, 3 min	25°C, 0.5 min; 8°C/min; 89°C, 0 min; 15°C/min; 150°C, 3 min
Total run time	13.5 min	15.7 min	15.56 min
Gases: Carrier (He) Detector make-up (N_2) Purge (He) Headspace (N_2)	3.9 mL/min 27 mL/min NA NA	1.5 mL/min NA 40 mL/min NA	1.5 mL/min 30 mL/min NA 10 mL
Autosampler/ sample volume	Varian model 8100 Injection volume: 1 µL Solvent plug volume: 0.5 µL Injection rate: 5.0 µL/sec Injection time: 0.05 min	Purge sample volume = 25 mL	Injection volume = 400 µL of headspace volume
Other		GC inlet temperature = 200°C Source temperature = 180°C MS resolution = 500 Multiplier volts = 1400 eV EI voltage = 70 eV	

NA = Not applicable

30-m length, 1.0-μm film thickness, 0.25-mm internal diameter; J&W Scientific)-- coupled to a medium-resolution, electron-impact, magnetic-sector MS (model TS-250; VG Tritech Ltd., Wythenshawe, Manchester, England). The MS was set up to monitor full-scan for CNCl as well as the internal standard and surrogate.

Headspace/GC-ECD Analysis. The method of Xie and Reckhow (5) was utilized with some modifications. Briefly, after the sample was brought up to room temperature, the internal standard (1,2-dibromopropane) was added to the sample. Then 10 mL of sample liquid was displaced by nitrogen gas to create a headspace. This was accomplished by first piercing the vial septum with a bare syringe needle and then forcing the needle of a syringe containing 10 mL of nitrogen through the same septum. As the nitrogen was forced into the vial, an equivalent amount of liquid was displaced through the bare needle, thus creating the desired headspace. The syringe and bare needle were removed after the required volume had been displaced. The vial was then shaken for 30 seconds. Next, using a 500-μL gas-tight syringe, 400 μL of the headspace volume was injected onto a split/splitless injector (in splitless mode). The analysis was conducted on a GC (Varian model 3600) with a [63]Ni ECD and a fused-silica capillary column (DB-1701, 30-m length, 1.0-μm film thickness, 0.25-mm internal diameter; J&W Scientific) to obtain baseline resolution of CNCl.

Water Samples. In order to directly compare each of the analytical techniques, samples were split and analyzed by the three methods. Several different sample matrix waters were obtained from various facilities nationwide, including two pilot plants and five full-scale water treatment plants. These locations were chosen on the basis of bromide ion concentration, total organic carbon concentration, pH, and chlorine-to-ammonia-nitrogen ratio (American Water Works Association Research Foundation [AWWARF], *Factors Affecting Disinfection By-Products Formation During Chloramination*, final report; AWWA and AWWARF, Denver, CO; in press). Table III outlines the range of water quality parameters for the samples used in this comparison. CNCl values ranged from 0.9 to 4.6 μg/L, with one pilot-plant sample representing an extreme condition (pH = 6) measuring as high as 12 μg/L.

Table III. Water Quality Parameters

Water Quality Parameter	Range
Bromide ion concentration	7 to 857 μg/L
Total organic carbon	1.4 to 8.9 mg/L
pH	6.7 to 9.2
Chlorine-to-ammonia nitrogen ratio	3:1 to 4.6:1
UV absorbance @ 254 nm	0.028 to 0.21

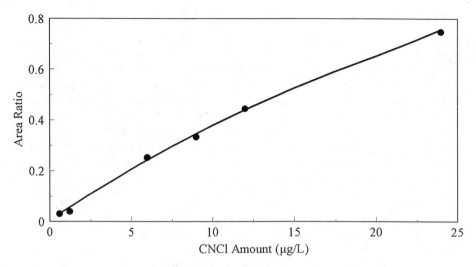

Figure 1. CNCl Calibration Curve for the P&T/GC-MS Method.
(Area ratio = [CNCl area/Internal standard area].)

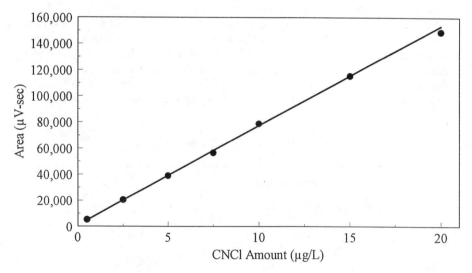

Figure 2. CNCl Calibration Curve for the Micro-LLE/GC-ECD Method.

Results and Discussion

Calibration and Quality Control. Calibration standards were analyzed in the same manner as the samples to compensate for less than 100-percent extraction or purge efficiency. The instruments were calibrated prior to sample analysis. Figures 1-3 show the calibration curves for the three analytical techniques. The P&T/GC-MS calibration curve (Figure 1) was curvilinear (second-order polynomial), whereas the micro-LLE/GC-ECD (Figure 2) and headspace/GC-ECD (Figure 3) techniques yielded linear calibration curves. Procedural standards led to accurate quantitation for these methods, each of which could detect CNCl concentrations as low as ~0.5 μg/L.

Quality control samples, including both accuracy (matrix spike) and precision (replicate) samples, were also analyzed in this comparison. Mean matrix-spike recoveries (±1 standard deviation) for CNCl were 98 ± 6.3 percent for the micro-LLE/GC-ECD method, 96 ± 18 percent for the P&T/GC-MS method, and 107 ± 14 percent for the headspace/GC-ECD method. The precision (±1 standard deviation) for CNCl was 3.7 ± 3.9 percent for the micro-LLE/GC-ECD method, 14.2 ± 9.5 percent for the P&T/GC-MS method, and 10.7 ± 7.3 percent for the headspace/GC-ECD method. The three methods demonstrated comparable accuracy; however, precision was best for the micro-LLE/GC-ECD technique.

Analytical Methods Comparison. The analyses were performed within the established 14-day holding period for CNCl. When the comparisons were performed, samples were analyzed within 24 h of each other.

Figure 4 compares the micro-LLE/GC-ECD and P&T/GC-MS methods. Typically, the micro-LLE/GC-ECD method produced CNCl results within ± 20 percent of the values determined by the P&T/GC-MS technique. In this limited data set, there somewhat higher results appear to have been produced by the micro-LLE/GC-ECD method. This difference may be partially a result of the calibration curves generated for each analysis. Figure 5 compares the headspace/GC-ECD and micro-LLE/GC-ECD methods. Once again, the micro-LLE/GC-ECD method produced CNCl results within ± 20 percent of the values determined by the headspace/GC-ECD method. Because of sampling limitations, no direct comparison was made between the headspace/GC-ECD and P&T/GC-MS methods. An indirect comparison could be made between these two methods based on the previous comparisons, as similar results were obtained for all three analytical techniques.

An advantage of the micro-LLE/GC-ECD method is that an expensive GC-MS system is not required. A comparison of the costs of the analyses in Southern California showed that a typical LLE/GC-ECD analysis (for other VOCs) ranged from $65 to $100, whereas the cost of a P&T/GC-MS analysis ranged from $150 to $260. Headspace analyses ranged from $150 to $300, in part because of uncommon usage. In addition, downtime on a GC-ECD system is significantly less than for the GC-MS.

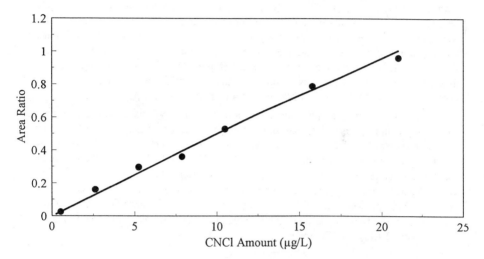

Figure 3. CNCl Calibration Curve for the Headspace/GC-ECD Method.
(Area ratio = [CNCl area/Internal standard area].)

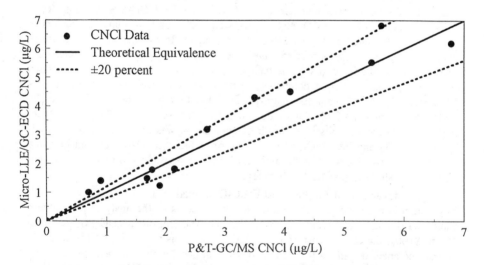

Figure 4. Analytical Methods Comparison Between the P&T/GC-MS and
Micro-LLE/GC-ECD Methods.

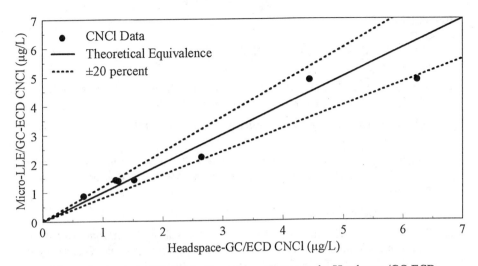

Figure 5. Analytical Methods Comparison Between the Headspace/GC-ECD and Micro-LLE/GC-ECD Methods.

Conclusions

As the comparisons show, the micro-LLE/GC-ECD, P&T/GC-MS, and headspace/GC-ECD methods are comparable analytical techniques for the determination of CNCl in drinking water. The accuracies of the three methods were comparable, and the micro-LLE method was the most precise. The results of split samples typically agreed to within ±20 percent. The cost comparison showed that the micro-LLE technique was the least expensive of the three analytical techniques. The comparative data presented here can provide a basis for acceptance of the micro-LLE/GC-ECD method as an alternative to the P&T/GC-MS and headspace/GC-ECD methods.

Acknowledgments

The authors thank Dr. James Symons of the University of Houston, Department of Civil and Environmental Engineering, and Dr. Gerald Speitel of the University of Texas at Austin, Department of Civil Engineering, for allowing the use of sample data related to the project "Factors Affecting Disinfection By-Products Formation During Chloramination," sponsored by AWWARF. Thanks are also extended to Alicia Diehl of the University of Texas at Austin for coordinating the sampling of the various utilities nationwide and to Peggy Kimball of Metropolitan for the final editing of this manuscript.

References

1. Hirose, Y.; Maeda, N.; Ohya, T.; Nojima, K.; Kanno, S. *Chemosphere* **1988**, *17*, 865-873.

2. Krasner, S. W.; McGuire, M. J.; Jacangelo, J. G.; Patania, N. L.; Reagan, K. M.; Aieta, E. M. *J. AWWA* **1989**, *81 (8)*, 41-53.

3. U.S. Environmental Protection Agency. *Fed. Reg.* **1994** (Feb. 10), *40CFR*, 6332-6444.

4. Flesch, J. J.; Fair, P. S. In *American Water Works Association Proceedings: 1988 Water Quality Technology Conference;* AWWA: Denver, CO, 1989; pp 465-474.

5. Xie, Y.; Reckhow, D. *Wat. Res.* **1993**, *27*, 507-511.

6. Sclimenti, M. J.; Hwang, C. J.; Speitel, G. E.; Diehl, A. C. In *American Water Works Association Proceedings: 1994 Water Quality Technology Conference;* AWWA: Denver, CO, 1995; pp 489-507.

NATURAL ORGANIC MATTER
RELATIONSHIPS AND CHARACTERIZATION

Chapter 10

Characterization of Natural Organic Matter and Its Reactivity with Chlorine

Gregory W. Harrington[1], Auguste Bruchet[2], Danielle Rybacki[2], and Philip C. Singer[1]

[1]Department of Environmental Sciences and Engineering, University of North Carolina, Chapel Hill, NC 27599–7400
[2]Centre International de Recherche Sur l'Eau et l'Environnement, Lyonnaise des Eaux Dumez, 38 rue du Président Wilson, 78230 Le Pecq, France

Hydrophobic extracts of natural organic matter (NOM) were isolated from five water supplies and characterized with pyrolysis gas chromatography/mass spectrometry (GC/MS) and carbon-13 nuclear magnetic resonance (^{13}C NMR) spectrometry. Carbon dioxide and phenol peak areas from pyrolysis GC/MS correlated with ^{13}C NMR estimates of carboxylic carbon content and aromatic carbon content, respectively. Phenol peak area from pyrolysis GC/MS was a qualitative indicator of chlorine consumption and disinfection byproduct (DBP) formation by the five hydrophobic NOM extracts. Phenolic carbon content, as estimated by ^{13}C NMR, gave a more quantitative indication of chlorine consumption and DBP formation by these five extracts.

The presence of NOM influences the use of chlorine in drinking water treatment for a number of reasons. NOM consumes chlorine, making chlorine unavailable for disinfection of pathogenic microorganisms or for oxidation of reduced metals such as ferrous iron. In addition, the reaction between chlorine and NOM produces a large number of halogenated and non-halogenated byproducts (*1-3*) that are commonly referred to as disinfection byproducts (DBPs). Several of these DBPs have been implicated as animal carcinogens and as possible human carcinogens (*4*).

The nature and extent of interaction between NOM components and aqueous chlorine species depend on numerous factors. Because pH influences electron distribution within NOM structures and the distribution of aqueous chlorine species, pH is an important factor in the interaction between NOM and aqueous chlorine species. The kinetic competition between hydrolysis, oxidation, and halogenation reactions also establishes pH as an important factor in determining the distribution of DBPs formed. In addition, some NOM components react with aqueous chlorine species at a much faster rate than other NOM components. Therefore, chlorine dose

0097–6156/96/0649–0138$15.25/0

and contact time are also important variables in determining the rate and extent of reaction between NOM and aqueous chlorine species. If bromide or iodide are present, aqueous chlorine species can oxidize these halides to reactive halogen species that are analogous to the aqueous chlorine species. Temperature is also an important variable in this interaction. Numerous models use one or more of these factors to simulate chlorine decay and DBP formation (*5-11*).

The aqueous chlorine species are electrophiles and tend to react with electron-rich sites in organic structures (*1,12-18*). Activated aromatic rings, aliphatic β-dicarbonyls, and amino nitrogen are examples of electron-rich organic structures that react strongly with chlorine. Because of this, the structural nature of the NOM species present in solution is also an important factor in determining the influence of NOM on the chlorination process. For aquatic humic substances, reactivity with chlorine is believed to increase with the activated aromatic carbon content of the humic substance (*19*). The electron distribution within NOM structures is influenced by pH, another reason why pH is an important variable in the interaction between NOM and chlorine.

Since the early 1980s, investigators have developed methods such as pyrolysis GC/MS and ^{13}C NMR spectrometry for evaluating the structural characteristics of NOM. Bruchet and his coworkers have used pyrolysis GC/MS methods for evaluating the relative distribution of four biopolymer classes in NOM samples (*20-22*). In addition, Maciel, Hatcher, Wilson, and their coworkers have used solid-state ^{13}C NMR techniques to yield quantitative estimates of carbon structural distribution in NOM samples (*23-32*). To date, no investigations have compared the results obtained by these two different methods of NOM characterization. Comparisons between chlorination results and pyrolysis GC/MS results, however, have been presented, but only in a qualitative manner (*20-22*). Qualitative comparisons have also been made between chlorination results and ^{13}C NMR results (*25*). Quantitative comparisons have been limited to analyzing the influence of aromatic carbon content, as estimated by ^{13}C NMR, on chlorine consumption by aquatic humic substances (*19*). One purpose of this research was to compare the results obtained by pyrolysis GC/MS and ^{13}C NMR spectrometry. Evaluating comparisons between chlorination behavior and results from the two NOM characterization methods was another objective of this research.

Experimental and Analytical Methods

General Approach. Hydrophobic NOM was extracted from five water supplies in the United States and lyophilized. Each of the freeze-dried extracts was analyzed with pyrolysis GC/MS and ^{13}C NMR to evaluate structural characteristics. Results obtained from pyrolysis GC/MS were compared with results obtained from ^{13}C NMR.

Each extract was reconstituted to make a synthetic water having a TOC concentration of 4 mg/L. The synthetic waters were chlorinated at pH 8.0 to produce a residual of 1.0 mg/L as Cl_2 after 24 hours. Chlorine consumption and DBP formation were evaluated for each synthetic water, and results were compared with structural characteristics of NOM extracts prior to chlorination.

Isolation of Hydrophobic NOM. A modified version of the Thurman and Malcolm (*33*) method was used to extract hydrophobic NOM from its source water. So that comparisons can be made with the published method, the following sections describe the details of the procedure used in this research.

    Resin Preparation. In this study, XAD-8 resin was obtained from the supplier (Rohm and Haas Corp., Philadelphia, PA) and cleaned to remove impurities left by the manufacturing process. The initial step involved the placement of resin in a freshly prepared solution of 0.1 N sodium hydroxide (NaOH). After 24 hours, the 0.1 N NaOH was decanted and another freshly prepared batch of 0.1 N NaOH was added to the resin. This procedure was repeated for five consecutive days, after which the resin was placed into a soxhlet extractor that had a capacity for 2 L of resin. The soxhlet extractor was used to clean the resin with a number of solvents, each solvent being used for 24 hours. The first solvent was methanol, followed by diethyl ether, acetonitrile, and a repeat of methanol. Upon completion of these steps, the resin was considered clean and was stored in methanol for future use.

    When the resin was needed, the methanol was decanted and a batch of deionized, organic-free water was added to the resin. After shaking for one minute, the water was decanted and replaced with a new batch of deionized, organic-free water. This step was repeated until the odor of methanol was no longer apparent. The resin was then packed to occupy an empty bed volume of 3 L in a glass column that was filled with deionized, organic-free water prior to resin addition. After packing the column, residual methanol was removed by running 50 bed volumes of deionized, organic-free water through the column at a rate of one bed volume per hour. The resin was then flushed at one bed volume per hour in sequential steps with 20 L of 0.1 N NaOH, 50 L of deionized, organic-free water, 20 L of 0.1 N hydrochloric acid (HCl), and 20 L of 0.01 N HCl. All NaOH and HCl solutions were prepared from deionized, organic-free water and reagent grade materials (Fisher Scientific Co., Pittsburgh, PA). Upon completion of these steps, the column was shipped to its desired location by overnight courier.

    Extraction Protocol. At each location, source water was pumped through a cartridge filter having a 1.0 µm nominal pore size. This filtered water was discharged into a 20 L overflow reservoir at a flow rate sufficient to maintain an overflow of filtered water. Filtered water was pumped from this reservoir, acidified to pH 2 with 1 N HCl, and applied to the resin-filled column. The 1 N HCl was stored in a 20 L reservoir and was prepared from reagent-grade, concentrated HCl on an as needed basis. Pumping of acid solution and of raw and filtered waters was performed with peristaltic pumps fitted with laboratory-grade Tygon tubing (pumps and tubing from Cole-Parmer Instrument Co., Niles, IL). At all locations, the extraction system was assembled and placed into operation by a member of the project team. Upon achieving satisfactory operation, the system was operated by members of the utility staff.

The flow rate through the column was two bed volumes per hour. This was monitored several times each day by collecting column effluent water with a graduated cylinder for a period of two minutes. Column effluent pH was also monitored at the same time as column effluent flow rate, with adjustments made to the acid feed rate if column effluent pH deviated from pH 2.0 by more than 0.1 pH unit. Influent and effluent TOC concentrations were monitored three times per week and the column was run until the requisite amount of organic carbon was collected. In all cases, shut-down criteria were based on the amount of organic carbon needed for experimentation and on an assumed carbon loss of 30 percent through elution and post-elution workup. At some utilities, column operation was terminated when 10 g of organic carbon were adsorbed to the column. The cutoff criterion was less than this at other sources.

The primary differences between the above approach and the Thurman and Malcolm (*33*) approach were the flow rate and the shut-down criteria. In this research, the flow rate was two bed volumes per hour as opposed to the 15 bed volumes per hour used by Thurman and Malcolm. The column was shut down in this research when the requisite amount of organic carbon was extracted from the water supply. Thurman and Malcolm established a column capacity factor of 100 as the criterion for ceasing column operation.

The primary reason for using the above operating protocol was to establish a method that could collect a large amount of hydrophobic NOM while limiting the burden placed on the water treatment plant operators who were monitoring the extraction system. For example, faster flow rates would have required more frequent replacement of the 1 N HCl solution or the use of a stronger HCl solution (e.g., 10 N). The frequent replacement of acid solution was not desirable because the operators had other, more important responsibilities than maintaining the extraction system. The use of a stronger acid solution would have posed safety concerns at all facilities. The use of Thurman and Malcolm's column capacity factor would have required frequent resin replacement, another burden the water plant operators would have been unwilling to bear. Although the isolation approach was different from that recommended by Thurman and Malcolm, the interpretations of the results obtained in this research were not influenced by the isolation procedure.

Elution and Post-Elution Protocol. Once extraction was complete, utility operators drained the column and returned it to the University of North Carolina by overnight courier. Upon receipt, the column was rinsed for one hour with one bed volume of deionized, organic-free water to remove chloride. This step was conducted with flow in the same direction as the adsorption step. Hydrophobic NOM was then eluted from the column in the reverse direction with 0.1 N NaOH at a flow rate of one bed volume per hour. The eluate was collected in 500 mL fractions and the TOC concentration of each fraction was analyzed. Depending on the source of hydrophobic NOM, six to ten of the most concentrated fractions were saved and combined for further preparation.

The next step involved removal of sodium and other cationic species from solution by cation exchange. Eluted solutions of hydrophobic NOM were passed through a column containing a 400 mL empty bed volume of AG-MP-50 cation exchange resin in hydrogen form (Bio-Rad Laboratories, Hercules, CA). Flow rate was maintained at one bed volume per hour and effluent pH was monitored to confirm the lack of breakthrough. The effluent was frozen in 300 mL batches by manually spinning a jar of sample in a bath of dry ice and isopropyl alcohol. Frozen samples were attached to a freeze-dryer (Labconco, Kansas City, MO) and lyophilized over a period of three to four days. Of the organic carbon estimated to have been adsorbed to the resin, 60 to 80 percent was eventually converted into freeze-dried form.

Sources of Hydrophobic NOM. Five water supplies were selected as sources for hydrophobic NOM. These sources are listed in Table I along with their location, general characteristics, and selected water quality parameters. This table shows the geographic diversity among the five sources, as well as the diversity of source type and organic water quality.

Of particular importance in Table I is the column that shows the percentage of organic carbon isolated from each source. With the Intracoastal Waterway, 71 percent of the organic carbon was isolated by the XAD-8 resin. In comparison, only 16 percent of the organic carbon was isolated from the Colorado River Aqueduct. The structural characteristics and chlorine reactivity of NOM that passed through the XAD-8 resin were not analyzed. Therefore, 84 percent of the organic carbon in the Colorado River Aqueduct was not characterized with respect to structural characteristics and chlorine reactivity. This was true of only 29 percent of the organic carbon in the Intracoastal Waterway. The relevance of this is discussed in the conclusions.

Characterization of Hydrophobic NOM Extracts. As noted above, the five hydrophobic NOM extracts were characterized with pyrolysis GC/MS and ^{13}C NMR spectrometry. The following sections describe the procedures used for these two methods.

Pyrolysis GC/MS. Approximately 10 mg of each freeze-dried hydrophobic NOM extract were shipped to the Centre International de Recherche Sur l'Eau et l'Environnement for pyrolysis GC/MS analysis. These samples were submitted directly to flash pyrolysis in a helium atmosphere using a Pyroprobe 100 filament pyrolyzer (Chemical Data Systems, Oxford, PA). Each sample was deposited into a 50 μL quartz tube and both ends of the quartz tube were plugged with quartz wool to prevent sample from escaping during introduction into the pyrolysis interface. Approximately 400 μg of sample were put into each quartz tube and the quartz tubes were then placed into the platinum filament of the pyrolysis probe. The probe was inserted into the pyrolysis oven, which was preheated at 200°C. Flash pyrolysis was performed by programming the platinum filament to a final temperature of 750°C at a rate of 20°C/msec with a final hold for 20 sec. The final temperature inside the quartz tube was 625°C ± 5°C, as determined by periodic checks with a type-K thermocouple minithermometer.

The pyrolysis oven was connected with the split/splitless injection port of a Carlo Erba 4160 GC interfaced with a NERMAG R-10-10 C quadrupole mass spectrometer. After pyrolysis, the gas-phase pyrolysis products were separated by gas chromatography on a 30 m DB wax fused silica capillary column. The temperature of this separation was programmed from 25°C to 220°C at a rate of 3°C/min. Upon chromatographic separation, the gas-phase pyrolysis products were identified by mass spectrometry. The mass spectrometer was operated at 70 eV and scanned from 20 to 400 amu at one scan per second.

For this paper, the peak area associated with the carbon dioxide pyrolysis fragment is reported as the fraction of the total peak area on the pyrochromatogram. All other peak areas are reported as the fraction of the total peak area minus the carbon dioxide peak area. For each hydrophobic NOM extract, these peak areas were used to calculate the relative contribution of four biopolymer classes to the overall nature of the extract. The calculation procedure and its application are summarized elsewhere (20-22).

^{13}C NMR Spectrometry. Freeze-dried extracts were submitted to solid-state ^{13}C NMR spectrometry and spectra were obtained with cross-polarization while spinning at the magic angle at 15 kHz. The spectrometer (Chemagnetics CMX-400) and magnet (Oxford Instruments) were rated with a ^{1}H frequency of 400 MHz and a ^{13}C frequency of 100 MHz. A repeat time of 1.01 sec was employed with a pulse width of 6.5 μsec, a contact time of 3 msec, an acquisition time of 8.193 msec, and a pulse delay of 1 sec. In addition, a sweep width of 125 kHz was used and spectra were generated from 10,000 to 13,775 scans with 1000 data points collected per scan. Also, line broadening was used at 200 Hz and the rotor was made of zirconium with vespel caps. Given the spin rate and ^{13}C frequency, spinning side bands could occur at chemical shift intervals of 149 ppm from the chemical shift of interest. This is consistent with the literature published by Hatcher and coworkers (23-25) and by Wilson and coworkers (26-28).

Once the spectra were obtained, several chemical shift ranges were integrated to estimate the structural distribution of carbon in each hydrophobic NOM extract. Chemical shifts of 0 ppm to 110 ppm were assumed to represent aliphatic carbon while chemical shifts of 110 ppm to 160 ppm were assumed to represent aromatic carbon. Carboxylic carbon and carbonyl carbon content were estimated from chemical shifts of 160 ppm to 190 ppm and 190 ppm to 220 ppm, respectively.

The aliphatic and aromatic regions of each spectrum were subdivided to evaluate the presence of oxygen in these types of carbon structures. The chemical shift range from 50 ppm to 90 ppm was assumed to represent aliphatic carbon that was singly bonded to one oxygen atom. For aliphatic carbon that was singly bonded to two oxygen atoms, the chemical shift range of 90 ppm to 110 ppm was assumed to be representative. Finally, the chemical shift range from 145 ppm to 160 ppm was assumed to represent aromatic carbon that was singly bonded to oxygen. For convenience, this latter type of carbon will be referred to as phenolic carbon for the remainder of this paper, even though phenoxy and amino aromatics are included in this portion of the spectrum.

It is emphasized that this approach yields an estimate of the structural distribution of carbon within a given sample. There are several reasons for making this point. First, the use of the above integration cutoff points is somewhat arbitrary because the chemical shifts of functional groups overlap. For example, some aliphatic carbons resonate at chemical shifts greater than 110 ppm while some aromatic carbons resonate at chemical shifts smaller than 110 ppm. Despite this, these cutoff points are at chemical shifts having minimal overlap and are consistent with the literature (23-32). The presence of spinning side bands can also diminish the ability to make quantitative estimates of structural distribution. Because all samples in this research were analyzed under the same NMR conditions, this effect is shared among all five samples. However, this issue becomes more significant if NMR spectra are obtained under different conditions.

Chlorination of Hydrophobic NOM Extracts. Freeze-dried extracts were reconstituted in deionized, organic-free water to produce a TOC concentration of 4 mg/L. Because each of these synthetic waters had the same TOC concentration, differences in reactivity with chlorine could be attributed to differences in the structural character of the hydrophobic NOM. Each of these synthetic waters also contained 2 mM sodium bicarbonate and 50 mM phosphate buffer. The latter was used to maintain pH 8.0 during chlorination. Sodium hypochlorite was added to each synthetic water at a dose that produced a free chlorine concentration of 1.0 mg/L as Cl_2 after a reaction time of 24 hours at 20°C. Reactions were performed in headspace-free containers and in the absence of light. The dose required for each of the synthetic waters was determined from initial tests with relatively small sample volumes. These initial tests indicated that the 1.0 mg/L target was achieved with a dose of 2.9 mg/L as Cl_2 for hydrophobic NOM from the Colorado River Aqueduct. However, hydrophobic NOM from the Intracoastal Waterway required a dose of 6.5 mg/L as Cl_2 to achieve the 1.0 mg/L target. Chlorine residuals were within 0.1 mg/L of the 1.0 mg/L target after 24 hours when the initially determined dose was applied for the detailed testing.

Following the 24 hour reaction period, free and total chlorine concentrations were determined by titration with ferrous ammonium sulfate using N,N-diethyl-p-phenylenediamine (DPD) as an indicator (34). Chlorine consumption was calculated as the difference between the chlorine dose and the chlorine concentration at 24 hours. The remaining sample was transferred to 40 mL vials, with each vial containing an appropriate quenching reagent for analysis of trihalomethanes and haloacetic acids. Concentrations of trihalomethanes and haloacetic acids were determined by procedures reported elsewhere (35).

Results and Discussion

NOM Characterization. A representative pyrochromatogram, produced from the Intracoastal Waterway extract, is shown in Figure 1. Tables II and III show the results obtained from analysis of this pyrochromatogram and those of the other four hydrophobic NOM extracts. Normalized peak areas of the five most abundant pyrolysis fragments are shown in Table II. Carbon dioxide was the most prominent peak in the pyrochromatograms for all five hydrophobic NOM extracts, a result that is typical of pyrolysis GC/MS in general. Hydrophobic NOM from the Turnpike Aquifer produced the most carbon dioxide while hydrophobic NOM from the

Table I. Locations and Characteristics of Hydrophobic NOM Sources

Source	Location	Source Type	Period of NOM Isolation	TOC (mg/L)	Hydrophobic TOC (%)[1]	Molar Absorbance at 254 nm (L/cm·mol C)
Intracoastal Waterway	Myrtle Beach, SC	swamp canal	5/19/92 - 5/25/92	17.4	71	501
Oradell Reservoir	Haworth, NJ	impoundment	6/8/92 - 6/25/92	5.4	43	299
Turnpike Aquifer	West Palm Beach, FL	limestone aquifer	9/10/92 - 9/17/92	14.4	48	331
Haggett's Pond	Andover, MA	natural lake	5/28/93 - 6/21/93	3.5	31	344
Colorado River Aqueduct	Lake Matthews, CA	impoundment	7/13/93 - 10/11/93	3.1	16	161

[1]Hydrophobic TOC is the amount of organic carbon removed by the XAD-8 resin over the time of NOM extraction (calculated from TOC concentrations that were measured three times per week).

Table II. Peak Areas of Selected Pyrolysis Fragments from Hydrophobic NOM Extracts[1]

Pyrolysis Fragment	Intracoastal Waterway	Oradell Reservoir	Turnpike Aquifer	Haggett's Pond	Colorado River Aqueduct
carbon dioxide[2]	0.726	0.680	0.802	0.621	0.546
phenol	0.190	0.153	0.197	0.154	0.115
toluene	0.092	0.019	0.079	0.093	0.102
acetic acid	0.075	0.096	0.059	0.045	0.107
p-cresol	0.085	0.034	0.089	0.081	0.077

[1]The five most abundant fragments are shown and, except for carbon dioxide, their peak areas are reported as (peak area)/[(total peak area) - (carbon dioxide peak area)].
[2]Carbon dioxide peak areas are reported as (peak area)/(total peak area).

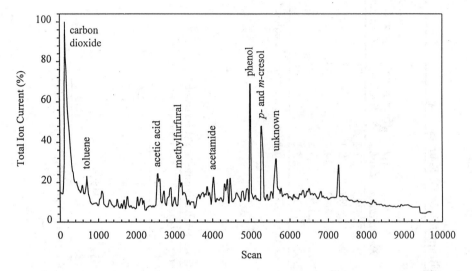

Figure 1. Pyrochromatogram produced by the pyrolysis of hydrophobic NOM from the Intracoastal Waterway.

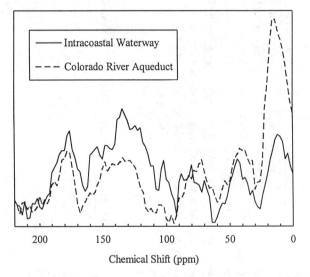

Figure 2. Normalized ^{13}C NMR spectra of hydrophobic NOM from the Intracoastal Waterway and the Colorado River Aqueduct.

Colorado River Aqueduct produced the least carbon dioxide. Normalized peak areas of the phenol and *p*-cresol pyrolysis fragments were greatest with hydrophobic NOM extracts from the Intracoastal Waterway and the Turnpike Aquifer. Table III indicates that hydrophobic NOM from the Colorado River Aqueduct had the greatest polysaccharide content. The other hydrophobic NOM extracts had significant concentrations of both polysaccharides and polyhydroxy aromatics.

Figure 2 shows representative ^{13}C NMR spectra for hydrophobic extracts from the Intracoastal Waterway and the Colorado River Aqueduct. The two spectra are normalized to the same total peak area. Table IV shows the results obtained from the integration of these spectra and from the spectra of the other three hydrophobic extracts. Aromatic carbon accounted for 21 to 33 percent of the carbon in the five hydrophobic NOM extracts, a range that is consistent with values published for other hydrophobic NOM extracts (*23-32*). Hydrophobic NOM from the Intracoastal Waterway and the Turnpike Aquifer had the largest aromatic carbon contents while hydrophobic NOM from the Colorado River had the smallest. Although the Intracoastal Waterway and Turnpike Aquifer extracts had similar aromatic carbon contents, the amount of oxygen-substituted aromatic carbon was significantly larger for the Intracoastal Waterway extract than for the Turnpike Aquifer extract. The Turnpike Aquifer extract was also lower in oxygen substituted aliphatic structures. The relative absence of oxygen-substituted carbon in Turnpike Aquifer extract is consistent with groundwater sources and may arise from the biodegradation of oxygenated organic structures (*23,30,36*).

For the five hydrophobic extracts studied, carboxylic carbon content correlated with aromatic carbon content (r = 0.933). This result is believed to be coincidental to these five extracts because published data do not indicate that such a correlation should exist as a general rule. Nevertheless, this correlation has some influence on data interpretation. Also, carbonyl contents were approximately the same for each of the five hydrophobic extracts. This implies that aliphatic carbon content should decrease among these five NOM extracts as aromatic and carboxylic carbon contents increase. In fact, inverse correlations were observed between aliphatic carbon content and aromatic carbon content (r = - 0.993) and between aliphatic carbon content and carboxylic carbon content (r = -0.965).

Comparison of Pyrolysis GC/MS and ^{13}C NMR. Correlation matrices were assembled to compare the results obtained by pyrolysis GC/MS with those obtained by ^{13}C NMR. Table V shows the correlations obtained when comparing normalized peak areas of pyrolysis fragments with chemical shift ranges from ^{13}C NMR spectrometry. A similar analysis is presented in Table VI for biopolymer classes.

The normalized peak area of the carbon dioxide pyrolysis fragment had correlation coefficients greater than 0.80 with aromatic carbon content and carboxylic carbon content. Because the samples were pyrolyzed in an inert atmosphere of helium, carbon dioxide production is expected to come from several sources including cleavage of carboxylic acid groups and release of intramolecular oxygen. Oxidation of carbon structures to carbon dioxide by carrier gas is not theoretically possible under these conditions. Therefore, the correlation between normalized carbon dioxide peak area and carboxylic carbon content can be expected. The

Table III. Relative Proportions of Biopolymers in Hydrophobic NOM Extracts

Biopolymer Class	Intracoastal Waterway	Oradell Reservoir	Turnpike Aquifer	Haggett's Pond	Colorado River Aqueduct
polysaccharides	0.46	0.48	0.36	0.36	0.61
proteins	0.07	0.10	0.09	0.13	0.08
amino sugars	0.08	0.07	0.03	0.10	0.15
polyhydroxy aromatics	0.39	0.35	0.53	0.40	0.15

Table IV. ^{13}C NMR Estimates of Carbon Structural Distribution in Hydrophobic NOM Extracts

Structural Type	Chemical Shift Range (ppm)	Intracoastal Waterway	Oradell Reservoir	Turnpike Aquifer	Haggett's Pond	Colorado River Aqueduct
aliphatic C (mmol/mmol C)	0 to 110	0.444	0.524	0.443	0.529	0.597
hydroxyl C (mmol/mmol C)[1]	50 to 90	0.134	0.171	0.084	0.166	0.154
anomeric C (mmol/mmol C)[2]	90 to 110	0.082	0.074	0.043	0.068	0.040
aromatic C (mmol/mmol C)	110 to 160	0.331	0.266	0.324	0.252	0.214
phenolic C (mmol/mmol C)[3]	145 to 160	0.086	0.065	0.068	0.066	0.058
carboxylic C (mmol/mmol C)[4]	160 to 190	0.155	0.133	0.157	0.143	0.119
carbonyl C (mmol/mmol C)[5]	190 to 220	0.071	0.076	0.076	0.076	0.069

[1] Includes aliphatic carbon that is singly bonded to one oxygen atom.
[2] Includes aliphatic carbon that is singly bonded to two oxygen atoms.
[3] Includes phenolic and phenoxy carbon.
[4] Includes carboxylic, ester, and quinone carbon.
[5] Includes aldehyde and ketone carbon structures.

Table V. Correlation Matrix for Pyrolysis Fragments and ^{13}C NMR Chemical Shifts

Pyrolysis Fragment	^{13}C NMR Chemical Shift Range						
	0 to 110 ppm (aliphatic C)	50 to 90 ppm (hydroxyl C)	90 to 110 ppm (anomeric C)	110 to 160 ppm (aromatic C)	145 to 160 ppm (phenolic C)	160 to 190 ppm (carboxylic C)	190 to 220 ppm (carbonyl C)
carbon dioxide	-0.944	-0.758	0.160	0.935	0.564	0.877	0.492
phenol	-0.996	-0.701	0.310	0.979	0.738	0.976	0.420
toluene	0.062	-0.261	-0.352	-0.065	0.116	0.089	-0.556
acetic acid	0.541	0.333	-0.168	-0.444	-0.307	-0.736	-0.603
p-cresol	-0.328	-0.606	-0.339	0.310	0.296	0.468	-0.285

Table VI. Correlation Matrix for Biopolymer Classes and ^{13}C NMR Chemical Shifts

Biopolymer Class	^{13}C NMR Chemical Shift Range (ppm)						
	0 to 110 ppm (aliphatic C)	50 to 90 ppm (hydroxyl C)	90 to 110 ppm (anomeric C)	110 to 160 ppm (aromatic C)	145 to 160 ppm (phenolic C)	160 to 190 ppm (carboxylic C)	190 to 220 ppm (carbonyl C)
polysaccharides	0.671	0.396	-0.228	-0.578	-0.311	-0.802	-0.807
proteins	0.234	0.398	0.111	-0.343	-0.398	-0.047	0.704
amino sugars	0.853	0.634	-0.173	-0.822	-0.397	-0.802	-0.708
polyhydroxy aromatics	-0.866	-0.615	0.205	0.805	0.450	0.909	0.740

correlation between normalized carbon dioxide peak area and aromatic carbon content was not expected. However, as noted earlier, ^{13}C NMR estimates showed a strong correlation between aromatic carbon content and carboxylic carbon content for these five hydrophobic NOM extracts. This autocorrelation is a possible explanation for the observed correlation between normalized carbon dioxide peak area and aromatic carbon content.

A similar result was observed for the normalized peak area of the phenol pyrolysis fragment. This peak area was also found to correlate with aromatic carbon content and with carboxylic carbon content, having correlation coefficients of 0.979 and 0.976, respectively. The phenol pyrolysis fragment would be expected to arise from aromatic structures. Therefore, the correlation between aromatic carbon content and the normalized peak area for the phenol fragment is consistent with expectations. As above, the autocorrelation between aromatic carbon content and carboxylic carbon content may explain the observed correlation between carboxylic carbon content and phenol peak area in pyrolysis GC/MS.

Table VI shows that polyhydroxy aromatics had correlation coefficients greater than 0.80 with both aromatic carbon content and carboxylic carbon content. This result is consistent with the results obtained for the normalized peak area of the phenol pyrolysis fragment. The amino sugars were observed to have a correlation coefficient of 0.853 with aliphatic carbon content. These biopolymers are likely to have structures that are mostly aliphatic, but polysaccharides would also be expected to have such structures. Because polysaccharide content was significantly larger than amino sugar content in these five extracts (see Table III), the observed correlation between amino sugar content and aliphatic carbon content was not necessarily expected.

The above results suggest that phenol peak area and polyhydroxy aromatic carbon content may be indicators of aromatic carbon content. Similarly, carbon dioxide may be an indicator of carboxylic carbon content. These pyrolysis GC/MS results had correlation coefficients greater than 0.80 with ^{13}C NMR results and these correlations were consistent with chemical explanations. The reader is cautioned, however, to remember that these results were obtained with a set of only five hydrophobic NOM extracts. This number of NOM extracts was insufficient to eliminate an autocorrelation between aromatic carbon content and carboxylic carbon content. This was also an insufficient number to eliminate an autocorrelation between the normalized peak areas of the phenol and carbon dioxide peaks in pyrolysis GC/MS. Furthermore, with the possible exception of phenol peak area as an indicator of aromaticity, the correlation coefficients were not sufficiently strong to indicate a quantitative relationship between pyrolysis GC/MS results and ^{13}C NMR results. More experiments with other extracts will be necessary to make conclusions in a definitive manner.

The reader is also cautioned to note that the above cases were the only cases where correlation coefficients greater than 0.80 were observed between pyrolysis GC/MS results and ^{13}C NMR results. From one perspective, this observation is acceptable because there were no significant correlations that were inconsistent with chemical expectations. For example, polysaccharide content from pyrolysis GC/MS did not correlate with aromatic carbon content from ^{13}C NMR, a result that was consistent with expectations. However, only a weak correlation (r = 0.671) was observed between polysaccharide content and aliphatic carbon content. The correlation between polysaccharide content and oxygen-substituted aliphatics (50 ppm to 90 ppm) was even weaker, with a correlation coefficient of 0.396. A stronger correlation was expected because polysaccharides produce signals in the 50 ppm to 90 ppm range on ^{13}C NMR spectra. The lack of a strong correlation for these cases is not consistent with chemical expectations. No explanation for this finding is apparent. Combining the data from this research with data from an additional set of NOM extracts may reveal more correlations.

NOM Characteristics and Reactivity with Chlorine. Correlations were evaluated between chlorination behavior and the normalized peak areas of pyrolysis fragments. A matrix of correlation coefficients is shown in Table VII, which shows that the normalized peak area of the phenol pyrolysis fragment was the best indicator of chlorine consumption. This was also true for the formation of chloroform, dichloroacetic acid, and trichloroacetic acid. These results are consistent with chemical expectations because the phenol pyrolysis fragment originates from activated aromatic structures. These structures are highly reactive with chlorine and others have shown correlations between chlorine consumption and the phenolic content of aquatic humic substances (*19*). Such structures are also known to produce significant quantities of the three chloro-organic byproducts shown in Table VII. Although the phenol pyrolysis fragment proved to be the best indicator of chlorine reactivity, the correlation coefficients were relatively weak (0.675 < r < 0.808) and suggest that this fragment is a qualitative indicator of chlorine reactivity.

A similar analysis was performed to compare chlorine reactivity with the biopolymer distributions determined from pyrolysis GC/MS (see Table VIII). The polyhydroxy aromatic class was the best indicator of chlorine reactivity, a result that was also consistent with previous observations (*20-22*). However, the correlation coefficients for this biopolymer class were not as good as those obtained with the phenol pyrolysis fragment. This suggests that numerous calculations, such as those required to estimate biopolymer distributions, may not be necessary if pyrolysis GC/MS is used as a tool to characterize the chlorine reactivity of NOM. The need to generate calibration curves for biopolymer standards such as dextran and chitin may also be unnecessary from this perspective. However, biopolymer distributions may serve as useful indicators of the other effects that NOM characteristics have on water treatment.

Correlation coefficients for relationships between chlorination behavior and ^{13}C NMR results are shown in Table IX. The strongest indicator of chlorine consumption was phenolic carbon content, with a correlation coefficient of 0.940. Phenolic carbon content was also the best indicator of chloroform, dichloroacetic acid, and trichloroacetic acid formation. For these cases, correlation coefficients ranged from 0.933 to 0.981. These correlation coefficients suggest that phenolic carbon content, as estimated by ^{13}C NMR, may show promise as a quantitative indicator of reactivity between NOM and chlorine. As noted earlier, phenolic carbon content was estimated from the chemical shift range of 145 ppm to 160 ppm. This chemical shift range includes phenolic, phenoxy, and amino aromatic carbon sites but

Table VII. Correlation Matrix for Pyrolysis Fragments and Chlorination Results

Pyrolysis Fragment	Chlorine Consumed	Chloroform Formed	Dichloroacetic Acid Formed	Trichloroacetic Acid Formed
carbon dioxide	0.651	0.548	0.504	0.507
phenol	0.808	0.754	0.708	0.675
toluene	-0.096	0.076	0.083	-0.029
acetic acid	-0.482	-0.572	-0.437	-0.252
p-cresol	0.140	0.275	0.248	0.122

Table VIII. Correlation Matrix for Biopolymer Classes and Chlorination Results

Biopolymer Class	Chlorine Consumed	Chloroform Formed	Dichloroacetic Acid Formed	Trichloroacetic Acid Formed
polysaccharides	-0.547	-0.567	-0.428	-0.284
proteins	-0.113	-0.042	-0.178	-0.325
amino sugars	-0.571	-0.457	-0.384	-0.375
polyhydroxy aromatics	0.638	0.596	0.490	0.406

Table IX. Correlation Matrix for ^{13}C NMR Chemical Shifts and Chlorination Results

^{13}C NMR Chemical Shift Range	Chlorine Consumed	Chloroform Formed	Dichloroacetic Acid Formed	Trichloroacetic Acid Formed
0 to 110 ppm (aliphatic C)	-0.833	-0.775	-0.743	-0.724
50 to 90 ppm (hydroxyl C)	-0.200	-0.163	-0.133	-0.110
90 to 110 ppm (anomeric C)	0.786	0.766	0.785	0.803
110 to 160 ppm (aromatic C)	0.830	0.763	0.750	0.752
145 to 160 ppm (phenolic C)	0.940	0.933	0.972	0.981
160 to 190 ppm (carboxylic C)	0.816	0.806	0.743	0.669
190 to 220 ppm (carbonyl C)	0.240	0.183	0.045	-0.035

does not measure the aromatic carbon sites that actually react with chlorine. In fact, these reactive sites are located at *ortho* and *para* positions from the carbon atom measured in this range of chemical shifts. Therefore, phenolic carbon content is used here as a surrogate measure of activated aromatic carbon sites.

These results are consistent with the observations of Reckhow, *et al.* (1990), who estimated phenolic carbon content from titration experiments rather than from ^{13}C NMR spectrometry (*19*). When estimated by titration, phenolic carbon content does not include phenoxy or amine-substituted aromatics. Direct comparisons of the two studies are shown in Figures 3 through 6, which plot measures of chlorine reactivity versus phenolic carbon content. Reckhow, *et al.*, obtained a good correlation between chlorine consumption and phenolic carbon content for a set of five fulvic acids (r = 0.977). Their fulvic acid data also yielded good correlations between DBP formation and phenolic carbon content, with correlation coefficients ranging from 0.933 to 0.991. Similar results were not obtained for the humic acids because the phenolic carbon content was not a good surrogate for activated aromatic carbon content.

Reckhow, *et al.*, used a chlorine dose of 20 mg/L as Cl_2 with a reaction time of 72 hours at pH 7.0. In this research, chlorine was dosed to generate a free chlorine concentration of 1.0 mg/L as Cl_2 after 24 hours at pH 8.0. Initial TOC concentrations were 5 mg/L and 4 mg/L for Reckhow, *et al.*, and for this research, respectively. Both studies used a temperature of 20°C for the reaction period. The slope of each line in Figures 3 through 6 is an indication of reaction stoichiometry. In each figure, the slopes of the two regression lines are not significantly different at a 95 percent confidence level. In Figure 3, the average slope of the two lines was 3.8 mmol of chlorine consumed per mmol phenolic C initially present. For chloroform formation, the average slope was 0.11 mmol of chloroform produced for each mmol of phenolic carbon initially present. The average slopes for dichloroacetic acid and trichloroacetic acid formation were, respectively, 0.04 mmol and 0.10 mmol per mmol of phenolic carbon initially present.

Conclusions

For the five hydrophobic NOM extracts studied in this research, the normalized peak area of the phenol pyrolysis fragment was a good indicator of aromatic carbon content as estimated by ^{13}C NMR. Also, a qualitative indication of carboxylic carbon content may be obtained from the carbon dioxide peak area of pyrolysis GC/MS. No other significant correlations were observed between pyrolysis GC/MS characteristics and ^{13}C NMR characteristics.

When NOM was characterized with pyrolysis GC/MS, the normalized peak area of the phenol pyrolysis fragment was the best indicator of reactivity between NOM and chlorine. When ^{13}C NMR was used to characterize NOM, phenolic carbon content was the best indicator of reactivity between NOM and chlorine. Therefore, results from both characterization methods suggest that phenolic structures are important sources of chlorine consumption by hydrophobic NOM. These structures are also important in the formation of halogenated DBPs such as chloroform and chloroacetic acids.

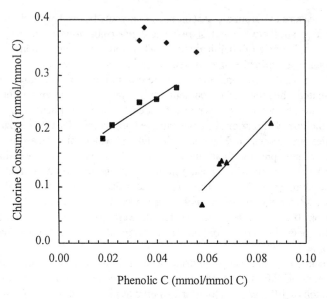

Figure 3. Relationship between chlorine consumption and initial concentration of phenolic carbon.

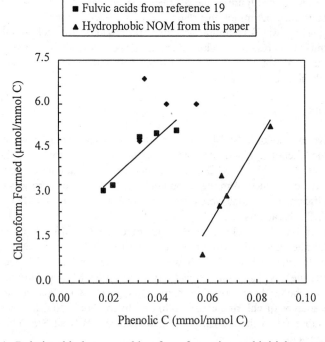

Figure 4. Relationship between chloroform formation and initial concentration of phenolic carbon.

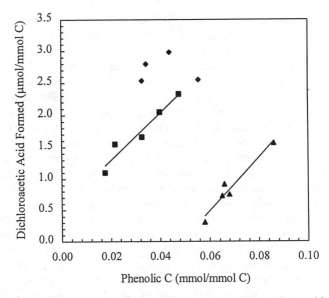

Figure 5. Relationship between dichloroacetic acid formation and initial concentration of phenolic carbon.

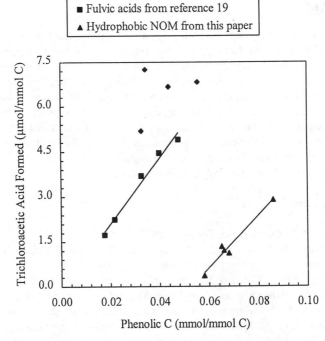

Figure 6. Relationship between trichloroacetic acid formation and initial concentration of phenolic carbon.

The chlorination results suggest that ^{13}C NMR may be a better method than pyrolysis GC/MS for characterizing NOM and its reactivity with chlorine. However, the amount of sample required for pyrolysis GC/MS analysis is considerably smaller than that required with more readily available NMR spectrometers, which operate with ^{13}C frequencies of approximately 25 MHz. Therefore, from a practical perspective, pyrolysis GC/MS may be the more attractive alternative. This is particularly true if a qualitative understanding of chlorine reactivity is sufficient. With the NMR spectrometer used in this study (100 MHz for ^{13}C), sample requirements for the two methods were approximately the same.

As noted earlier, the reader is cautioned that the above conclusions were gained from a limited set of five hydrophobic NOM extracts. This set of extracts has a relatively narrow range of aromatic carbon content (21 to 33 percent). Aromatic carbon content was also correlated with carboxylic carbon content. Additional experiments will be needed with other NOM extracts to make the above conclusions in a more definitive manner. Furthermore, hydrophobic NOM is representative of only a portion of the NOM present in a water. In the sources used for this study, 16 to 71 percent of the NOM was isolated as hydrophobic NOM. Other fractions of NOM are known to contain phenolic structures (*29,30*) and similar relationships between chlorine reactivity and phenolic content may also be expected for these other NOM fractions. However, these other NOM fractions may also contain other structures that are reactive with chlorine. Research is needed to determine whether this is indeed the case.

Acknowledgments

The contribution of personnel from the Town of Andover, MA; the City of Myrtle Beach, SC; Palm Beach County, FL; the Hackensack Water Company; and the Metropolitan Water District of Southern California is greatly appreciated. Many experiments and laboratory analyses were performed by Gretchen Cowman, Lori Harrington, Dan Schechter, and Marla Smith. Howard Weinberg provided guidance for the DBP analyses. The ^{13}C NMR experiments were performed by Weiguo Sun with the assistance of Yue Wu and Chi-Duen Poon. Financial support for this work was provided by the American Water Works Association Research Foundation and by in-kind contributions from the Centre International de Recherche Sur l'Eau et l'Environnement and from the five participating utilities.

Literature Cited

1.   Rook, J. J. *Environ. Sci. Technol.* **1977**, *11*, 478-482.
2.   Norwood, D. L.; Johnson, J. D.; Christman, R. F.; Millington, D. S. In *Water Chlorination: Environmental Impact and Health Effects*; Jolley, R. L.; Brungs, W. A.; Cotruvo, J. A.; Cumming, R. B.; Mattice, J. S.; Jacobs, V. A., Eds.; Ann Arbor Science: Ann Arbor, MI, 1981, Vol. 4; pp 191-200.
3.   Stevens, A. A.; Moore, L. A.; Slocum, C. J.; Smith, B. L.; Seeger, D. R.; Ireland, J. C. In *Aquatic Humic Substances: Influence on Fate and Treatment*

of *Pollutants*; Suffet, I. H.; MacCarthy, P., Eds.; Advances in Chemistry Series 219; American Chemical Society: Washington, DC, 1989; pp 681-695.

4. Bull, R. J.; Kopfler, F. C. In *Health Effects of Disinfectants and Disinfection By-Products*; Amer. Water Works Assoc. Research Foundation: Denver, CO, 1991.

5. Moore, G. S.; Tuthill, R. W.; Polakoff, D. W. *Jour. Amer. Water Works Assoc.* **1979**; *71*, 37-39.

6. Kavanaugh, M. C.; Trussell, A. R.; Cromer, J.; Trussell, R. R. *Jour. Amer. Water Works Assoc.* **1980**; *72*, 578-582.

7. Urano, K.; Wada, H.; Takemasa, T. *Water Res.* **1983**, *17*, 1797-1802.

8. Engerholm, B. A.; Amy, G. L. *Jour. Amer. Water Works Assoc.* **1983**; *75*, 418-423.

9. Qualls, R. G.; Johnson, J. D. *Environ. Sci. Technol.* **1983**, *17*, 692-698.

10. Amy, G. L.; Chadik, P. A.; Chowdhury, Z. K. *Jour. Amer. Water Works Assoc.* **1987**; *79*, 89-97.

11. Harrington, G. W.; Chowdhury, Z. K.; Owen, D. M. *Jour. Amer. Water Works Assoc.* **1992**; *84*, 78-87.

12. Rook, J. J. *Jour. Amer. Water Works Assoc.* **1976**; *68*, 168-172.

13. Larson, R. A.; Rockwell, A. L. *Environ. Sci. Technol.* **1979**, *13*, 325-329.

14. Norwood, D. L.; Johnson, J. D.; Christman, R. F.; Hass, J. R.; Bobenreith, M. J. *Environ. Sci. Technol.* **1980**, *14*, 187-190.

15. de Laat, J.; Merlet, N.; Dore, M. *Water Res.* **1982**, *16*, 1437-1450.

16. Boyce, S. D.; Hornig, J. F. *Environ. Sci. Technol.* **1983**, *17*, 202-211.

17. Reckhow, D. A.; Singer, P. C. In *Water Chlorination: Chemistry, Environmental Impact and Health Effects*; Jolley, R. L.; Bull, R. J.; Davis, W. P.; Katz, S.; Roberts, M. H.; Jacobs, V. A., Eds.; Lewis Publishers: Chelsea, MI, 1985, Vol. 5; pp 1229-1257.

18. Scully, F. E.; Howell, G. D.; Kravitz, R.; Jewell, J. T.; Hahn, V.; Speed, M. *Environ. Sci. Technol.* **1988**, *22*, 537-542.

19. Reckhow, D. A.; Singer, P. C.; Malcolm, R. L. *Environ. Sci. Technol.* **1990**, *24*, 1655-1664.

20. Bruchet, A.; Rousseau, C.; Mallevialle, J. *Jour. Amer. Water Works Assoc.* **1990**; *82*, 66-74.

21. Bruchet, A.; Anselme, C; Duguet, J. P.; Mallevialle, J. In *Aquatic Humic Substances: Influence on Fate and Treatment of Pollutants*; Suffet, I. H.; MacCarthy, P., Eds.; Advances in Chemistry Series 219; American Chemical Society: Washington, DC, 1989; pp 93-105.

22. Mallevialle, J.; Anselme, C; Marsigny, O. In *Aquatic Humic Substances: Influence on Fate and Treatment of Pollutants*; Suffet, I. H.; MacCarthy, P., Eds.; Advances in Chemistry Series 219; American Chemical Society: Washington, DC, 1989; pp 749-767.

23. Kögel-Knabner, I.; Hatcher, P. G.; Zech, W. *Soil Sci. Soc. Am. J.* **1991**, *55*, 241-247.

24. Benner, R.; Hatcher, P. G.; Hedges, J. I. *Geochim. Cosmochim. Acta.* **1990**, *54*, 2003-2013.

25. Norwood, D. L.; Christman, R. F.; Hatcher, P. G. *Environ. Sci. Technol.* **1987**, *21*, 791-798.
26. Baldock, J. A.; Oades, J. M.; Waters, A. G.; Peng, X.; Vassallo, A. M.; Wilson, M. A. *Biogeochem.* **1992**, *16*, 1-42.
27. Leenheer, J. A.; Wilson, M. A.; Malcolm, R. L. *Org. Geochem.* **1987**, *11*, 273-280.
28. Gillam, A. H.; Wilson, M. A. *Org. Geochem.* **1985**, *8*, 15-25.
29. McKnight, D. M.; Bencala, K. E.; Zellweger, G. W.; Aiken, G. R.; Feder, G. L.; Thorn, K. A. *Environ. Sci. Technol.* **1992**, *26*, 1388-1396.
30. Aiken, G. R.; McKnight, D. M.; Thorn, K. A.; Thurman, E. M. *Org. Geochem.* **1992**, *18*, 567-573.
31. Malcolm, R. L.; MacCarthy, P. *Environ. Sci. Technol.* **1986**, *20*, 904-911.
32. Malcolm, R. L. In *Humic Substances in Soil, Sediment, and Water: Geochemistry, Isolation, and Characterization*; Aiken, G. R.; McKnight, D. M.; Wershaw, R. L.; MacCarthy, P., Eds.; Wiley-Interscience: New York, NY, 1985; pp 181-209.
33. Thurman, E. M.; Malcolm, R. L. *Environ. Sci. Technol.* **1981**, *15*, 463-466.
34. *Standard Methods for the Examination of Water and Wastewater.* Clesceri, L. S.; Greenberg, A. E.; Trussell, R. R., eds. 17th edition; American Public Health Association, American Water Works Association, Water Pollution Control Federation: Washington, DC, 1989.
35. McGuire, M. J.; Krasner, S. W.; Reagan, K. M.; Aieta, E. M.; Jacangelo, J. G.; Patania, N. L.; Gramith, K. M. In *Disinfection By-Products in United States Drinking Waters*; United States Environmental Protection Agency and Association of Metropolitan Water Agencies: Washington, DC, 1989; Vol. 2.
36. Thurman, E. M. In *Humic Substances in Soil, Sediment, and Water: Geochemistry, Isolation, and Characterization*; Aiken, G. R.; McKnight, D. M.; Wershaw, R. L.; MacCarthy, P., Eds.; Wiley-Interscience: New York, NY, 1985; pp 87-103.

Chapter 11

Use of Pyrolysis Gas Chromatography–Mass Spectrometry to Study the Nature and Behavior of Natural Organic Matter in Water Treatment

K. A. Gray[1], A. H. Simpson[2], and K. S. McAuliffe[2,3]

[1]Department of Civil Engineering, Northwestern University,
2145 Sheridan Road, Evanston, IL 60208–3109
[2]Department of Civil Engineering and Geological Sciences,
University of Notre Dame, Notre Dame, IN 46556

Pyrolysis(PY)-GC-MS has been used to characterize the chemical nature of the organic matrix of three surface waters derived from diverse locations in North America. The chemical fingerprints of water samples treated by conventional and enhanced coagulation are presented and are statistically analyzed relative to their disinfectant byproduct (DBP) yield. This approach seeks to establish quantitative structure-function relationships between chemical components of the organic matrix and DBP formation potential, and illustrates a number of insightful trends among the pyrolysis fragments of these waters. The aromatic signature of these waters was poorly correlated to trihalomethane formation potential (THMFP). For a set of critical pyrolysis fragments identified for these samples, a general pattern was found for the precursor material of THMFP, whereas the chemical nature of precursors for haloacetic acid (HAA) appeared to be more source specific. Furthermore, the precursors to THM and HAA showed distinct chemical characteristics as revealed by PY-GC-MS.

Pyrolysis-GC-MS is an analytical technique which has been used extensively in industry, soil science and geochemistry to study the structure of complex, non-volatile, organic macromolecules and recently has been applied to the study of natural organic material (NOM) occurring in systems of interest to environmental engineers. Pyrolysis (PY) is a method that thermally cleaves an organic molecule into volatile fragments which are then separated by gas chromatography (GC) and identified by mass spectrometry (MS). Under controlled conditions this technique yields a reproducible fragmentation pattern, or fingerprint, which is highly characteristic of the parent organic material.

In this research PY-GC-MS has been used to monitor the changes in organic quality produced by the coagulation of surface waters from different locations. The primary goal of this work was to characterize by PY-GC-MS the chemical nature of disinfection byproduct (DBP) precursors in different types of waters treated by conventional and optimized coagulation. Specifically, the objectives of this work were to: 1) Chemically fingerprint by PY-GC-MS the organic matrix of raw and treated waters from 3 locations and describe qualitatively the chemical nature of NOM removed under various conditions of coagulation; 2) Evaluate statistically PY-GC-MS data

[3]Current address: 16488 NW Argyle Way, Portland, OR 97229

relative to DBP yields for each sample in order to identify chemical markers for precursors; 3) Compare the PY-GC-MS data of different waters to determine the general or site specific nature of the relationships established for DBP precursors.

At the present time it is neither possible to predict how NOM will influence treatment effectiveness or efficiency, nor is it known *a priori* to what extent DBPs will be formed in a particular treatment scenario. One of the reasons for this is that NOM structure is not known. A second factor is that NOM quality varies in time and space and the ability to monitor critical changes in NOM quality is lacking. In light of new regulatory initiatives such as the Disinfectant-Disinfection Byproduct Rule, it is critical that a better understanding of these relationships and processes be obtained.

Although PY-GC-MS is an established technique in a number of disciplines and is employed for many different applications, it has been used primarily as a method to acquire a qualitative picture of the organic character of a sample. In the analysis of well defined polymers, relatively simple PY-GC-MS fingerprints are obtained which can be evaluated to determine parent structure. This is not the case for NOM which is so polymorphous, heterogeneous and ill-defined that the elucidation of a parent structure is virtually impossible. It may be possible, however, to identify by PY-GC-MS salient structural features that can be correlated with a particular behavior in water treatment.

It is hypothesized that if PY-GC-MS data are combined with data on treatment performance and other measures of water quality, distinctive structural features (i.e., chemical markers) can be identified which may serve to predict the performance or function of NOM in treatment. Water samples have been collected from three surface waters: East Fork Lake (Cincinnati, OH), Buffalo Pound Lake (Regina, Saskatchewan, Canada), Colorado River (Metropolitan Water District of Southern California). The organic quality as characterized by PY-GC-MS has been determined for raw and treated samples of each water. These data were then analyzed statistically relative to TOC removal and DBP formation data in order to establish structure-function relationships for NOM under various scenarios.

Methods

All water samples were collected in amber glass bottles, stored at 4 °C, and sent to the University of Notre Dame via overnight delivery. Measurement of DBP formation potential (DBPFP) was conducted under uniform formation conditions (pH 8 ± 0.1, 20 °C $\pm$ 1 °C, 24 ± 1 h, and 1 ± 0.3 mg/L free chlorine residual) for the East Fork Lake samples (5) and for the Buffalo Pound Lake and Colorado River samples 7 day DBPFP values were determined at pH 7 and a 2-3 mg/L free Cl_2 residual (9). The haloacetic acid formation potential (HAAFP) was determined by summing the concentrations of the following 6 species: mono-, di-, and trichloroacetic acids, mono- and dibromoacetic acids, and bromochloroacetic acid.

The PY-GC-MS method employed in this work is described in detail elsewhere (1) and is briefly explained here. A modified version of the procedure developed by Bruchet et al. (2) has been used. Sample preparation involved filtering samples through 0.45 μm glass fiber filters and concentrating the DOC to a level of 1 mg/100 μL by a combination of rotary vacuum evaporation and room temperature evaporation under N_2. DOC losses in the concentration steps due to precipitation of inorganic salts were typically less than 10%. If salt precipitation occurred and the solid was colored, it was rinsed in an attempt to minimize DOC loss. Pyrolysis samples were prepared in quartz capillary tubes from this final concentrate by coating the tubes with 20 μL of the concentrate (200 μg DOC) and allowing them to dry overnight at room temperature. This process was repeated until a final amount of 1 mg organic carbon was reached. The specifications for each component of the analysis are shown in Table I. The quartz tubes of samples were placed inside a coiled platinum filament of a CDS Pyroprobe 2000 at an initial interface temperature of 70 °C. The interface was then ramped from 70 to 250 °C at 30 °C/sec simultaneously with the temperature ramp of the pyrolysis filament.

Table I. Specifications of PY-GC-MS method.

Analytical Step	Instrument	Conditions
Pyrolysis	Chemical Data Systems Pyroprobe 2000	Heated filament Temp ramp: 100-725 °C @ 20 °C/msec; Hold for 60 sec. Internal temperature 625 ± 5°C
Gas Chromatography	Fisons 8030	Supelcowax 10, 60 m column Splitless injection Temperature gradient: 45 °C hold for 15 min; 45-260 °C @ 2 °C/min; 260°C hold 45 min.
Mass Spectrometry	Fisons MD 800	Operated at 70 eV; Scanning 20-400 amu @ 1 scan/sec
Data Analysis	Digital 433dxLp	NIST Library

The pyrolysis temperature was ramped resistively by a booster current from an initial temperature of 70 to 725 °C at 20 °C/msec and held for 60 seconds at a final, internal temperature of 625 °C +/- 5 °C. The pyrolysis fragments were quickly swept onto a polar column of the Fisons 8030 GC. Splitless injection onto a Supelcowax 10, 60 meter column was used and compound separation and elution were achieved with the following temperature gradient: 45 °C held for 15 min, 45-260 °C ramp at 2 °C /min and 260 °C held for 45 min.

A Fisons MD 800 Mass Spectrometer operated at 70 eV served as the detector and typically scanned from 20-400 amu at 1 scan/sec. A Digital 433dxLp data system and National Institute of Standards and Technology Library were used. A combination of techniques was employed for identification of the separated pyrolysis fragments: 1) Analysis of the mass spectra; 2) Consideration of GC elution times; 3) Library match (matches with goodness of fit less than 850/1000 were rejected).

This method produced a pyrochromatogram which plotted the percentage of total ion current against the elution time. In addition to fragment identity a number of parameters described the peaks of the pyrochromatogram such as scan number, retention time, peak height and peak area. These data were read to output files in Dbase format in order to create a database allowing easy data manipulation. A database program, Microsoft Access, was used to facilitate storage of the data files generated by PY-GC-MS and Microsoft Query was used to select and sort particular data for testing various hypotheses.

A modified version of the semi-quantitative interpretation of pyrolysis data proposed by Bruchet (2) was also applied in this work. All pyrolysis fragments were considered and grouped into one of 5 classes based on the following hierarchical order: halo-substituted, nitrogen-containing, aromatic, aliphatic, unknown. For the data obtained in this study, the general chemical nature of a sample was characterized by classifying all the identified peaks into one of the above chemical classes and the percentage of total peak height was calculated for each category. This work was focused on assessing basic chemical functionality, rather than determining the biopolymeric sources of NOM. Yet, there is general agreement between the Bruchet interpretation and the approach described here, i.e. polysaccharide≈aliphatic; nitrogen containing≈protein; aromatic≈polyhydroxyaromatic.

Neither classification approach, however, is sufficiently sensitive to reveal other than major changes in the matrix due to treatment. For this reason, statistical analyses have been employed. Analysis of variance (ANOVA) has been used to compare between-sample variance and within-sample variance to determine if significant differences existed between samples taken at different points in a treatment train. While ANOVA reveals differences among samples and treatments, it fails to indicate how changes in NOM nature were related to such functions as DBPFP. Therefore, principal component analysis with and without varimax rotation and multiple regression have been used to determine how a critical set of pyrolysis peaks can be used to predict the behavior of a water, i.e., DBPFP. Complete discussions of these approaches and results are provided elsewhere (3,4). In this paper, a correlation matrix was determined for the three waters to identify the pyrolysis fragments that were most strongly correlated with a particular function such as trihalomethane formation potential (THMFP) or haloacetic acid formation potential (HAAFP). Comparison of these correlation coefficients allowed assessment of how general the structure-function relationships are for different types of waters undergoing similar treatment strategies. It is postulated that such an approach may reveal the chemical nature of precursors to DBPs as well as the chemical nature of NOM removed by a particular treatment process.

Results and Discussion

PY-GC-MS results and specific DBP yields are presented for each water. These data are then compared to identify chemical markers for DBPFP and to determine if a general relationship exists among the pyrolysis fingerprints of the samples, the patterns of organic removal produced by coagulation and the production of DBPs.

East Fork Lake. East Fork Lake is a reservoir in suburban Cincinnati which had a TOC concentration of 4.8 mg /L, a turbidity of 3.4 ntu and pH=8.2 at the time of sampling in Sept., 1993. Organic removals, THMFP and HAAFP were determined for this water at various points in a pilot scale experiment of conventional and enhanced coagulation. Details of this experiment are provided elsewhere (5). In the conventional mode, a coagulant dose of 44 ppm alum was used producing a post-coagulation pH of 7.34. Enhanced coagulation conditions were created by increasing the alum dose to 152 ppm which caused a decrease in pH to 6.57. Samples of the raw, settled, and filtered waters from each pilot treatment train have been analyzed for TOC, THMFP and HAAFP, and by PY-GC-MS.
Table II summarizes the percent removals of TOC, THMFP and HAAFP for filtered samples from the conventional and enhanced coagulation treatment trains. Thirty-five percent of the initial TOC was removed under the conditions of conventional alum coagulation resulting in a reduction of 30% in THMFP and 43% HAAFP. Under enhanced coagulation conditions a greater TOC reduction of 50% was achieved which corresponded to a 46% decrease in THMFP and 80% reduction in HAAFP.
The PY-GC-MS fingerprint of the raw water from East Fork Lake is shown in Figure 1. This water displayed a pronounced biological signature based on strong peaks of acetonitrile, toluene, acetic acid, acetamide and phenol. Substituted cyclopentenones, and propanoic acid are pyrolysis fragments typically associated with secondary polysaccharide sources (6). In Table III the classification of this pyrolysis profile into broad chemical classes is given. This autumn sample of East Fork Lake was primarily aliphatic in nature with a moderate aromatic and minor nitrogenous signature. Based on these features, the NOM quality in this lake was probably influenced by algal dynamics.
The PY-GC-MS fingerprints of the conventional and enhanced coagulation samples appear in Figures 2 and 3, respectively. These profiles were very similar to one another showing a general decrease in the peak height of all fragments relative to the peaks of acetic acid and acetamide. In other words, the organic material not removed by coagulation, the residual DOC, was characterized by strong pyrolysis peaks of acetic acid and acetamide, and displayed a weak aromatic signature relative to the raw water. By difference, then, the NOM removed by either the conventional or enhanced coagulation of this water appeared to be characterized by acetone, acetonitrile, and aromatic pyrolysis fragments.

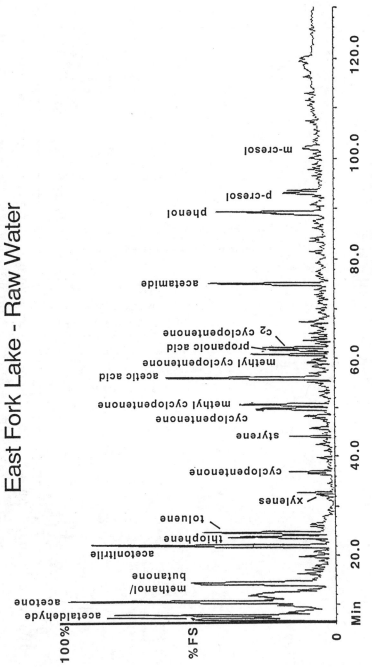

Figure 1: PY-GC-MS Fingerprint of Raw Water of East Fork Lake.

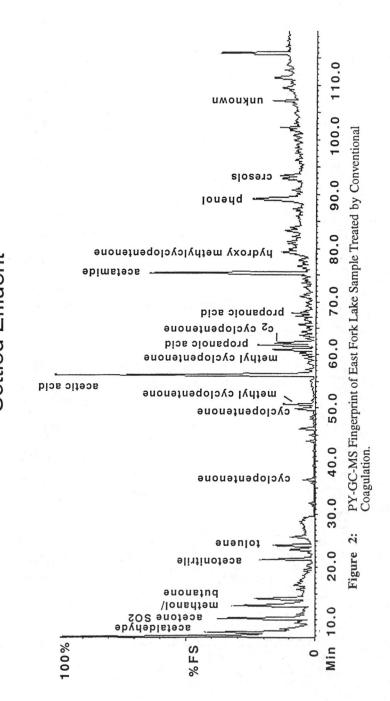

Figure 2:   PY-GC-MS Fingerprint of East Fork Lake Sample Treated by Conventional Coagulation.

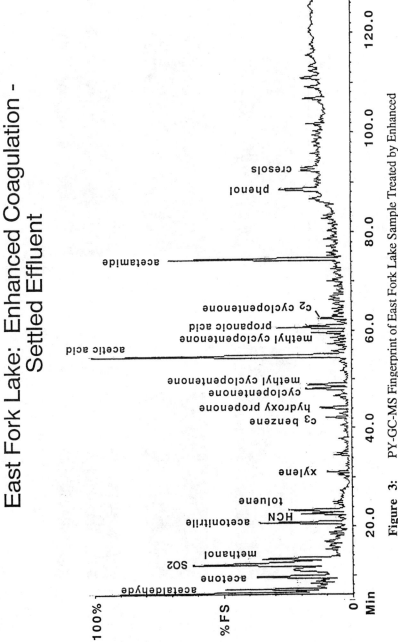

Figure 3: PY-GC-MS Fingerprint of East Fork Lake Sample Treated by Enhanced Coagulation.

East Fork Lake

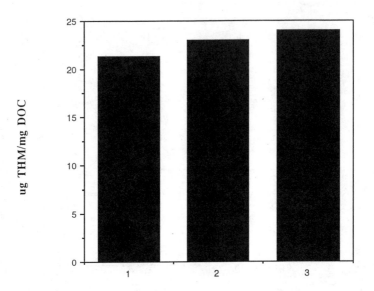

1=Raw, 2=Conventional, 3=Enhanced

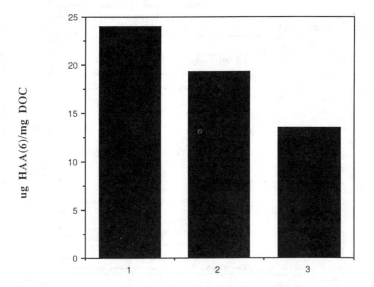

1=Raw, 2=Conventional, 3=Enhanced

Figure 4:   Normalized DBP Yields of East Fork Lake.

Table II: Percent removals of TOC, THMFP and HAAFP.

Sample	TOC	THMFP	HAAFP
East Fork Lake			
Conventional Coagulation	35	30	43
Enhanced Coagulation	50	46	80
Buffalo Pound Lake			
Conventional Coagulation	23	3	22
Enhanced Coagulation	45	20	44
Colorado River-MWDSC			
Conventional Coagulation	11	9	40
Enhanced Coagulation	50	30	65

Yet, when the pyrolysis fragments of these pyrochromatograms are categorized into the chemical classes shown in Table III, only minor differences are apparent in their general chemical nature in comparison to the raw water. In fact, when the DBP yields are normalized to the residual TOC concentration as shown in Figure 4, a similar trend is observed. Little difference was observed between the specific THM yield of the raw or treated effluents. No selective removal of THM precursors was produced by the coagulation of these waters and THMFP reduction closely followed the pattern of TOC removal. These results are rather surprising given the somewhat diminished aromatic nature of the treated effluents and the general belief that the aromatic fraction of NOM has a higher THMFP (7).

An opposite trend was observed for the specific HAA yields where, based on these data, it appeared that HAA precursors were selectively removed in the coagulation of this water. These results, however, must be considered with caution, because it has been proposed that at lower DOC levels and higher Br:DOC ratios, the kinetics of HAA formation are shifted to favor the formation of 3 additional HAAs (tribromoacetic acid, dichlorobromoacetic acid and dibromochloroacetic acid) that cannot be quantified at this time. An apparent decrease in HAA yield, then, may only be a shift in speciation. Although there are no bromide data available for these samples, if these results are valid, the precursors to HAA in this water may have a significant aromatic nature and are removed well by coagulation.

Buffalo Pound Lake. Buffalo Pound Lake is a shallow lake located in Regina, Saskatchewan, Canada having moderate levels of alkalinity and hardness and moderate to high levels of TOC. The higher TOC levels are associated with algal and plant activity that plague the lake during periods of high productivity. At the time of these samples, Nov., 1993, the turbidity was relatively low (3.5 NTU) and the TOC was moderate at approximately 4.0 mg/L (DOC=3.6 mg/L).

Samples were obtained in a pilot facility which compared the performance of conventional alum coagulation (40 ppm @ pH 7.1) to enhanced or optimized coagulation (30 ppm alum @ pH 5.5). Table II shows the TOC removals achieved by these two treatments. Conventional or baseline coagulation produced a 23% reduction in TOC, whereas optimizing the conditions (decrease in pH and alum dose) resulted in increasing the TOC removal to 45%. The enhanced coagulation mode also produced greater reductions in THMFP and HAAFP in comparison to the baseline treatment. As shown in Table II,

Table III: Chemical Classification of East Fork Lake Samples, where AL=aliphatic, AR=aromatic, N=nitrogen-containing, and UN=unknown. Values are percentages of total peak height. See text for details.

East Fork Lake	AL	AR	N	UN
Raw Water	58	24	15	3
Conventional Coagulation	48	16	23	13
Enhanced Coagulation	56	20	18	6

conventional treatment reduced the THMFP by 3% and HAAFP by 22% in comparison to a 20% THMFP and 44% HAAFP reduction in the enhanced mode.

Figures 5-7 show the PY-GC-MS fingerprints for the Buffalo Pound Lake raw and treated waters.   In general, this water was characterized by a very large number of pyrolysis fragments (> 300) many of which were substituted phenols and naphthalenes. The raw water, shown in Figure 5, exhibited major peaks of acetone, acetonitrile, phenol and various cresols, butanone/methanol, acetamide, cyclopentanones and toluene.   Based on this fragmentation pattern the classification of the chemical nature of this water (Table IV) shows that the aliphatic (AL) and aromatic (AR) fractions were equivalent at approximately 40% each of the total peak height.   The nitrogen containing peaks (N) comprised 14% and about 6% of the total peak height was unknown. This raw water also displayed a strong biological signature, but, in addition to the algal-derived material which is thought to promote the aliphatic signature, the source of the aromatic signature was probably decomposition of detrital material produced by higher plants.

The 23% TOC removal produced by conventional coagulation resulted in relatively minor changes to the general nature of the PY-GC-MS fingerprint of this water, as illustrated in Figure 6. The aromatic fragments of phenol, substituted phenols, substituted naphthalenes and toluene remained as prominent features of the profile.   The heights of these aromatic peaks relative to acetone, acetonitrile and acetamide, though, were reduced and this slight reduction in aromaticity was also reflected in the chemical classification shown in Table IV.   In contrast, enhanced coagulation resulted in not only a higher degree of TOC removal (45%), but also a dramatic change in the PY-GC-MS fingerprint.   As shown in Figure 7 acetone and acetic acid were the major peaks and there was a dramatic reduction in the aromaticity of the sample (Table IV).

In comparison to the behavior of East Fork Lake, a different relationship between TOC removal and DBPFP reduction was observed for this surface water under these coagulation conditions.   This is illustrated more clearly in Figure 8 which shows DBP yields normalized to TOC concentration.   The specific yield of THM was greater in the treated effluents than it was in the raw water. This comparison indicates that while treatment diminished the absolute amount of THMs due to the decrease in TOC, THM precursors were not preferentially removed in either treatment mode.   Rather, for this water and these treatment strategies, the precursors comprised a higher percentage of the residual TOC in the treated effluents than in the raw water. In fact, enhanced coagulation produced a greater enrichment of THM precursors. The normalized data for HAAs differed slightly, in that, although there was not selective precursor removal with treatment, there did not appear to be enrichment either, and the specific yields for each water were similar. These data must be approached with caution, however, because, as explained in the previous section, the kinetics of HAA formation may be altered at lower TOC concentrations in the presence of bromide, or at higher Br:DOC ratios, to form the 3 additional HAAs that, as yet, cannot be quantified. The bromide concentration of the raw water, however, was less than the detection limit (9). Nevertheless, while these data suggest that the specific HAA

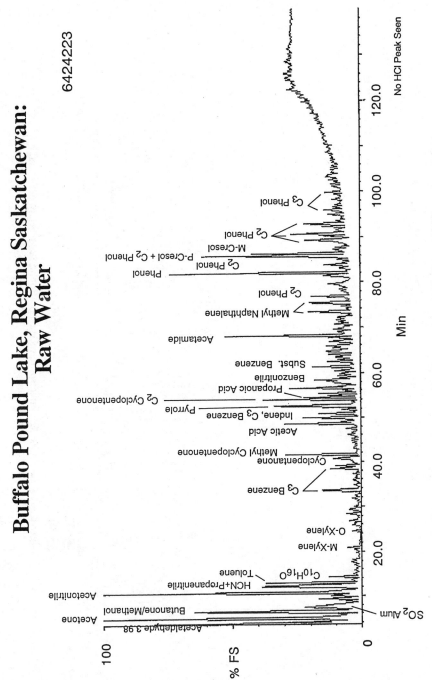

Figure 5: PY-GC-MS Fingerprint of Raw Water of Buffalo Pound Lake.

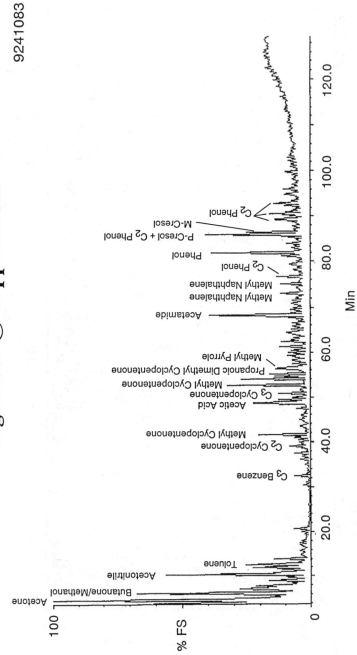

9241083

Figure 6: PY-GC-MS Fingerprint of Buffalo Pound Lake Sample Treated by Conventional Coagulation.

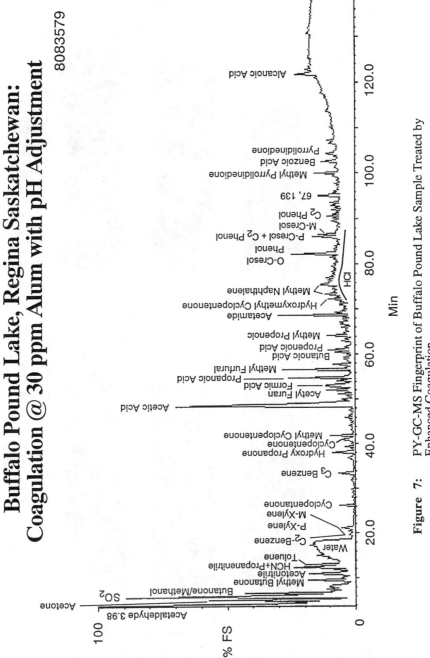

Figure 7: PY-GC-MS Fingerprint of Buffalo Pound Lake Sample Treated by Enhanced Coagulation.

Buffalo Pound Lake

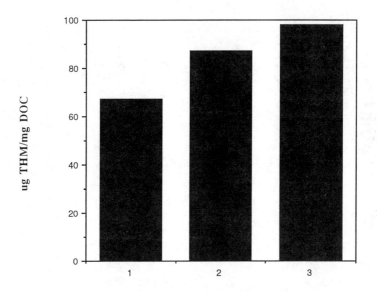

1=Raw, 2=Conventional, 3=Enhanced

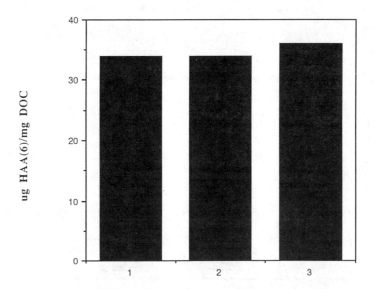

1=Raw, 2=Conventional, 3=Enhanced

Figure 8:   Normalized DBP Yields of Buffalo Pound Lake.

yield did not change with treatment, the total HAA formation potential was not measured and it is possible that higher concentrations of the 3 unmeasured HAAs were formed.

It is interesting, though, that in this water, too, the reduction in aromaticity as measured by the PY-GC-MS analysis of the sample treated by enhanced coagulation did not correspond to greater reductions in DBPFP. Instead, in this case the diminished aromatic signature was associated with an increase in the specific THM yield. These data suggest that the precursors to THM formation in this water were poorly removed by coagulation and once again, contrary to the conventional wisdom, decreases in aromatic compounds did not result in a selective removal of DBP precursors. These results suggest that the precursors to DBPs in this water either have a significant aliphatic nature, or that the precursors, especially of THMs, are unique and possibly not well characterized by PY-GC-MS.

Table IV: Chemical Classification of Buffalo Pound Lake Samples, where AL=aliphatic, AR=aromatic, N=nitrogen-containing, and UN=unknown. Values are percentages of total peak height. See text for details.

Buffalo Pound Lake	AL	AR	N	UN
Raw Water	40	40	14	6
Conventional Coagulation	46	31	13	-
Enhanced Coagulation	63	18	10	9

Metropolitan Water District of Southern California (MWDSC). Water from the Colorado River was used in a demonstration plant comparison of conventional and optimized, or enhanced, coagulation using 25 mg/L $FeCl_3$ in Feb., 1994. As shown in Table II, a TOC reduction of 11% was achieved under conventional conditions at pH 7.2, whereas a TOC reduction of 50% was produced under enhanced conditions at pH 5.7. In the first case, the TOC reduction corresponded to a 9% and 40% decrease in THMFP and HAAFP, respectively. Under optimized conditions, TOC removal produced a 30% decrease in THMFP and 65% reduction in HAAFP.

Figures 9-11 show the PY-GC-MS fingerprints of the raw, conventional and enhanced waters, respectively. As seen in Figures 9, acetamide and acetic acid were the dominant pyrolysis fragments of the influent water to this demonstration plant which was taken from the Colorado River. Various substituted phenols, naphthalenes, propanoic acids and pyrrolidinediones comprised secondary peaks. The chemical nature of this water as categorized in Table V was primarily aliphatic (50%) and its aromatic nature was 30% of the total peak height. Based on this pattern of pyrolysis fragments, it is thought that the NOM of this water was derived from both anthropogenic and natural biological sources. Although there were peaks present that can be attributed to biopolymeric sources (i.e., acetamide, methyl furfural), the biological signature seemed weaker than the anthropogenic signature due to the dominance of a few peaks and the absence of the variety of fragments usually accompanying biological productivity, i.e., typical polysaccharide and protein fragments (8).

For the most part, the raw water's chemical nature, as characterized by PY-GC-MS, was not modified to a large extent by conventional coagulation, although there appeared to be a slightly greater reduction in the parent material promoting aliphatic fragments (Figure 10). Enhanced coagulation, however, did result in more significant changes in chemical nature. As observed in Figure 11 all fragments relative to acetic acid

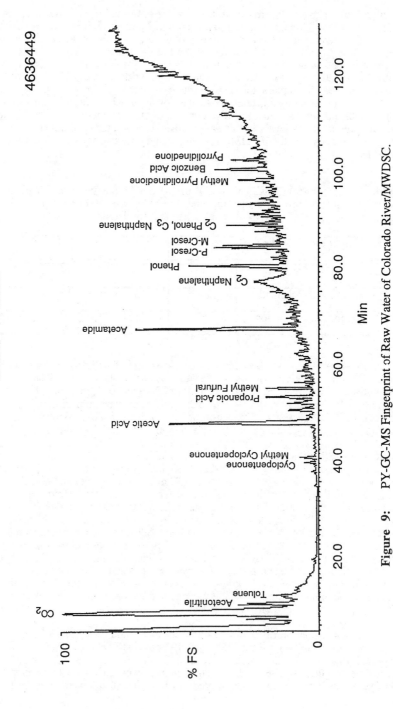

Figure 9:   PY-GC-MS Fingerprint of Raw Water of Colorado River/MWDSC.

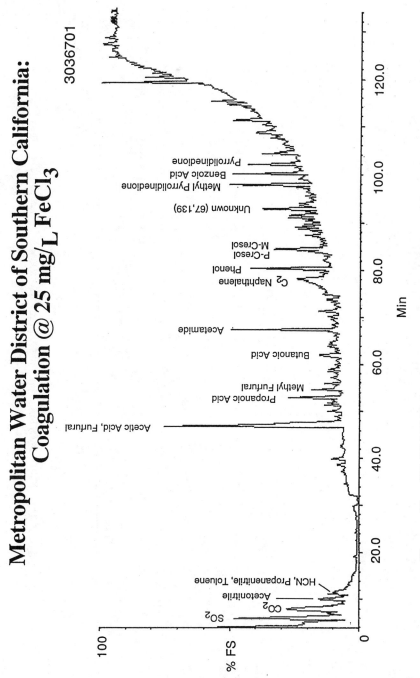

Figure 10: PY-GC-MS Fingerprint of Colorado River/MWDSC Sample Treated by Conventional Coagulation.

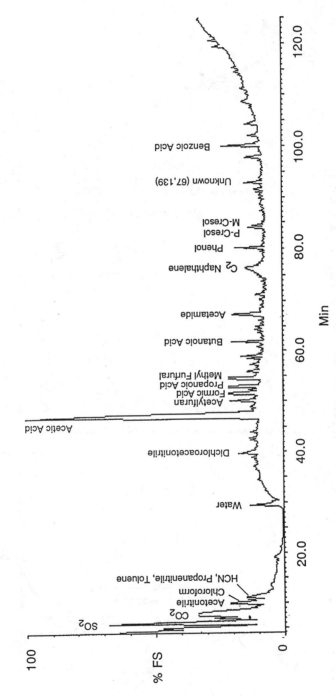

Figure 11:   PY-GC-MS Fingerprint of Colorado River/MWDSC Samples Treated by Enhanced Coagulation.

were substantially reduced resulting in a relative increase in the aliphatic fraction and reduction in the aromatic fraction (Table V).

The specific THM and HAA yields are shown in Figure 12 and the pattern of normalized DBPFP among these samples was very similar to that observed for East Fork Lake samples. Little change in specific THM yield was achieved by either treatment, but the specific HAA yield produced by both conventional and enhanced coagulation was less than that of the raw water suggesting that selective removal of HAA precursors occurred in coagulation. Although the bromide ion concentration was less than the detection limit in the raw water, these results need to be taken with caution, due to the uncertain value of Br:DOC for these waters and the unquantifiable concentrations of the 3 additional HAAs. If changes in chemical nature as characterized by PY-GC-MS are compared in a qualitative sense to the normalized DBP data shown in Figure 12, it is expected that in this water the predominant nature of the THM precursors would be aliphatic, while for HAAs it would be aromatic. This behavior is similar to that of East Fork Lake, but differs from what was observed for the Buffalo Pound Lake samples.

Table V: Chemical Classification of Colorado River, Metropolitan Water District of Southern California, where AL=aliphatic, AR=aromatic, N=nitrogen-containing, and UN=unknown. Values are percentages of total peak height. See text for details.

Colorado River-MWDSC	AL	AR	N	UN
Raw Water	50	30	13	7
Conventional Coagulation	40	31	20	8
Enhanced Coagulation	63	13	10	14

Statistical   Analysis. A single regression for each of the chemical fragments of a water's organic fingerprint against THMFP and HAAFP at each sampling point was performed. The correlation coefficients for those pyrolysis fragments that were either common to each set of samples or showed a strong relationship are listed in Table VI. These results are consistent with the qualitative observations discussed in previous sections and illustrate some general relationships between features of the organic profiles and DBPFP, as well as some site specific patterns.

In most cases, the magnitude and sign of the correlation coefficients for specific THM yield were very similar. In general, the aliphatic fragments, except the isomers of methyl cyclopentenone, were positively correlated, whereas the aromatic fragments and the isomers of methyl cyclopentenone were strongly, negatively correlated to THMFP, especially in the case of MWDSC. Among the three waters, very high negative correlations were found for m-cresol. These similarities are striking given the differences in chemical nature, NOM source, geography and treatment of these waters. Only the pyrolysis fragments of acetic acid and propanoic acid showed relatively strong positive correlation to THMFP. These fragments are not highly specific for a particular parent structure, but are typically observed in waters influenced by algal productivity and may be the pyrolysis products of an aliphatic backbone highly substituted by oxygen. In this comparison, then, THMFP was associated with very general aliphatic features, but strongly, negatively correlated to the pyrolysis fragments of phenol and the isomers of methyl cyclopentenone.

Very different trends were observed for specific HAA yield. First, the magnitude and sign of the correlation coefficients for specific HAA yield differed among these waters. While similar trends were followed by East Fork Lake and MWDSC, the pattern of correlation coefficients for Buffalo Pound Lake was directly opposite. Secondly, in the

MWDSC

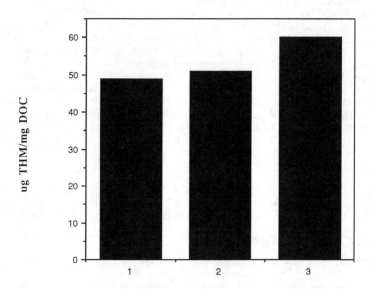

1=Raw, 2=Conventional, 3=Enhanced

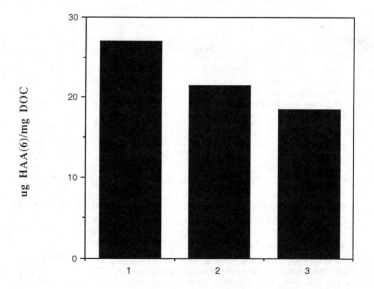

1=Raw, 2=Conventional, 3=Enhanced

Figure 12: Normalized DBP Yields of Colorado River/MWDSC.

Table VI: Comparison of Correlation Coefficients between Various Pyrolysis Fragments and Specific DBP Yields.

Fragment	East Fork Lake	Buffalo Pound Lake	Colorado River-MWDSC
	THM/TOC	*THM/TOC*	*THM/TOC*
HAA(6)/TOC	-0.76	0.57	-0.86
Acetaldhyde	0.44	0.55	-
Acetamide	0.63	0.49	-0.98
Acetic Acid	0.46	0.78	1.00
Acetone	0.16	-0.38	-
HCN	-0.50	-0.72	-
m-Cresol	-0.86	-0.99	-0.96
Methyl-cyclopen (1)	-0.86	-0.84	-0.99
Methyl-cyclopen (2)	-0.58	-0.56	-0.90
o-Cresol	-	-	-0.98
Phenol	-0.45	-0.92	-0.94
Propanoic Acid	0.60	0.57	0.96
	HAA/TOC	*HAA/TOC*	*HAA/TOC*
Acetaldhyde	-0.49	0.31	-
Acetamide	-0.97	0.14	0.94
Acetic Acid	-0.75	0.71	-0.88
Acetone	-0.44	-0.31	-
HCN	0.84	0.15	-
m-Cresol	0.69	-0.62	0.72
Methyl-cyclopen (1)	0.70	-0.89	0.91
Methyl-cyclopen (2)	0.56	-0.88	0.73
o-Cresol	-	-	0.92
Phenol	0.81	-0.56	0.65
Propanoic Acid	-0.72	0.61	-0.76

case of East Fork Lake and MWDSC, those fragments strongly and positively correlated to specific THM yield showed somewhat strong negative correlation to specific HAA yield, whereas the aromatic fragments and the isomers of methyl cyclopentenone were positively correlated with HAA yield. Although the pattern of positive and negative correlation for specific THM and HAA yields in Buffalo Pound Lake samples was similar, the values of the correlation coefficients were lower in the case of HAA in comparison to THM yield. This comparison suggests that the organic characteristics of the precursors to HAA may be less general than was the case for specific THM yield and more dependent on the type of water.

It is interesting to note that those waters (East Fork Lake and Colorado River/MWDSC) which showed similar organic fingerprints and similar removals under very different conditions of optimized coagulation, also displayed very similar trends with respect to their correlation coefficients as seen in Table VI. The specific THM yields of the MWDSC samples, however, were more than double those measured for the East Fork

Lake samples because 7-day DBPFP values have been compared to one day measurements. Therefore, while PY-GC-MS revealed similarities and differences in organic quality and illustrated trends in behavior, PY-GC-MS analysis cannot, as yet, provide a basis for making quantitative predictions about DBP yields. The method does, however, provide insight into the chemical nature of DBP precursors and in that light, this study has characterized the aliphatic nature of DBP precursors.

Conclusions

In this comparison of three diverse surface waters, a variety of optimized conditions of coagulation was found to produce high removals in TOC and varying degrees of reduction in DBPFP. In two cases (Buffalo Pound Lake and MWDSC) significant decreases in the aromatic signatures as characterized by PY-GC-MS were produced by enhanced coagulation. Yet, such dramatic reductions in aromatic nature did not consistently result in a corresponding decrease in specific DBP yield. In fact, in all three of these waters, two common aliphatic pyrolysis fragments (acetic acid and propanoic acid) were found to be positively correlated to specific THM yield and very strong negative correlations were found for phenolic fragments.

Enhanced coagulation reduced TOC concentrations and correspondingly, DBPFP. Specific DBP yield, however, was diminished in only a limited number of instances, indicating that selective removal of DBP precursors did not occur in most cases. A general aliphatic nature was found in each of these waters for THM precursors which were not preferentially removed under optimized conditions of coagulation. Preferential removal of the aromatic precursors to HAA may have occurred, but this is an uncertain trend given the unknown yields of the 3 additional and unquantified brominated HAAs. This comparison suggests that in most cases distinct precursor pools exist for THM and HAA formation and PY-GC-MS can identify waters expected to have similar structure-function relationships between NOM and DBPFP.

Acknowledgements

This work was sponsored by the U.S. EPA (Cooperative Agreement No. 821825011) and American Water Works Research Foundation (Contract No. 814-92). The authors gratefully acknowledge the contributions of Dick Miltner, Thomas Speth, Steve Randtke and Bob Hoehn.

References

1. Gray, K.A., K.S. McAuliffe, R. M. Bornick, A.H. Simpson, A.J. Horne and P. Bachand. Evaluation of Organic Quality in Prado Wetlands and Santa Ana River by PY-GC-MS. Final Report to Orange County Water District, April, 1996.
2. Bruchet, A., C. Rousseau and J. Mallevialle. *Jour.AWWA*, 1990,**8 2**:9:66-74.
3. Gray, K.A., K.S. McAuliffe and A.H. Simpson. Third and Fifth Quarterly Reports to U.S. Environmental Protection Agency, Cooperative Agreement No. 821825011, 1994.
4. Gray, K.A., K.S. McAuliffe and A.H. Simpson, "NOM Characterization of Four Surface Waters by PY-GC-MS," Final Report to American Water Works Research Foundation, May, 1996.
5. Miltner, R.J., S.A. Nolan, M.J. Dryfuse, R.S. Summers. "Evaluation of Enhanced Coagulation for DBP Control," in the Proceedings of the 1994 National Conference on Environmental Engineering (American Society of Civil Engineers, New York) 484-491.
6. Irwin, W. J. 1982. Analytical Pyrolysis (Marcel Dekker, New York).
7. Reckhow, D. A., P. C. Singer, and R. L. Malcolm. *Environ.Sci. &Technol.* 1990, **2 4**:1655-1664.
8. Gray, K.A., K. S. McAuliffe, "Use of PY-GC-MS to Fingerprint the Influences of Algal Material on NOM," presented in the seminar entitled "Natural Organics and Drinking

Water-From Ecology to Engineering," at the 1994 Annual Meeting of the American Water Works Assoc., New York, NY, June, 1994.
9. Smith, L.A., A. M. Dietrich, P.D. Mann, P. H. Hargette, W.R. Knocke, R.C. Hoehn, and S.J Randtke, "Effects of Enhanced Coagulation on Halogenated Disinfection Byproduct Formation Potentials," presented at the 1994 Annual Meeting of the American Water Works Assoc., New York, NY, June, 1994.

Chapter 12

Use of UV Spectroscopy To Study Chlorination of Natural Organic Matter

Gregory V. Korshin, Chi-Wang Li, and Mark M. Benjamin

Department of Civil Engineering, Box 352700,
University of Washington, Seattle, WA 98195–2700

Effects of chlorination on the width and intensity of absorbance of the electron-transfer band in the UV spectra of NOM are studied. For $\lambda > 250$ nm, chlorination causes the absorbance to decrease. For a wide range of chlorine doses and chlorination times, the decrease in absorbance is linearly related to the amount of $CHCl_3$ generated. The contraction of the ET band caused by chlorination is suggested to correspond to selective removal of activated aromatic rings and breakdown of NOM molecules into smaller fragments. For conditions typically encountered in drinking water treatment, chlorination of NOM substantially increases the intensity of fluorescence. This result is consistent with the oxidation of the NOM into smaller fragments. It is proposed that the width of the ET band is an indicator of the concentration of aromatic carbon in NOM, in both unaltered and chlorinated samples. This hypothesis is supported by data from ^{13}C solid-state NMR and UV absorbance measurements. It is concluded that the UV spectroscopy has significant capabilities to probe the structure and reactions of NOM.

Natural organic matter (NOM) dissolved in water both absorbs light and fluoresces (1). The bulk of NOM (typically, from <70% to >90% of the total organic carbon) is comprised of polydisperse, polymeric humic substances. These substances may be subdivided into several sub-fractions, the most important of which are humic, fulvic and hydrophilic acids (2). The rest of the organic carbon is comprised of carbohydrates, simple organic acids, proteins and amino acids. The contribution of NOM fractions other than humic substances to the absorbance and fluorescence of NOM is generally negligible (3). Since this communication deals primarily with optical properties of NOM, references to NOM in this paper refer specifically to the humic fraction of the NOM. In the UV region, and especially at $\lambda > 250$, light absorption is

0097–6156/96/0649–0182$15.00/0

due predominantly to aromatic units in the NOM structure (*4,5*). To date, no quantitative theory of UV absorbance of NOM has been proposed, and the relationships among molecular conformation and reactivity and the spectral parameters of NOM have not been adequately explored. Most researchers have limited their data collection to monitoring the absorbance at 254 nm (A_{254}), using these values as a rough indicator of the overall NOM concentration. This wavelength has been chosen primarily because it is the wavelength of the brightest line in the emission spectrum of low-pressure emission mercury lamps (*6*), and at this wavelength the absorbance of NOM normally is high enough to be measured reliably.

A_{254} is often a good surrogate parameter both for DOC and for the trihalomethane formation potential (THMFP) (*7-9*). The value of A_{254} prior to chlorination has also been used in multi-parametric statistical models used to predict THM formation (*10,11*), but the term has not shown to provide any direct insight into the chemistry of interaction between chlorine-based oxidants and NOM. Also, although chlorination of NOM dramatically decreases light absorption (*12-15*), no formal analysis of the change in A_{254} induced by chlorination has been offered. Semi-empirical kinetic models of DBP formation (e.g., *16-19*) generally make no attempt made to use spectral information, other than using the initial value of A_{254} as noted above.

There is a tremendous, unrealized potential for using absorption spectroscopy to understand the chemistry of NOM. The technology for measuring absorbance is both highly sensitive and experimentally simple. Measurements usually do not require sample preparation (except, generally, filtration to remove particles), and the range of linearity between absorbance and the concentration of NOM (measured as DOC) typically extends from below 0.1 mg/L to several tens or even hundreds of mg/L, depending of the chosen wavelength (*20-21*). This paper represents an attempt to partially bridge the gap from general statements acknowledging the utility of UV spectrophotometry to quantitative measurements describing the relationship between the electronic spectra of NOM and the properties of NOM, including its behavior in reactions with halogens.

A Hypothesis of the Compound Electron-Transfer Band in the UV spectra of Natural Organic Matter

The absorbance of light by NOM is due primarily to the presence of aromatic structures incorporated into the molecules of humic substances. Depending on the NOM's origin, aromatic carbon may constitute from <10% to >30% of the total organic carbon (*4, 5, 15, 22, 23*). Carboxylic groups are also prominent in the structure of NOM and define its largely acidic character. The aromatic rings found in NOM are not uniform and may be substituted with a variety of activating and non-activating functional groups. Given the range of average molecular weight typical for NOM (depending on the origin and experimental techniques, the estimates vary from <1000K to >5000K (*20, 21, 23-26*)), there is a virtually countless number of possible combinations of aromatic rings substitution patterns in the molecules.

In the electronic spectra of aromatic compounds, there is always an absorption band (often referred to as the electron-transfer (ET) band) centered at 240<λ<260 nm. For the simplest of aromatic compounds, benzene, the maximum of this band is found near 253 nm (27). The intensity of the ET band in the UV spectrum of benzene is relatively low because of very strong quantum-mechanical prohibition. However, the presence on the ring of polar functional groups such as -OH, -COOH, -OCH$_3$ and -O-CO- can increase the molar extinction coefficient (ε) from ca. 200 cm^2/mol for unsubstituted benzene to as high as several thousand cm^2/mol. By contrast, non-polar aliphatic groups attached to the ring do not increase the intensity of the ET transition significantly, with ε generally remaining below 300 cm^2/mol.

The shape of the ET band in any pure aromatic compound dissolved in water can be fairly well described as a Gauss function of energy. Similarly to the electronic transitions in benzene, each aromatic chromophore in more complicated NOM molecules is expected to have an ET band. The exact wavelength (or, alternatively, the energy of light quanta) of its maximum and the molar absorptivity of the band for a given chromophore reflect the chromophore's structure and environment. The light absorption by a mixture of chromophores is simply the sum of their individual contributions:

$$A_{ET}(E) = \sum_{\text{all ET bands}} \varepsilon_i c_i \exp\left[-\frac{4\ln(2)\,(E - E_{o,i})^2}{\Delta_i^2} \right] \tag{1}$$

where ε_i is the molar extinction coefficient for the ET band from a particular chromophore i, c_i is the concentration of that type of chromophore, and $E_{o,i}$ and Δ_i are the position of the maximum and the width (measured at wavelengths where the absorbance is 50% of the maximum) of the corresponding band of light absorption. Since the energy of light quanta is related to the wavelength of the light as follows:

$$E = \frac{1240}{\lambda} \tag{2}$$

where E is expressed in electron-volts (eV) and λ in nanometers, Equation 1 could be modified to express absorbance as a function of wavelength, but the function would not be Gaussian.

We have postulated that, when the absorbances from all the aromatic chromophores in NOM are superimposed, the spectrum can be modeled as having a composite ET band, which also has a Gaussian dependence on energy, i.e., we have postulated that the composite ET band may be described as:

$$A_{ET}(E) = \varepsilon_o c_o \exp\left[-\frac{4\ln(2)(E - E_{o,ET})^2}{\Delta_{ET}^2} \right] \tag{3}$$

where A_{ET}, E_o, and Δ_{ET} are parameters analogous to the individual ET bands, but calculated as the best-fit for the composite band. Similarly, ε_o and c_o are best-fit values based on the chromophore mixture in the sample. The values of c_o and ε_o cannot be determined independently by analysis of UV spectra; only their product $A_o = c_o \varepsilon_o$ may be found. Thus, characterization of the composite ET band requires assigning values of A_o, E_o and Δ_{ET} to the band. (We have postulated that the overall spectrum of NOM includes other composite bands as well. In this paper, we discuss only the ET band, and so the subscript 'ET' is omitted in all future reference to E_o and Δ.)

The use of the Gauss function is common for modeling the electronic spectra of species in solution (28). Alternative representations of spectral band shapes such as Lorenz and Voigt profiles or combinations of the Gauss and Lorentz profiles are possible as well (28). An option of using the Lorentz function or combinations of the Lorentz and Gauss functions is incorporated into some software packages (29). In our experience, however, the use of mathematical functions other than Gaussian has not been successful in modeling the UV spectra of NOM. The applicability of this band shape may be directly tested through a simple mathematical procedure. In a range of energies where bands from different electronic transitions do not overlap (for the ET band, we estimate this to be the case at ca. E<4.8 eV, or λ>260 nm), the logarithm of the absorbance is expected to be a relatively simple function of the energy of light quanta, as shown in Equation 4, which is derived directly from the Gauss formula:

$$\ln\left(\frac{A(E)}{A(E_{ref})}\right) = \frac{4\ln(2)}{\Delta^2}\left((E_{ref} - E_{o,max})^2 - (E - E_{o,max})^2\right) \qquad (4)$$

E_{ref} is an arbitrary energy chosen as the reference point for the calculation. The quadratic expression in brackets on the right side of the equation will be called the modified Gauss coordinate. According to Equation 4, the logarithm of the ratio of absorbance at any energy to that at the reference energy is proportional to the value of the modified Gauss coordinate, with a slope equal to $4(ln2)/\Delta^2$. For a given experimental spectrum, the values of all the parameters in Equation 4 are known, so the hypothesis that the spectrum is Gaussian and that it can be characterized by a single set of Gaussian parameters can be tested.

In this communication, an attempt is made to apply this conceptual approach to interpreting the changes in the values of A_o and Δ caused by chlorination of a water sample containing NOM, as well as to demonstrate possible practical application of UV spectrophotometry in the studies of NOM chlorination.

Changes of UV Spectra of NOM upon Chlorination

Water used in the experiments was from a water supply reservoir in Mt. Vernon, WA, or was NOM that was concentrated from this water by reverse osmosis or by adsorption on the surface of iron oxide (30). DOC concentrations in the raw water ranged from 3.2 to 4.0 mg/L. Reverse osmosis retained virtually 100% of the UV-absorbing species; adsorption by iron oxides retained, on average, 95% of the UV absorption. DOC retention efficiency by the two techniques was approximately 90% and 85%, respectively. The samples were buffered at pH 7.0 with phosphate (concentration 0.03 M), dosed with 0.5 to 40 mg/L chlorine, and incubated in the dark at 25°C for up to 168 hours. Chlorine consumption, DOC, UV absorbance, fluorescence, and the concentration of $CHCl_3$ were monitored. A high precision, high dynamic range Perkin-Elmer Lambda-18 spectrophotometer was used to measure the absorbance of light (typically, a 5-cm quartz cell was used). Fluorescence spectra were recorded using a Perkin-Elmer LS-50B fluorometer. The concentration of $CHCl_3$

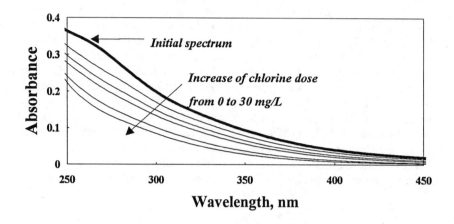

Figure 1. UV spectra of Mt. Vernon NOM subjected to chlorination. Reaction time 24 hours. Initial DOC 10 mg/L, pH 7.00. Chlorine dose varies from 0 to 30 mg/L.

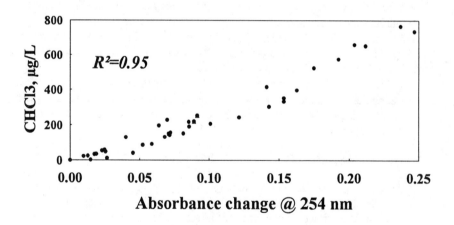

Figure 2. Relationship between the decrease of UV absorbance at 254 nm caused by chlorination and the concentration of $CHCl_3$ generated. NOM was from untreated or pre-concentrated Mt. Vernon water (initial DOC 3.6 and 10 mg/L, respectively). Reaction times were 2 to 168 hours, chlorine doses from 0 to 40 mg/L. pH 7.00, phosphate buffer.

was measured either by the purge-and-trap or the extraction method (*31*). Volatile DBPs other than $CHCl_3$ were not detected, apparently because of the absence of bromide in this water (analysis for bromide was done by ionic chromatography, detection limit was *ca.* 25 µg/L). Prior to recording the UV and fluorescence spectra and measuring the concentration of $CHCl_3$, free chlorine was quenched with sodium sulfite at a dose approximately 20% in excess of the stoichiometric requirement.

A set of UV spectra of the chlorinated NOM is presented in Figure 1. At $\lambda>230$ nm, addition of chlorine always causes UV absorbance to decrease. The relative decrease of absorbance becomes more pronounced with increasing λ. As a result, both the intensity of the band at its maximum (A_o), *i.e.* at an energy close to 4.90 eV and the bandwidth (Δ) decrease.

The decrease of absorbance is not surprising *per se*, since it is widely accepted that chlorine attacks activated aromatic rings which, in turn, constitute the predominant UV-absorbing chromophore in NOM (*32, 33*). Because of the relationship between UV absorbance and the concentration of aromatic carbon in NOM and because aromatic groups are widely believed to be DBP precursors, the change of the UV absorbance caused by chlorination might be correlated with the formation of DBPs (exemplified by $CHCl_3$ in this communication). For the sake of conformity with previous research, this hypothesis is evaluated here using absorbance data at a wavelength of 254 nm (4.88 eV), though other wavelengths (for example, 270 and 300 nm) have been used as successfully.

Based on the premise that chlorine attacks activated aromatic rings, and DBPs such as $CHCl_3$ are formed as a result, it follows that the more aromatic rings are destroyed, the more DBPs accumulate in the solution. As a first approximation, we use ΔA_{254}, defined as the decrease in absorbance at 254 nm in the chlorinated sample compared to the unchlorinated one, as a measure of the destruction of such rings. If the reactions between chlorine and the aromatic rings it attacks have a constant yield for a particular type of DBP, then a given decrease in A_{254} should correspond to a fixed amount of that DBP generated, regardless of whether the decrease in A_{254} occurs in a short period of time in a solution containing high concentrations of NOM and Cl or a much longer time in a less concentrated solution.

Figure 2 shows values of $[CHCl_3]$ for the Mt. Vernon NOM as a function of ΔA_{254} for chlorine doses from 0.5 to 40 mg/L, reaction times from 2 to 168 hours, and for two DOC concentrations. Over this wide range of parameters, a very good linear correlation exists between ΔA_{254} and the concentration of $CHCl_3$ ($R^2=0.95$). The best-fit line does not pass through the origin: at $\Delta A_{254}<0.02$, negligible amounts of $CHCl_3$ were produced even though the change of absorbance was significant. This result, which was observed in several experiments, may be explained by the presence of rapidly reacting species in solution or sites in NOM molecules that do not generate $CHCl_3$ immediately (if at all). The presence of such sites on NOM molecules has been reported (*14, 34*). Some of the reactions with NOM might produce intermediates that subsequently react to form $CHCl_3$, as suggested in (*33, 35*). When these intermediates form, incorporation of chlorine could destroy the aromaticity of the functional group (while possibly leaving the cyclical structure) and eventually cleave the oxidized group from the molecule. A decrease in UV absorbance would be detected as soon as the aromaticity is lost, before $CHCl_3$ are released, even if this DBP is ultimately formed.

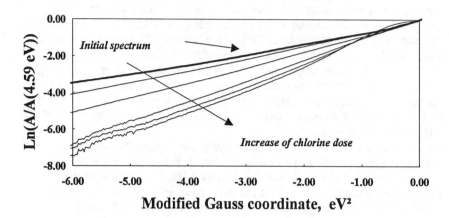

Figure 3. The spectra shown on Figure 1, represented in modified Gauss coordinates (see text). The maximum of the ET absorbance band is presumed to be located at 4.95 eV (251 nm), and the absorbance at that energy is used as a reference value in the calculations.

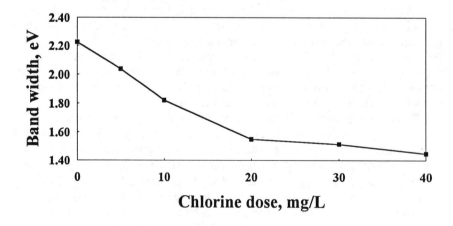

Figure 4. Dependence of the width of the ET band in the spectra of Mt. Vernon NOM on chlorine dose. Reaction time 24 hours.

Given the wide variations of reaction time and chlorine dose, the set of the spectral data is remarkably consistent. This seems to confirm that in the range of experimental conditions explored, the nature of reaction sites and the yield of $CHCl_3$ remain largely constant. Parameters such as the concentration of bromide, pH and ambient temperature may cause major changes of the speciation, kinetics and yield of the haloform reaction (*33-37*). Studies of effects of these parameters on the relationship between ΔA_λ and $[CHCl_3]$ are being conducted in our laboratory.

Correlations Between the Parameters of the ET Band and Properties of NOM

The fact that the relative decrease of absorbance becomes more prevalent with the increasing λ apparently corresponds to a contraction of the ET band caused by chlorination. A more consistent evaluation of the contraction of the ET band may be done for the absorbance spectra represented in the modified Gauss coordinates (Figure 3). In these calculations, 4.59 eV (270 nm) was selected as the reference energy. Independent numerical experiments showed that at 4.59 eV, other absorption bands typical of aromatic compounds contribute negligibly to absorbance. Furthermore, we have found that the maximum of the ET band for a wide variety of NOM sources and under varying conditions of NOM alteration by oxidation, coagulation and adsorption is in the range 4.80 to 5.0 eV (248 to 258 nm). These data will be discussed in more detail elsewhere.

The experimental data represented in Figure 3 can be fitted with a linear function with R^2 from 0.985 to 0.999. The near-linearity applies over a dynamic range of absorbances of three orders of magnitude. The best-fit values of Δ for the data in Figure 3 are shown in Figure 4. Chlorination substantially decreases the width of the ET band. From an initial value of 2.20 eV, Δ decreases almost proportionally to the chlorine dose, to 2.04 eV and 1.82 eV for 5 and 10 mg/L chlorine, respectively. Further increases in chlorine dose do not cause very significant changes, and the value of Δ stabilizes at ca. 1.5-1.6 eV at chlorine doses above 20 mg/L. Numerical experiments show that the change in Δ values far exceeds the uncertainty caused by the *a priori* estimate of the E_0 value.

It is also possible that the width of the ET band in the raw water is increased by interactions among the individual chromophores (*27, 38*). If inter-chromophore interactions broaden the spectral bands, the broadening would be most apparent in NOM molecules with a high concentration of aromatic carbon and/or high molecular weight.

Chlorination destroys many chemically-active chromophores, but many others must remain unaffected (*15*). In terms of NOM chemistry, this leads to two concurrent processes: NOM molecules in the raw water break down into smaller fragments, and aromatic units found in these smaller fragments are less activated. Furthermore, there is much less inter-chromophore interaction because of the selective oxidation of the reactive chromophores and the breakdown of the molecules.

Fluorescence spectroscopy confirms the breakdown of NOM molecules caused by chlorination. Fluorescence spectra of chlorinated Mt. Vernon NOM are shown in Figure 5. The intensity of fluorescence spectra was corrected for inner filter effect as recommended in (*39*). In contrast to the trend in UV absorbance, intensity of fluorescence increases when the sample is chlorinated. The incongruent direction of

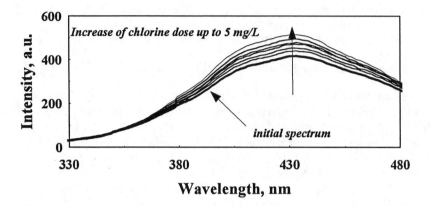

Figure 5. Emission fluorescence spectra of chlorinated Mt. Vernon NOM. 24 hours contact time, pH 7.0, DOC 10 mg/L. Excitation at 251 nm. Intensity of emission is corrected for inner filter effect.

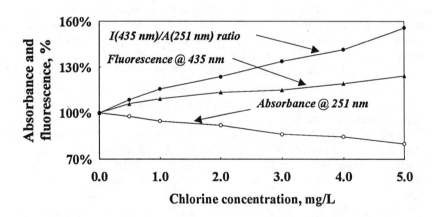

Figure 6. Relative changes of absorbance, fluorescence intensity, and the ratio of fluorescence intensity to absorbance for low doses of chlorine. Fluorescence intensity is recorded at 435 nm, UV absorbance and excitation of fluorescence at 251 nm. Mt. Vernon NOM. 24 hours contact time, pH 7.0, DOC 10 mg/L.

the changes of absorbance and fluorescence is apparent in Figure 6, in which the relative changes of absorbance and fluorescence are represented. For instance, for a reaction time of 24 hours and a chlorine dose of 5 mg/L, the absorbance at 251 nm (the wavelength used for excitation of fluorescence) decreases ca. 20% from the initial value, while fluorescence increases ca. 24%. The ratio of fluorescence intensity to the absorbance of light used for its excitation, which may be considered a surrogate parameter for the quantum yield of fluorescence, increases ca. 56%. This means that the fluorophores present in the molecules of chlorinated NOM are more likely to emit light than the fluorophores present in the initial non-chlorinated NOM, i.e. the yield of fluorescence increases. Since both the fluorophores and chromophores in NOM molecules are thought to be aromatic functional groups but exhibit completely different trends caused by NOM oxidation by chlorine, this may correspond to different types of local topochemistry within the large NOM molecules in which these functional groups are found. However, the discussion on this matter requires much more experimental data and goes beyond the scope of this communication.

At this point, the authors hypothesize that the breakdown of NOM molecules into smaller units can explain the changes in both absorbance and fluorescence. Literature reports are unanimous in confirming that decreases in the average molecular weight of NOM are always accompanied by increases in fluorescence intensity (*20, 21, 25, 40-43*). Mechanistically, this observation can be attributed to the decreased probability of radiationless losses caused by vibrations in smaller NOM molecules, which tend to have a more rigid structure than larger molecules.

The increase of fluorescence intensity caused by chlorination is not monotonous. At high chlorine doses and extended reaction times, the intensity of fluorescence may decrease rather than increase, though the ratio of fluorescence intensity to absorbance is always higher than in the initial sample. The non-monotonous change of fluorescence during chlorination may be a result of different kinetics of the reaction of the oxidant with the chromophores and fluorophores, though both types of group are aromatic units. The kinetic inequality of the reactivities of chromophores and fluorophores seems to correspond to their different topochemisty. It is our opinion that the combination of the fluorescence and absorption spectroscopy may eventually yield very important information regarding reactions of NOM, but a complete discussion on this matter is beyond the scope of current communication.

Based on the results of the absorbance and fluorescence analyses, it is concluded that chlorination of NOM is accompanied by selective oxidation of activated aromatic rings and reduction of the molecular weight of NOM molecules. These changes manifest themselves as decreases in the absorbance and contraction of the ET band, and in increases in the intensity of fluorescence of chlorinated NOM. The total aromaticity in the chlorinated NOM is decreased (as confirmed by (*15*)).

Though these effects were observed in a series of chlorinated NOM samples, it is possible that the correlation between the concentration of aromatic carbon and the value of Δ may still apply. This hypothesis is supported by independent data. For a series of independent NOM samples, the correlation between the amount of aromatic carbon and the corresponding parameters of the UV spectra may be evaluated directly if both the UV and NMR spectra of the corresponding samples are known. Reckhow et al. (*44*) presented the percentage of total and activated aromatic carbon (calculated

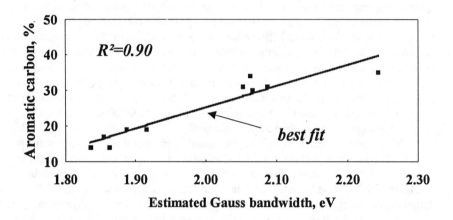

Figure 7. Estimated width of the ET band in the spectra of humic and fulvic acids from five water sources. UV and NMR data presented by Reckhow et al. [44] are used for calculations.

by integration of the NMR spectra), absorbance at 254 and 400 for fulvic and humic acids derived from five independent sources. This set of data was used to correlate between Δ with the aromaticity of NOM.

Presuming that the ET band is centered exactly at 254 nm (4.88 eV), and assuming that no other bands contribute to the absorbance at 254 and 400 nm, the value of Δ was calculated based on equation 4. The results of these calculations are a first approximation since a more precise calculation of Δ requires more extensive set of data. The estimated position of maximum of the ET band is not critical for the calculations of Δ as long as the value of E_o is between 4.85 and 4.95 eV. The calculated values of Δ are most likely less than the real number, since the reported values of A_{254} certainly contain some contributions from bands other than the ET band. The interference of bands other that the ET band is not anticipated at 400 nm.

The estimated values of Δ correlated well with the aromaticity of NOM (Figure 7). For the set of 10 samples, the correlation between aromaticity and Δ is R^2=0.90. For the ratio of A_{400}/A_{254}, which may be used as a surrogate for the value of Δ, the correlation is somewhat weaker (R^2=0.88). The correlation between the value of Δ and the percentage of activated aromatic carbon was worse than for the total aromatic carbon (R^2=0.53) though the precision in this latter case may be hampered by a limited accuracy of the data for activated aromatic carbon, the impact of which into the total organic carbon was in the range 1.2 to 4.0% for the set of samples discussed in (*44*). For these low percentages, the integration of the NMR spectra is imprecise, whereas the estimate of the total aromatic carbon was expected to be 14 to 35%.

The agreement in the interpretation of the data obtained in completely independent series of experiments confirms that the UV spectroscopy of NOM has a significant potential in probing the properties of NOM. Based on the analysis of the shape and intensity of the ET band, it seems possible to predict the total aromaticity of NOM and the rate and yield of its reactions with chlorine and other oxidants, and to explore the correlation between the UV spectra and the molecular weight of the NOM fractions as, for example, in size-exclusion chromatography.

Conclusions

The concept of representing the electron transfer band in the UV spectra of NOM by a Gauss-shaped band is introduced. The relationship between the width and intensity of absorbance of the ET band for a series of chlorinated NOM samples is explored. For λ>250 nm, chlorination causes the absorbance to decrease. The decrease of absorbance is in a very good linear correlation with the generation of $CHCl_3$. It is concluded that direct observation of absorbance changes permits to estimate the generation of DBPs and to monitor chlorination reactions *in situ*. Chlorination causes the ET band to contract. This effect is suggested to correspond to selective removal of activated aromatic rings, breakdown of NOM molecules into smaller fragments and decrease of inter-chromophore interactions. Fluorescence spectra provide direct confirmation of this hypothesis: the intensity of emission in chlorinated NOM samples is considerably increased which typically takes place for low molecular weight NOM fractions. Based on the results for chlorination, it is proposed that the width of the ET band is sensitive to the concentration of aromatic carbon in all NOM samples including those unaffected by oxidation with chlorine. Calculations based on the independent data of ^{13}C solid-state NMR and UV absorbance confirm this. It is concluded that the UV spectroscopy has significant capabilities to probe the structure and reactions of NOM.

Acknowledgments

This work has been funded by AWWA Research Foundation (grant # 159-94). Support from the engineering companies HDR Engineering, Inc. (Omaha, NE), SAUR (Maurepas, France) and DYNAMCO (West Sussex, UK) is greatly appreciated. Personal thanks to Steve Reiber (HDR Engineering) for keen interest and encouragement of this study.

References

1. MacCarthy, P.; Rice, J.A. In *Humic Substances in Soil, Sediment and Water;* Aiken, G.R. et al., Eds.; John Wiley & Sons: New York, NY, 1985.
2. Leenheer J.A. *Env.Sci. Technol.*, **1984**, *15*, 578-587.
3. Laane, R.W.P.M., Koole, L.. *Neth. J. Sea Res.*, **1982**,*15*, 217-227.
4. Traina, S.J., Novak, J., Smeck, N.E. *J. Environ. Qual.*, **1990**, *19*, 151-153.
5. Novak, J.M., Mills, G.L., Bertsch, P.M. *J. Environ. Qual.*, **1992**, *21*, 144-147.
6. *Handbook of Chemistry and Physics.* Lide, D.P., Ed. 71st Edition. CRC Press: Boca Raton, 1990.
7. Edzwald, J.K., Becker, W.C., Wattier K.L. *J. Amer. Water Works Assoc.*, **1985**, *77*, April, 122-132.
8. Singer, P.C.; Chang, S.D. *J. Amer. Water Works Assoc.*, **1989**, *81*, August, 61-65.
9. Reckhow, D.A.; Singer P.C. *J. Amer. Water Works Assoc.*, **1990**, *82*, April, 173-180.
10. Engelholm, B.A.; Amy, G.L. *J. Amer. Water Works Assoc.*, **1983**, *75*, August, 418-423.
11. Amy, G.L,. Chadik, P.A., Chowdhury, Z. *J. Amer. Water Works Assoc.*, **1987**, *79*, July, 89-97.
12. Gjessing, E.T. *Physical and Chemical Characteristics of Aquatic Humus.* Ann Arbor Science Publishers, Inc.: Ann Arbor, MI, 1976.
13. Van Breemen, A.N.; Nieuwstad, T.J.; van der Meent-Olieman, G.C. *Water Res.*, **1979**, *13*, 771-779.
14. Jensen, J.N.; Johnson, J.D.; St.Aubin, J.; Christman R.F. *Org. Geochem.*, **1985**, *8*, 71-76.
15. Hanna, J.V.; Johnson, W.D.; Quezada, R.A.; Wilson M.A.; Xiao-Qiao, L. *Environ. Sci. Technol.*, **1991**, *25*, 1160-1164.
16. Urano, K.;Wada, H.; Takemasa, T. *Water Res.*, **1983**, *17*, 1797-1802.
17. Adin , A.; Katzhendler, J.; Alkaslassy, D.; Rav-Acha, Ch. *Water Res.*, **1991**, *25*, 797-805.
18. Kavanaugh, M.C.; Trussell, A.R.; Cromer, J.; Trussell, R.R. *J. Amer. Water Works Assoc.*, **1980**, *72*, October, 579-582.
19. Peters, C.J.; Young, R.J.; Perry, R. *Env. Sci. Technol.*, **1980**, *14*, 1391-1395.

20. Levesque, M. *Soil Science,* **1972,** *113,* 346-353.
21. Smart, P.L.; Finlayson, B.L.; Rylands, W.D.; Ball C.M. *Water Res.,* **1976,** *10,* 805-811.
22. Wilson, M.A., Vassalo, A.M., Perdue, E.M., Reuter, J.H. *Anal. Chem.,* **1987,** *59,* 551-558.
23. Swift, R.S., Leonard, R.L., Newman, R.H. *Sci. Tot. Env.,* **1992,** *117/118,* 53-63.
24. Malcolm, R.L., P.MacCarthy. *Env. Sci. Technol.,* **1986,** *20,* 904-908.
25. Hall,K.J.; Lee, G.F. *Water Res.,* **1974,** *8,* 239-251.
26. Dycus, P.J.M., Healy, K.D., Wells, M.J.M. *Separat. Sci. Technol.,* **1995,** *30,* 1435-1444.
27. Jaffe, H.H.; Orchin, M. *Theory and Applications of Ultraviolet Spectroscopy.* John Wiley and Sons: New York, NY, 1962.
28. Pelikan, P.; Ceppan, M.; Liska, M. *Applications of Numerical Methods in Molecular Spectroscopy;* CRC Press: Boca Raton, 1994.
29. PeakSolve. Peak Fitting for Windows. Galactic Industries Corp., 395 Main Street, Salem, NH, 03079.
30. Benjamin, M.M.; Chang, Y.-J.; Li, C.-W.; Korshin, G.V. *NOM Adsorption onto Iron-Oxide-Coated Sand;* AWWA Research Foundation: Denver, CO, 1993.
31. *Standard Methods for the Examination of Water and Wastewater.* Clesceri, L.S.; Greenburg, A.E.; Rhodes Trussell R., Eds; 17th Edition. American Public Health Association, American Water Works Association, Water Pollution Control Federation: Washington, D.C, 1989.
32. Rook, J.J. *Water Treatment. Exam.,* **1974,** *23,* 234-243.
33. Norwood, D.L.; Cristman R.F. *Environ. Sci. Technol.,* **1987,** *21,* 791-798.
34. Qualls, R.G.; Johnson, J.D. *Env. Sci. Technol.,* **1983,** *17,* 692-698.
35. Tretyakova, N.Yu.; Lebedev, A.T.; Petrosyan V.S. *Env. Sci. Technol.,* **1994,** *28,* 606-613.
36. Stevens, I.A.; Slocum, C.J.; Seeger, D.R; Robeck, G.G. *J. Amer. Water Works Assoc.,* **1976,** *68,* November, 615-620.
37. Symons, J.M.; Fu, P.L.K.; Dressman, R.C.; Stevens, A.A. *J. Amer. Water Works Assoc.,* **1987,** *79,* September, 114-118.
38. Scott, A.I. *Interpretation of the Ultraviolet Spectra of Natural Products.* Pergamon Press: New York, NY, 1964.
39. Lakowicz, J.R. *Principles of Fluorescence Spectroscopy.* Plenum Press: New York, 1983.
40. Stewart, A.J.; Wetzel, R.G. *Limnol. Oceanogr.,* **1981,** *26,* 590-597.
41. Stewart, A.J.; Wetzel, R.G. *Limnol. Oceanogr.,* **1980,** *25,* 559-564.
42. Visser, S.A. In *Aquatic and Terrestrial Humic Materials;* Christman R.F.; Gjessing, T.E., Eds.; Ann Arbor Science: Ann Arbor, MI, 1983, p.183-202.
43. Green, S.A.; Morel, F.M.M.; Blough, N.V. *Environ. Sci.Technol.,* **1992,** *26,* 294-302.
44. Reckhow; D.A., Singer, P.C.; Malcolm, R.L. *Environ. Sci.Technol.,* **1990,** *24,* 1655-1664.

Chapter 13

Effect of Ozonation and Biotreatment on Molecular Size and Hydrophilic Fractions of Natural Organic Matter

Margarete T. Koechling[1], Hiba M. Shukairy[2], and R. Scott Summers[1]

[1]Department of Civil and Environmental Engineering, P.O. Box 210071, University of Cincinnati, Cincinnati, OH 45221–0071
[2]Post Graduate Research Program, Oak Ridge Institute for Science and Education, U.S. Environmental Protection Agency, 26 West Martin Luther King Drive, Cincinnati, OH 45268

The impact of ozonation and ozonation combined with biotreatment on the molecular size (MS) distribution, hydrophilicity, chlorine reactivity and chemical composition of a groundwater natural organic matter was investigated. Chlorination of isolated fractions was conducted under constant precursor concentrations and chlorination conditions. In the untreated solution, the <1K MS fraction was the most reactive to chlorination, and the nonhumic fraction showed the highest bromine incorporation. Ozonation caused a shift to smaller MS and hydrophilic fractions, and from a polyaromatic nature to a polysaccharidal and proteinaecous nature, as measured by pyrolysis-GC/MS analyses, but did not affect the disinfection by-product (DBP) specific yield in the larger MS fractions. It did cause a shift towards more bromosubstituted DBPs. Subsequent biotreatment was most effective in dissolved organic carbon removal for the <1K MS fraction, but did not exhibit selective DBP precursor removal in any fraction.

The use of ozonation for primary disinfection is increasing in the US (*1*). To effectively control ozonation by-products and provide a biologically stable water to the consumers, biological filtration should be used after ozonation. Ozonation oxidizes natural organic matter (NOM), resulting in an increase in polarity and hydrophilicity, and the breakdown of larger molecules into smaller ones. Ozone reacts with unsaturated and aromatic functionalities in NOM to form aldehydes, ketones and carboxylic acids (*2-5*).

In general, low molecular weight NOM that has a low ultraviolet (UV) absorbance to dissolved organic carbon (DOC) ratio (specific UV absorbance), is expected to be more easily biodegradable because of lower mass transfer resistance and perhaps due to greater accessibility for enzymatic attack. Also, saturated organic compounds seem to be more easily utilized by microorganisms leading to greater

0097–6156/96/0649–0196$15.00/0

biodegradability of NOM sources with low specific UV absorbance (*6*).   Some ozonation by-products, such as low molecular weight aldehydes and keto acids, are relatively small molecules and very biodegradable (*3,7,8*).   Therefore, ozonation enhances biodegradability which results in increased DOC removal by subsequent biofiltration (*3,6,9-12*).   For biotreatment, Collins and Vaughan (*13*) reported the largest molecular size (MS) fraction (>10K) to be more easily removed during biofiltration, whereas the overall mass in the smallest fraction (< 0.5K MS) did not change.

The use of ozonation results in some control of the halogenated disinfection by-products (DBPs) (*14-16*).   The combination of ozonation and biotreatment is very effective for DBP precursor removal as assessed by formation potential (*11,12,15,16*).   Andrews and Huck (*17*) reported that lower MS fractions exerted greater chlorine demand per mole carbon, but did not always produce more halogenated DBPs.   Owen et al. (*9*) showed increased trihalomethane formation potential (THMFP) specific yield after ozonation (1 mg O_3/mg DOC) in the <1K MS fraction, while that in the unfractionated and nonhumic fractions remained unchanged.   Jackson (*18*) reported a decrease in specific yields of THMFP and total organic halide formation potential (TOXFP) in the large and small MS fractions after ozonation.

Ozonation and biotreatment affect the speciation of the halogenated organic DBPs.   The speciation is dependent on the ratios of bromide (Br^-) to DOC, Br^- to chlorine, and the chlorination holding time (*19-22*).   Higher ratios of Br^- to DOC and Br^- to chlorine result in a shift towards more bromosubstituted DBPs, while longer holding times favor chlorosubstitution.   In general, bromosubstituted DBPs have a higher mass than similar chlorosubstituted compounds, and are of more health concern (*23*).   One way to assess DBP speciation is to use bromine incorporation factors.   THM bromine incorporation (*n*) is defined by Gould et al. (*24*) as

$$n = TTHM\text{-}Br/TTHM \ (\mu mol/\mu mol) \tag{1}$$

where TTHM-Br is the sum of the molar concentrations of the bromosubstituted THM species and total trihalomethane (TTHM) represents the sum of the molar concentrations of all THM species formed.   *n* varies between 0 and 3, where *n* = 0, if only chloroform is formed, and *n* = 3, if bromoform is the sole THM species.   Similarly for the haloacetic acids (HAAs), the HAA bromine incorporation (*n'*) is defined as the ratio of the sum of the molar concentrations of bromosubstituted HAAs to the sum of the molar concentrations of all analyzed HAA species (*14,15*).   In this study, only six of the nine HAA species were quantified and reported as HAA6; three of them are bromosubstituted (monobromoacetic acid, dibromoacetic acid, and bromochloroacetic acid).   Therefore, in this case *n'* varies from 0 to 2.

Ozonation oxidizes bromide to bromate, resulting in a decrease in the bromide concentration.   The DOC concentration normally remains almost unchanged after ozonation at doses typically used for inactivation in drinking water treatment.   The DOC concentration decreases after biotreatment, while the bromide concentration is not affected.   Thus, the bromide to DOC ratio changes with these treatment processes resulting in changes in DBP speciation (*14,15*).

Objectives

The objectives of this study were to investigate the impact of ozonation and ozonation followed by biotreatment on the MS distribution and hydrophilicity of NOM with regards to DOC distribution and DBP formation under constant precursor concentrations and chlorination conditions. Only under these constant reaction conditions can the selectivity of ozonation and ozonation combined with biotreatment be assessed. The specific yield (DBP formation per mg DOC ($\mu g/mg$)) for each MS fraction and for the hydrophilic fraction, before and after each treatment scheme, was determined. The impact of treatment on DBP speciation in each fraction was evaluated under constant precursor and chlorination conditions.

Experimental Approach

The experimental approach, presented in Figure 1, was to treat a NOM solution at high concentrations, separate it into MS and hydrophilic fractions, adjust all fractions to constant concentrations and chlorinate under constant conditions. Three treatment schemes were investigated: 1) no treatment; 2) ozonation; and 3) ozonation combined with biotreatment. Ultrafiltration and XAD-8 resin fractionation were conducted after each of the three treatment levels. The resultant MS and nonhumic fractions were chlorinated under constant organic (DOC) and inorganic (Br^-) precursor concentrations. This was done by treating and fractionating NOM solutions at high concentrations, then diluting each of the treated and untreated fractions to a constant DOC, and spiking to a constant bromide concentration prior to chlorination under uniform formation conditions (UFC) (*25*). DBP precursors were assessed by measuring dissolved organic halide (DOX), THM, and HAA6 concentrations after a 24 hour holding time and a free chlorine residual of 1.0 ± 0.4 mg/L.

　　　At least 20 percent of the samples were replicated. The error bars shown in the figures were calculated as relative percent differences for duplicates and standard deviation for triplicate samples.

Materials and Methods

Natural Organic Matter. The organic matter used in this study, Fuhrberg NOM (FuNOM), was isolated by a strong basic anion exchange resin (Lewatit MP 500 A, Bayer Chemical Co.) used in the treatment of groundwater with a high NOM content (7 mg/L DOC) at Fuhrberg (Hannover), Germany. The resin was regenerated with a solution containing 10 percent sodium chloride and 2 percent sodium hydroxide. The regenerate contained 20,000 mg/L DOC and a wide range of inorganic anions. This solution was diluted to the desired DOC concentration with laboratory clean water, adjusted to neutral pH with sulfuric acid, and filtered through a 0.22 μm membrane prior to further use. (Laboratory clean water is tap water which has been passed through reverse osmosis, ion exchange and an activated carbon column.)

Treatment. Ozonation of a concentrated NOM solution (80 mg/L DOC) was conducted at neutral pH in a bench-scale batch reactor. A transferred ozone to DOC ratio of 0.75 mg/mg was used. Liquid phase ozone residual was in the range of 0.7 to 0.9 mg O_3/L. During ozonation, the DOC concentration decreased about 10 percent. Biotreatment was performed by recirculating the diluted ozonated solution (20 mg/L DOC) for 14 hours in a bioacclimated sand filter. This time was chosen to limit DOC removal to the easily biodegradable fraction. The DOC concentration decreased by 15 percent, which corresponds to the easily biodegradable fraction for this particular NOM source (26).

Molecular Size Fractionation. Three NOM molecular size fractions, >3000 (>3K), 1000-3000 (1-3K), and <1000 (<1K), were isolated. YM3 and YM1 membranes (Amicon Corp.) with stated molecular weight cutoffs of 3000 and 1000 Daltons were used. Typically, other studies (13,27) that attempted to evaluate intermediate MS fractions have used mass balances between the influent solutions and permeates to calculate water quality parameters for the intermediate MS fractions. In this study, isolated MS fractions were used for chlorination experiments. The retentate of the YM3 membrane was used for the >3K MS fraction, while the permeate was filtered through a YM1 membrane to yield a retentate, the 1-3K MS fraction, and the <1K MS permeate. Prior to use, the retentates were diafiltered with three cell volumes of laboratory clean water, which should result in at least 95 % removal of molecules smaller than the MS cutoff of that membrane (28). Thus, this procedure yields isolated MS fractions of NOM. To minimize any biodegradation during ultra-filtration of the ozonated solutions, the UF system and all necessary glassware were sterilized before usage.

Humic/Nonhumic Fractionation. The hydrophobic/hydrophilic (humic/nonhumic) fractionation was conducted based on the method proposed by Leenheer (29). All treated and untreated samples were adjusted to 10 mg/L DOC, pH of 2.0, and run through the XAD-8 resin (Supelco) column at a flowrate of 4 mL/min. The empty bed contact time of the resin column was 2.2 min. The resin effluent was collected until a k' value of 120 was achieved, and was termed the nonhumic (hydrophilic) fraction. The k' value is the ratio of NOM adsorbed onto the resin to the NOM dissolved in the liquid phase. It is approximately computed as the ratio of treated sample volume, V, to the void volume of the resin column, V_0: k' = V/V_0 - 1 (29). Higher k' values indicate a more hydrophobic character of the NOM in the effluent, while lower values refer to a more hydrophilic effluent.

Chlorination. Prior to chlorination, all fractionated and unfractionated solutions were adjusted to constant precursor concentrations: 2 mg/L DOC and 100 μg/L Br$^-$. Chlorination was done under UFC: pH = 8.0 ± 0.2, incubation temperature = 20 ± 1°C, holding time = 24 ± 1 hours, free Cl_2-residual = 1.0 ± 0.4 mg/L (25). At the end of the holding time in the dark, the residual chlorine was quenched with anhydrous sodium sulfite, and the DBPs (DOX, THMs, HAA6) were preserved and measured according to the recommended methods.

Analytical Methods. A description of the analytical methods used in this study and their minimum detection limit (MDL) is given in Table I.

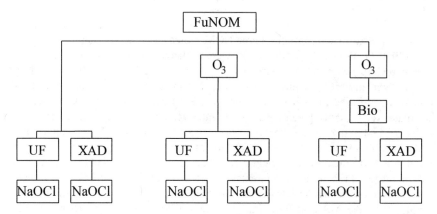

Figure 1. Experimental approach

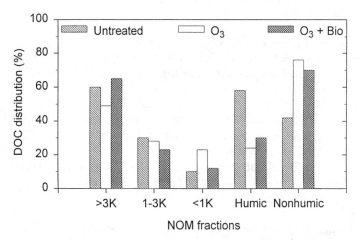

Figure 2. Effect of treatment on DOC distribution

Table I. Analytical Methods

Parameter	Method	MDL	Reference
DOC	5310 C	0.1 mg/L	30
DOX	5320 B	3 µg Cl⁻/L	30
THM	502.2, Revision 2.1	2 µg/L	31
HAA	6251 B	0.1 µg/L for DCAA, TCAA, BCAA, DBAA 1.4 µg/L for CAA and BAA	30
Cl_2-residual	4500-Cl D	0.1 mg/L	30
Cl_2-strength	4500-Cl B	0.17 mg/mL	30
O_3-residual	4500-O_3 B	0.02 mg/L	30
O_3-applied dose and off-gas	422	30 µg/L	32
UV	5910 B	0.01/m	30
Bromide	IC, Method 300	8.3 µg/L	33

Results and Discussion

DOC Distribution. Figure 2 shows the impact of ozonation and ozonation combined with biotreatment on the MS and humic/nonhumic DOC distribution of the FuNOM. For the untreated solution, 60 percent of the total DOC was found in the >3K MS fraction. The 1-3K and <1K MS fractions accounted for 30 and 10 percent, respectively. Ozonation of the FuNOM solution decreased the DOC in the >3K MS fraction to 50 percent with a corresponding increase in the <1K MS fraction to 25 percent, indicating the breakdown of larger NOM molecules into smaller ones. Subsequent biotreatment of the ozonated FuNOM solution was most effective for the DOC removal in the <1K MS fraction, with some decrease observed in the 1-3K MS fraction. This decrease in the DOC concentration in the smaller MS fractions yielded an increase in the relative amount in the >3K MS fraction.

The untreated FuNOM solution was composed of 58 percent humic and 42 percent nonhumic DOC. Ozonation increased the nonhumic fraction to 75 percent, indicating that the NOM changed in character from a hydrophobic to a dominant hydrophilic nature. Subsequent biotreatment decreased the nonhumic fraction to 70 percent of the total DOC, suggesting some biodegradability in this fraction. The DOC distribution of the humic fraction was calculated as the difference between the influent and the nonhumic DOC concentrations.

The shift towards smaller MS fractions and more hydrophilic compounds after ozonation observed in this study is in accord with other studies (9,10,34). The >3K MS and humic fractions were found to be the least biodegradable. The biotreatment contact time in this study was 14 hours, which removed only the easily biodegradable fraction of the NOM, which seemed to be contained in the <1K MS and the nonhumic

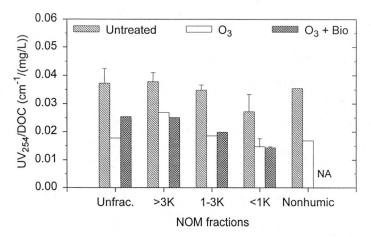

Figure 3. Effect of treatment on UV absorbance
NA: not analyzed

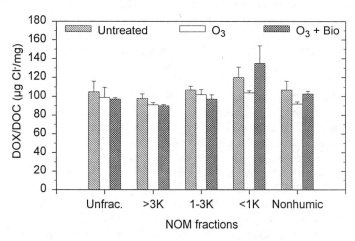

Figure 4. Effect of treatment on DOX formation

fractions.   Collins and Vaughan (*13*) utilized a biotreatment contact time between 5 and 7 days, and found biodegradation of other fractions.

UV Absorbance.   The effect of treatment on the MS and hydrophobic/hydrophilic distribution of FuNOM as assessed by specific UV absorbance is shown in Figure 3. The results are plotted as the specific UV absorbance (ratio of UV absorbance at a wavelength of 254 nm to the DOC concentration).   Prior to treatment all but the <1K MS fraction had nearly the same specific UV absorbance.   There seemed to be a slight decrease in specific UV absorbance with decreasing molecular size.   Ozonation caused a large decrease in all examined fractions and in the unfractionated solution, indicating selective oxidation of the UV absorbing functional groups.   These results are in agreement with other studies (*9,10,35*).   Subsequent biotreatment showed no further selectivity for UV absorbing compounds, but removed UV absorbance and DOC to the same extent.   In another study for the same unfractionated NOM source, biological oxidation resulted in equivalent removal of DOC and UV absorbing compounds (*22*).

DBP Formation.   Figures 4 and 5 show the DOX and TTHM specific yields, respectively, for the investigated NOM fractions.   The NOM was fractionated before and after treatment at DOC concentrations of about 16 mg/L to allow the fractionated solutions to be diluted to a constant DOC concentration of 2 mg/L prior to chlorination.   Therefore, any differences in the specific yields can be attributed to changes in NOM characteristics by treatment and not to differences in precursor concentrations and/or chlorination conditions.

DOX and TTHM specific yields behaved similarly.   Overall, the specific yield of all fractions in the untreated samples were similar, with an average of 107 μg Cl⁻ /mg DOC for DOX and 36 μg/mg DOC for TTHMs, while a trend of increasing yields with decreasing MS was indicated.   The <1K MS fraction exhibited the highest formation.   Similar results were reported by Owen et al. (*9*) for THM formation potential of different waters from southwestern US rivers.   In those studies, however, the precursor concentrations changed with treatment.

Treatment did not significantly affect the DOX or the TTHM specific yields of the unfractionated solution or the larger MS fractions.   The specific yield of the <1K MS fraction decreased upon ozonation, indicating the possibility of selective oxidation of DOX and THM precursors in this fraction.   After subsequent biotreatment the specific yield in this fraction increased, suggesting that the less reactive compounds created by ozonation may be more easily biodegradable, and thus are removed by biotreatment to a higher extent than the more reactive precursors. Thus, biotreatment results in NOM having a higher percentage of reactive precursors which could not be utilized by the microorganisms at this contact time.   Therefore, the specific yield increased.

Although ozonation increased the fraction of nonhumic NOM (Figure 2), the specific yields of DOX and TTHM were not affected by ozonation or subsequent biotreatment, indicating that neither of the two treatment processes selectively removed DOX or THM precursors.

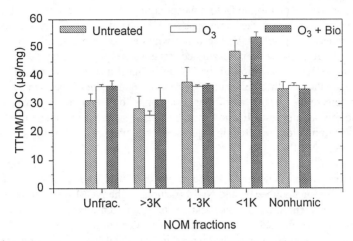

Figure 5. Effect of treatment on TTHM formation

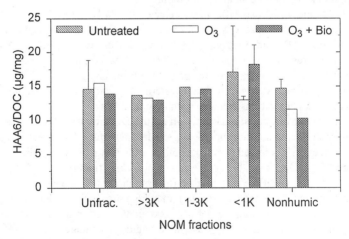

Figure 6. Effect of treatment on HAA6 formation

The HAA6 specific yields are shown in Figure 6. The results of the untreated solutions are similar to those found for DOX and TTHM specific yields. The <1K MS fraction seemed to have a slightly higher HAA6 specific yield compared to the average of 15 µg/mg DOC. The unfractionated solution and the two larger MS fractions were not significantly affected by either of the two investigated treatment processes. The largest impact was observed for the <1K MS and the nonhumic fraction. For the <1K MS fraction, the HAA6 specific yield decreased after ozonation, indicating the possibility for selective oxidation of the precursor compounds. After biotreatment, the specific yield increased, which can be attributed to the behavior as discussed for the TTHM formation.

For the nonhumic fraction, the specific yield decreased after both ozonation and ozonation coupled with biotreatment, indicating selective removal of hydrophilic HAA6 precursors. However, care should be taken in drawing definite conclusions about the impact of treatment on total HAA precursors, as three of the nine HAA species were not quantified.

Figure 7 demonstrates the effect of treatment on the THM bromine incorporation factor, n. An average n value of 0.4 was found for the untreated solutions, except for the nonhumic fraction which exhibited a 50 percent higher THM bromine incorporation. Under the constant bromide, DOC and chlorination conditions of this study, ozonation resulted in significant increases, 50 to 100 percent, in bromosubstitution, for the unfractionated and the MS fractions. Similar behavior was seen for n', the HAA6 bromine incorporation factor, as shown in Figure 8. Ozonation converts fast reacting precursors to slower ones, oxidizing aromatic and unsaturated functional groups to yield aliphatic ketones, aldehydes and carboxylic acids (5). Activated sites, such as α-protons to carbonyl groups, may be prone to bromosubstitution because of the faster reaction kinetics of bromine compared to chlorine. Bromosubstituted groups are bulky and have the tendency to leave the parent compound more easily because of steric effects (22). Merlet (36) has shown a higher yield of bromoform formation from acetyl-function compounds. The results from this study indicate that ozonation changes the characteristics of NOM to a form that may be more susceptible to bromosubstitution, while destroying other sites that may have become chlorosubstituted. Therefore, n and n' increased upon ozonation. Subsequent biotreatment had little impact on DBP speciation.

Owen et al. (9) reported a slightly higher THMFP specific yield for the <1K MS fraction compared to the >1K MS fraction for different untreated water sources. The THM specific yield of the <1K MS fraction increased after ozonation. The percentage brominated THMs were significantly greater in the nonhumic and <1K MS fractions than in the unfractionated bulk waters. In another study, Jackson (18) found that THM and TOX specific yields in large and small MS fractions decreased with increasing ozone doses for Ohio River water that was spiked to high bromide concentrations. Andrews and Huck (17) reported that ozonation increased THM yield for lake water and its humic fraction and HAA yield for river water, while it decreased HAA yield for lake water fractions, indicating a source dependency for the formation of these halogenated DBPs. Shukairy et al. (35) investigated ozonation (0.9 mg O_3/mg DOC) and biotreatment (5 day contact time) for DBP precursor removal in MS fractions for the same groundwater source used in this study. TOXFP

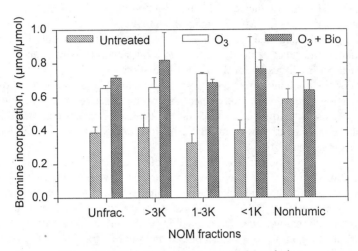

Figure 7. Effect of treatment on THM speciation

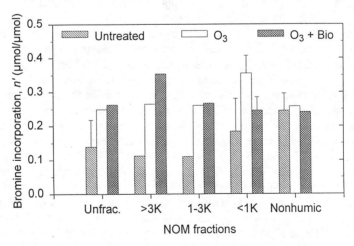

Figure 8. Effect of treatment on HAA6 speciation

specific yield decreased slightly in each MS fraction after both ozonation and subsequent biotreatment, indicating some selective TOX precursor oxidation. A slight increase in THMFP specific yield was observed after biotreatment in all fractions, with and without preozonation. Also, a shift to more brominated THMs was found after ozonation. However, in all of the above cited studies, DOC and bromide concentrations were not adjusted before chlorination. Therefore, the changes in the Br⁻ to DOC ratio with treatment and fractionation may have controlled DBP formation and speciation.

Pyrolysis - Gas Chromatography/Mass Spectrometry Results. Because chlorination was conducted under constant organic (DOC) and inorganic (Br⁻) precursor concentrations and chlorination conditions, the effects seen with treatment were caused solely by changes in the characteristics of the organic matter. In an attempt to investigate any chemical composition differences, the different NOM fractions were analyzed by pyrolysis - gas chromatography/mass spectrometry (PY-GC/MS) and reported according to Bruchet's biopolymer groupings (*37*). The results are presented as percent composition based on four polymer groupings in Table II. FuNOM was found to be composed of polysaccharides, proteins, and polyhydroxyaromatic groupings. No aminosugars were detected in the unfractionated solution or in any of the fractions, with or without treatment. The main trend that can be deduced from these results is that the largest percentage of polymer groupings in FuNOM (including fractions) is of polyhydroxyaromatic character. Also, the

Table II. Pyrolysis - gas chromatography/mass spectrometry analyses (% composition)

Sample	Polysaccharides	Proteins	Polyhydroxy-aromatics
Unfractionated			
untreated	23	14	63
after O_3	36	17	48
after O_3 + bio	33	17	51
>3K MS fraction			
untreated	22	11	67
after O_3	38	17	45
after O_3 + bio	38	13	49
1-3K MS fraction			
untreated	32	11	56
after O_3	39	19	43
after O_3 + bio	39	19	42
<1K MS fraction			
untreated	10	15	75
after O_3	24	29	47
after O_3 + bio	19	30	51

polysaccharide percent composition in the untreated solutions is slightly lower in the <1K MS fraction and slightly higher in the 1-3K MS fraction in comparison to the >3K MS fraction and the unfractionated solution. Ozonation yielded a decrease in the percentage of polyhydroxyaromatic composition and a slight increase in the percentage of polysaccharidal and proteinaceous composition. The <1K MS fraction appeared to be the most affected by ozonation. Biotreatment following ozonation did not seem to change the organic matter composition more than ozonation alone, which is similar to the results deduced from chlorination studies for the same NOM as shown by the DBP results in this and other studies (22). The decrease in UV absorbance due to ozonation observed for the unfractionated FuNOM and all the fractions, Figure 3, may well reflect the observed decrease in the percentage of the polyhydroxyaromatic composition. However, no correlations between DBP formation and PY-GC/MS results can be attempted at this stage. Fromme et al. (38) have presented data that suggest that these biopolymer groupings do not correlate well to DBP formation. PY-GC/MS results as reported herein are only one step in the direction of trying to describe the nature of NOM and its behavior with treatment.

Summary

The NOM investigated in this study had a DOC MS distribution of 60 percent >3K, 30 percent 1-3K, and 10 percent <1K. About 60 percent of the NOM was found to be hydrophobic (humic), and 40 percent to be hydrophilic (nonhumic) in nature. Characterization by pyrolysis-GC/MS showed that the percent composition in the polyhydroxyaromatic group dominated the unfractionated solution as well as all MS fractions.

Ozonation caused a shift to the smaller MS and hydrophilic fractions, indicating the breakdown of larger molecules and the creation of more polar compounds. Ozonation also decreased the specific UV absorbance, indicating selective reaction with UV absorbing functional groups in all fractions. This was also reflected in a decrease in the percent composition of the polyhydroxyaromatic group and an increase in the polysaccharide group as shown by the PY-GC/MS results. Biotreatment was most effective in removing DOC in the <1K MS fraction, but did not change the nature of the NOM as assessed by the above measures.

The <1K MS fraction had a higher reactivity to chlorination compared to the other fractions and the unfractionated solution, as seen by the specific DBP yields. Little selectivity for DBP precursor removal by ozonation or subsequent biotreatment was found, except in the <1K MS fraction which showed a slightly lower specific DBP yield after ozonation and a slightly higher one after subsequent biotreatment.

The nonhumic fraction showed higher bromine incorporation in the untreated NOM solution compared to the MS fractions. An increase in THM and HAA6 bromine incorporation factors, n and n', respectively, was observed upon ozonation in all fractions, indicating a shift towards more bromosubstituted compounds.

It is important to note that the DBP formation in this study was investigated under constant organic (DOC) and inorganic (Br⁻) precursor concentrations and chlorination conditions before and after treatment. Thus, any changes in the specific yield could be attributed to changes in the NOM characteristics by treatment. Little selectivity was found for DBP precursor removal, indicating that neither ozonation alone nor ozonation coupled with biotreatment preferentially removed DBP precursors. This leads to the conclusion that lower DBP formation after ozonation and ozonation combined with biotreatment reported in other studies occurred because of an overall removal of organic and inorganic precursors. Thus, the use of these processes will be helpful for utilities, as they lower the DOC concentration and subsequent DBP formation.

Acknowledgments

The authors would like to acknowledge Thomas Speth, USEPA - Water Supply and Water Resources Division, and William Fromme, now Cincinnati Water Works, for the pyrolysis - GC/MS analyses and their help in the data interpretation. This research was funded by the USEPA - Water Supply and Water Resources Division through cooperative agreement CR 821891, but the views expressed here are those of the authors and do not necessarily reflect the views of the USEPA. Mention of trade names does not constitute endorsement or recommendation for use by the USEPA.

Literature Cited

1. Rice, R.G. *Ozone Sci. & Eng.*, **1995**, 17 (5)
2. Glaze, W.H.; Koga, M.; Cancilla, D. *Environm. Sci. & Techn.* **1989**, 23 (7), 838
3. Schechter, D.S. ; Singer, P.C. *Ozone Sci. & Eng.* **1995**, 17 (1), 53
4. Killops, S.D. *Water Research* **1986**, 20 (2), 153
5. Langlais, B.; Reckhow, D.A.; Brink, D.R. (Eds.) *Ozone in Water Treatment: Application and Engineering,* Cooperative Research Report, AWWA Research Foundation and Compagnie Generale des Eaux, Lewis Publisher, Chelsea, Michigan, 1991
6. Goel, S.; Hozalski, R.M.; Bouwer, E.J. *J. AWWA* **1995**, 87 (1), 90
7. Miltner, R.J.; Summers, R.S. Proc. of *AWWA - Annual Conf.* **1992**, Vancouver, BC, p. 181
8. Shukairy, H.M. Ph.D. dissertation, **1994**, University of Cincinnati, Cincinnati, Ohio
9. Owen, D.M.; Amy, G.L.; Chowdury, Z.K.; Paode, R.; McCoy, G.; Viscosil, K. *J. AWWA* **1995**, 87 (1), 46
10. Takahashi, N.; Naki, T.; Satoh, Y.; Katoh, Y. *Ozone Sci. & Eng.* **1995**, 17(5), 511
11. Wang, J.Z.; Summers, R.S.; Miltner, R.J. *J. AWWA* **1995**, 87 (12), 55
12. Shukairy, H.M.; Miltner, R.J.; Summers, R.S. *J. AWWA* **1995**, 87 (10) 71
13. Collins, M.R.; Vaughan, C.W. Proc. of *AWWA-WQT Conf.* **1993**, Miami, Florida, p. 1249
14. Shukairy, H.M.; Miltner, R.J.; Summers, R.S. *J. AWWA* **1994a**, 86 (6) 72
15. Symons, J.M.; Speitel, G.E.; Diehl, A.C.; Sorensen, Jr., H.W. *J. AWWA* **1994**, 86 (6) 48
16. Speitel, G.E.. Jr.; Symons, J.M.; Diehl, A.C.; Sorenson, H.W.; Cipparone, L.A. *J. AWWA* **1993**, 85 (5) 86

17. Andrews, S.A.; Huck, P.M. Proc. of *AWWA - WQT Conf.* **1993**, Miami, Florida, p. 177

18. Jackson, J.L. **1993**, M.S. Thesis, University of Cincinnati, Cincinnati, Ohio

19. Amy, G.L.; Tan, L.; Marshall, K.D. *Water Research* **1991**, *25* (2), 191

20. Summers, R.S.; Benz, M.A.; Shukairy, H.M.; Cummings, L. *J. AWWA* **1993**, *85* (1), 88

21. Symons, J.M.; Krasner, S.M.; Sclimenti, M.J.; Simms, L.A.; Sorensen, H.W., Jr.; Speitel, G.R., Jr.; Diehl, A.C. In *Disinfection By-Products in Water Treatment: The Chemistry of Their Formation and Control;* Minear, R.A.; Amy, G.L., Eds.; CRC Lewis Publishers, Boca Raton, FL, 1996; pp 91-130

22. Shukairy, H.M.; Summers, R.S. In *Disinfection By-Products in Water Treatment: The Chemistry of Their Formation and Control;* Minear, R.A.; Amy, G.L., Eds.; CRC Lewis Publishers, Boca Raton, Florida, 1996; pp 311-335

23. Bull, R.J.; Kopfler, F.C. *Health Effects of Disinfectants and Disinfection By-Products,* AWWA Research Foundation, Denver, CO, 1991

24. Gould, J.P.; Fitchorn, L.E.; Urheim, E. In *Water Chlorination: Environmental Impact and Health Effects;* Jolley, R.L. et al., Ed.; Ann Arbor Science Publishers; Ann Arbor, MI, 1983; Vol. 4

25. Summers, R.S.; Hooper, S.M.; Shukairy, H.M.; Solarik, G.; Owen, D.M. *J. AWWA* **1996**, *88* (6)

26. Wang, J.Z.; Summers, R.S. Proc. of Nat. *ASCE Conf.* **1994**, Boulder, Colorado, p. 452

27. Jackson, J.L.; Hong, S.; Summers, R.S. Proc. of *AWWA - WQT Conf.* **1993**, Miami, Florida, p. 93

28. Ingham, K.C.; Busby, T.F.; Sahlestrom, Y.; Castino, F. In *Ultrafiltration Membranes and Applications;* Cooper, A.R., Ed.; Plenum Press, New York, NY, 1980, pp 141-158

29. Leenheer, J.A. *Environm. Sci. & Techn.* **1981**, *15* (5), 578

30. *Standard Methods for the Examination of Water and Wastewater,* APHA, AWWA and WEF, Washington, D.C., 19th ed., 1995

31. USEPA *Methods for the Determination of Organic Compounds in Drinking Water*, 1988, EPA/600/4-88/039

32. *Standard Methods for the Examination of Water and Wastewater,* APHA, AWWA and WEF, Washington, D.C., 16th ed., 1985

33. USEPA *Methods for the Determination of Inorganic Substances in Environmental Samples*, 1993, EPA/600/R/93/100 - Draft

34. (Solarik, G.; Hooper, S.M.; Summers, R.S.; Owen, D.M. accepted for presentation at the Annual *AWWA Conference* in Toronto, Canada, 1996)

35. Shukairy, H.M.; Koechling, M.T.; Summers, R.S. Proc. of Annual *AWWA Conf.* **1994b**, New York, NY, p. 941

36. Merlet, N. Ph.D. diss., **1986**, Université de Poitiers, Poitiers, France

37. Bruchet, A.; Rousseau, C.; Mallevialle, J. *J. AWWA* **1990**, *82* (9), 66

38. Fromme, W.R.; Speth, T.F.; Summers, R.S. In *Preprints of Papers* presented at the 210th ACS National Meeting, August 20-24, 1995, Chicago, IL, *35* (2), 631

Chapter 14

Natural Organic Matter Characterization and Treatability by Biological Activated Carbon Filtration

Croton Reservoir Case Study

C. M. Klevens[1], M. R. Collins[1,6], R. Negm[2], M. F. Farrar[3], G. P. Fulton[4], and R. Mastronardi[5]

[1]Department of Civil Engineering, College of Engineering and Physical Sciences, Kingsbury Hall, and [2]Department of Microbiology, College of Life Science and Agriculture, Rudman Hall, University of New Hampshire, Durham, NH 03824
[3]Metcalf & Eddy of New York, Inc., 603 42nd Street, New York, NY 10165
[4]Hazen & Sawyer, P.C., 730 Broadway, New York, NY 10003
[5]New York City Department of Environmental Protection, Jerome Park Demonstration Plant, Bronx, NY 10468

Raw water filtration of the Croton Reservoir water by GAC (10 min EBCT) was compared to treatment by ozone/biological activated carbon (BAC) filtration (10 min and 5 min EBCT) which may precede the diatomaceous earth filters proposed for this water supply for New York City. Average results showed 44-57% reduction in trihalomethanes and haloacetic acids, 28-30% reduction in chlorine demand and 21-23% reduction in TOC after BAC treatment compared to removals of 14%, 8% and 14%, respectively, achieved by GAC filtration alone. The removal of BDOC was strongly related to the removal of the 'fast' biodegradable fraction. Substrate in BAC treated effluents was comprised of more slowly biodegradable compounds than were present in the raw water. The changes in dissolved organic carbon characteristics from ozonation (i.e., to more hydrophilic, smaller, less reactive compounds) were confirmed using standard NOM fractionation techniques. In general, biomass levels were heterogeneously distributed with the highest levels at the filter surface.

In 1971 the City of New York instituted a research program specifically directed to its Croton Reservoir supply, in anticipation of the eventual need to provide some treatment beyond chlorination for this watershed. Metcalf & Eddy of N.Y. and Hazen & Sawyer P.C. were engaged at that time, in joint venture, to conduct the corresponding investigations (Principe et al. 1994). Croton is the oldest and smallest

[6]Corresponding author

0097–6156/96/0649–0211$20.00/0

of three reservoir systems serving New York City; it was placed in service in 1852 and provides roughly 10% or 140-240 mgd water yield compared to 470-600 mgd (40%) from the Catskills system and 580-750 mgd (50%) from the Delaware system (Bunch and Kerr 1995). Construction of a new, low profile, 450 mgd filtration facility for treatment of Croton Reservoir source water has been proposed (Principe et al. 1994).

Commitment to providing filtration of the Croton supply was formalized in a report in 1979 wherein the Consultants recommended ozone and diatamaceous earth (DE) filtration to provide water quality comparable to rapid sand filtration, at a lower cost (Principe et al. 1994). Operating experience with DE filtration showed that only particulate matter was removed, and treated effluent contained essentially the same levels of dissolved organic carbon (DOC) as the raw water (Mastronardi et al. 1993). Since ozone at the low doses applied (0.5-1.5 mg/L or ≈ 0.5 mgO$_3$/mg DOC) accomplished little mineralization of DOC, many of the large molecules of NOM were converted to more readily biodegradable dissolved organic carbon (BDOC). The potential for problems with regrowth in the water distribution system was of growing concern.

Thus an additional recommendation resulting from the Demonstration Plant experience was to incorporate granular activated carbon (GAC) filter-adsorbers between ozone and DE, whereby additional DOC and BDOC could be removed, along with reductions of chlorine demand and disinfection by-product (DBP) formation. Biofilm development is promoted in the adsorber by the increased biodegradability of ozonated organics which, together with the GAC surface for bacterial growth, results in biological "activation" of the carbon. The combination of ozone followed by activated carbon is commonly referred to as biological activated carbon (BAC).

Research questions to be addressed in this research program were to evaluate BAC treatment of NOM in ozonated Croton Reservoir water. Specific objectives for evaluation of the pilot treatment facility were to:

- evaluate seasonal variability on raw water NOM characteristics including relative biodegradability, hydrophilic/hydrophobic content and apparent molecular weight (AMW) distributions.
- analyze and evaluate treatment performance with respect to seasonal variability and removal of total and dissolved organic carbon, UV$_{254}$ absorbance, chlorine demand, DBP formation with respect to THM and HAA byproducts, and BDOC stability,
- evaluate NOM treatability by characterization of NOM fractions after GAC, ozone, and ozone/BAC, and to
- quantify levels of biomass and bioactivity with depth in mature, steady state GAC and BAC biofilters.

Methodology

Pilot Facility Description. The Consultants constructed a 38 L/min pilot facility at the Croton Reservoir Intake Gatehouse Facility which began continuous operation in Sept 1993. The pilot plant included initially two parallel trains: a control 'GAC' adsorber alongside a preozonated or 'BAC' adsorber, each maintained at a flowrate of 9 L/min (7.3 m/h) to attain 10 min empty bed contact time (EBCT).   In May 1994, a second BAC adsorber was installed to evaluate performance with a higher flowrate, 18 L/min (15 m/h) or 5 min EBCT (BAC5).   The plant continued continuous operation until Feb 1995, except for regular shutdowns for backwashing and periodic repairs.   The present monitoring program was initiated in Jan 1994 after the 10 min GAC and BAC filters had been exhausted with respect to adsorption, and biodegradation was the primary mode of treatment.   However, the initial maturation period for the 5 min BAC filter was included in this program, therefore, water quality parameters have been corrected to exclude the initial 30 days (May - June 1994) of operation of this column where organic carbon removals were higher due to adsorption.

Raw water influent to the pilot facility was bled from the main Croton aqueduct intake prior to chlorination.   A 1 kg/day air feed ozone generator (Griffin Technics, Lodi, NJ) was used to apply 1.5 ± 0.5 mgO$_3$/L or roughly 0.5 mgO$_3$/mg DOC.   Ozone bubbles were dissipated through a disk diffuser at the bottom of a 5.5m x 10cm diameter countercurrent flow column.

GAC media Type BPL 4x10 (Calgon Carbon Corp., Pittsburgh, PA), as selected by The Joint Venture, was packed to a bed height of 1.2 m in each of the three adsorbers.   This media type was selected to allow high surface area (1100 m^2/g) with low headloss (< 1cm H$_2$O at 18 m/h, 3.7 mm mean particle diameter).   The filters beds were housed in 30 cm diameter Schedule 40 PVC pipe of 5.5 m height, to allow for bed expansion of roughly 50% during backwash.   Columns were equipped with side sampling ports spaced at 30 cm distances along the height of the bed for extracting water and media samples, with additional ports every 5 cm near the top.

Water level above the filter beds was allowed to accumulate to approximately 1.2 m head before backwashing, but backwashing was conducted weekly whether or not terminal headloss had been reached.   Sampling was usually conducted several days later, to allow reacclimation of the columns following this disturbance.   Influent raw water was diverted for use as backwash, with waste solids returned to the reservoir.   Some filter run lengths were cut short (1 day) due to problems with air binding (Farrar 1994).   Excessive headloss caused by air bubbles entrenched within the filters was experienced when the temperature difference between the raw water and adsorbers was significant, as in the spring season.   This temperature differential caused dissolved gases, i.e., oxygen, to come out of solution upon contacting the warm media.

Sampling and Analysis Program.   Sampling and analysis efforts were conducted between 5 Jan 94 and 6 Feb 95.   Water and media samples were collected from the pilot plant for analysis of various parameters as summarized in Table I.

Biweekly or monthly aqueous monitoring samples were collected from five sampling locations, the raw water, the GAC column effluent, the ozonated water, the BAC 10 effluent, and the BAC 5 effluent.   Samples were shipped overnight in ice-packed coolers to the University of New Hampshire, where they were refrigerated at 4°C and analyzed for parameters including total and dissolved organic carbon (TOC/DOC).   UV-absorbance at 254nm wavelenght (UV$_{254}$), chlorine demand, total trihalomethanes - simulated distribution system test (TTHMSDS), and

Table I. Sampling and Analyses Summary

WATER ANALYSIS	Sampling Frequency	Preservation, Max. Holding	Method	Reference
Nonpurgeable Total Organic Carbon (TOC)	Biweekly	2 wks, H3PO4 4°C	UV-promoted persulfate oxidation	Std Methods 1992
Dissolved Org. Carbon (DOC)	Seasonal	20 hours	0.7 um GF/F Whatman	
UV-absorbance @ 254nm	Seasonal	2 wks, 4°C	Spectroscopy	
Chlorine Demand - Simulated Distrib System	Monthly	2 wks, 4°C headspace free	Colorimetric DPD Pillows	
Trihalomethanes[a]	Monthly	2 wks, 4°C headspace free	Liquid-liquid Extraction GC-EDC	
Haloacetic Acids (6)[a]	Seasonal	2 wks, 4°C headfree, NH4Cl	Method 552-GC	El
Biodegradable Organic Carbon (BDOC)	Monthly	2 wks, 4°C	Recirculating Sand Biofilters	Mogren et 1990
Hydrophilic/Hydrophobic Separations on XAD8	Seasonal	2 wks, 4°C	Resin adsorption and ion exchange	Leenheer 1981
Apparent Molecular Weight	Seasonal	2 wks, 4°C	Ultrafiltration	Collins et al. 1986
Iron, Magnesium, Calcium, Manganese cations	Seasonal	1 month, 4°C HNO$_3$	ICP Emiss. Spectroscopy	Std Methods 1992
GAC/BAC MEDIA				
Bacteria Extraction	Summer, Fall	1d, 4°C	Pyrophosphate, pH7	Balkwill 1985
Dry Weight Determination	Summer, Fall	1d, 4°C	103°C dried 24 hr	Std Meth 1992
Biomass	Summer, Fall Summer, Fall	Immediate	Formaldehyde fixation/Vacuum Dessication	Chesbro et al. 1994
Protein Assay	Summer, Fall Summer, Fall	Immediate	Dodecyl Sulfate/NaOH Bicinchroninic Acid	Chesbro et al. 1994
DAPI Direct Counts	Summer, Fall	5 d, 4°C	Epifluorescent Microscopy	Std Meth 1992
HPC on R2A Agar	Summer, Fall	Immediate	5.10d Incubation, 15°C	Std Meth 1992
Phospholipid Analyses (University of Cincinnati)	Summer, Fall	d, 4°C	Attached organisms extract w/CHC13/CH3OH	Wang 1995
Activity - Incorporation of [32]Phosphate in Lipids	Fall	Immediate	[32]Phosphate labeling extract w/CHCl$_3$/CH$_3$OH	Chesbro et al. 1994

[a]Bromide levels were <0.1 mg/L so bromated DBP species were generally less than 5 percent of the total.

biodegradable organic carbon (BOC).   Additional parameters were analyzed onsite by the Consultants, including pH, dissolved oxygen, temperature, color, turbidity, and UV_{254}.   SDS conditions included chlorination at 0.8-1.0 mg Cl_2/mg TOC and buffering to pH 7.5 for incubation at 20°C.   A chlorine residual of 0.5 ± 0.3 mg/L was targeted following 24h sample incubation.   Chlorine demand was based on these same conditions.

Five seasonal sampling events were also undertaken, to evaluate the treatability of specific NOM fractions under the variability introduced by seasonal changes.   Sampling was conducted on 23 Mar 94 and 6 Feb 95 for characterization of the Winter season, 8 Jun 94 for the spring samples, 16 Aug 94 for Summer samples, and 17 Oct 94 for Fall season characterization.   Both water and media samples were collected during the Summer and Fall sampling events.   Seasonal aqueous samples were analyzed for bulk parameters similarly as biweekly monitoring samples, but were also fractionated into hydrophilic/hydrophobic DOC, apparent molecular weight distributions (AMW), and biodegradable fractions.   TOC/DOC, UV_{254}, chlorine demand and TTHMSDS were quantified in most NOM fractions derived from seasonal samples.   Haloacetic Acids SDS (six compounds) and metals (iron, manganese, magnesium and calcium) were analyzed in unfractionated seasonal samples.   Winter 1995 water samples were processed for all but molecular weight (MW) analyses.

Media samples were obtained from 3 depths within the filters; samples were extracted from the filter housings through the side sampling ports.   'Top' was defined as the sample obtained 5 cm from the bed surface, 'Middle' was that obtained at 60 cm depth, and 'Bottom' media was taken from 120 cm depth or just above the filter gravel base.   After collecting the water samples in the Summer and Fall sampling trips, the columns were allowed to drain to facilitate the collection of media samples.   Media was gathered using a hollow metal rod, flame/ethanol sterilized between samples, to extract media from the center of the filter beds at the different depths.   Split duplicate samples were collected from each depth in sterile Whirlpak (Nasco West Inc., Modesto, CA) plastic bags or in organic-free glass vials, for biomass (as microbial lipid phosphates) analysis by the University of Cincinnati, Civil and Environmental Engineering Department, and for additional biomass and bioactivity analyses by the University of New Hampshire Microbiology Department.   A small volume of pore water was also collected from the columns to ensure bacteria remained moist over the overnight shipping period.

Analytical Program.   Details on specific analytical methods may be found elsewhere (Klevens 1995).   Special emphasis was given to NOM fractionations which are summarized below.

Standard techniques for NOM characterization were used to evaluate molecular weight distribution (AMW), hydrophilic/hydrophobic fractions, and biodegradable organic carbon (Klevens 1995).   The characteristics of organic matter in molecular weight categories <0.5k, 0.5-3k, 3-10k, and >10k apparent molecular weight units (amu) were determined by processing samples in parallel through the respective ultrafiltration membranes (YM and YC types, Amicon Corp., Danvers, MA).

Hydrophobic DOC was extracted by adsorption on nonionic XAD-8 resin (Rohm and Haas, Philadelphia, PA) following acidification of the samples to pH 2. Direct measurement of DOC, UV_{254}, chlorine demand and TTHM on both the total and hydrophilic portions was used to determine, by difference, the contribution corresponding to the hydrophobic fraction.

BOC/BDOC was evaluated from samples recirculation through acclimated biofilters over a 7 day period (adapted from Mogren et al., 1990). Between analyses, the biofilters were maintained active by recirculating Croton raw water; 5 gallon raw water carboys were refreshed about every 3 weeks for maintaining adequate feed supply. Estimates of BDOC fractions were obtained for seasonal samples by analysis of the DOC curve vs. time, where the slope of the curve changed from a sharp (presumably exponential) decrease after depletion of the fast biodegradable organic compounds to a more graduate decrease, indicative of assimilation of the more slowly biodegradable compounds. As outlined in Figure 1, the operationally defined distinction between fast and slow BOC was determined by bisecting the extreme tangents of the DOC curve. In almost all cases, the curve bisection occurred near the 24 h mark of incubation. Non-biodegradable or refractory DOC was defined as the average concentration remaining after 5-7 days contact time through the biofilters. BOC content was calculated as the difference in initial and final DOC concentrations. End value DOC concentrations on any particular sample were variable, probably confounded by releases of microbial byproducts. Refractory DOC concentration was thus estimated where the DOC curve versus time flattened to a steady state end value (Figure 1).

Because DOC measurements were based on filtrations performed 6-20 h after sample collection (upon return to the lab), it is likely that DOC values did not include some part of the fast biodegradable substrate that decayed during this shipping time. The fraction lost was confounded with particulate organic carbon (POC), since TOC measures were based on samples acidified onsite upon collection, whereas DOC samples were obtained after the initial shipping time. This overlap affected mostly raw water and ozonated samples, and particularly those from the Fall season (highest TOC), since GAC and BAC effluents were routinely found to contain negligible particulate matter. To distinguish between particulate and fast biodegradable organic carbon, selected samples which include preserved and unpreserved aliquots were used. The unpreserved aliquot (e.g., intended for BDOC or UV_{254} analysis) was sampled and preserved upon receipt for 'lab-TOC', and also was filtered for preserving a parallel aliquot for DOC quantification. These results were compared to the same samples that had been preserved onsite for TOC. This analysis showed between 0.1-0.2 mg/L POC (n=3) in Fall 1994 raw water samples, which was accounted for when estimating BDOC for similar Fall raw water samples. The difference between total and dissolved samples for other sites (GAC, Ozonated and BAC effluents) was negligible (0.017 $\pm$ 0.019 mg/L, n=12), thus their respective BDOC concentrations were based on the difference between TOC and refractory TOC.

Field and analytical quality control procedures were applied to properly qualify the results obtained from this research. Statistical tests for comparison of variances and means were based on 95% confidence levels ($\alpha = 5\%$). Limits of detection (LOD) or method detection limits (MDL) and limits of quantitation (LOQ) were estimated for analytical procedures based on methods prescribed by Standard Methods (1989 and 1992) or Miller and Miller (1993). Blanks were used in all procedures to guard against contamination from reagents, glassware, and biased instrumentation drift. Duplicates and some triplicates were used for assessment of precision at different levels of processing/analyses. Accuracy controls included comparison to standards or spikes, where appropriate. Table II includes a summary of QA/QC results for the analytical program.

Biomass Analyses. Samples of GAC media were collected from pilot filter columns and characterized with the assistance of University of New Hampshire Microbiology and University of Cincinnati Civil and Environmental Engineering Departments. Table I lists the analyses performed. A brief summary of the analyses follows.

Bacterial Extraction and Dry Weight Determination. Sterile technique was used in transferring 4-6g wet weight GAC or sand media coated with biofilm to a preweighted 25 mL Erlenmayer flask. Fifteen mL of sterile, pH 7, 1% sodium pyrophosphate (NaP_2O_7) buffer was transferred to each flask; samples were then agitated by mechanical vortex mixer for two 30s bursts with an intermediate 30s rest (Balkwill and Ghiorse 1985). Extraction was conducted using two sequential rinses (7.5 mL each) of pyrophosphate solution to maximize recovery of bacteria, though the efficiency of extraction was not quantified. The extracted bacteria and buffer were transferred to labeled test tubes and delivered immediately to University of New Hampshire Microbiology department for additional processing.

Sand or carbon media aliquots (5-8 g wet weight) processed for bacterial extraction were dried at 103°C and weighed after 24, 36, and 48 h to confirm a constant dry weight was reached. Sample dry weights ranged 1-2 g with exact weights being noted for standardizing all bacterial analyses on a dry weight basis.

Biomass and Protein Assay Determinations. The procedures included gross biomass quantitation from addition of dilute formaldehyde to extracted samples, sedimentation (centrifugation) of the biomass, and vacuum dessication of solids for determination of dry weight total biomass. Protein determination was conducted by solubilizing the extracted bacteria by addition of NaOH (0.2M) and sodium dodecyl sulfate (1%), incubating the mixture 5 min at 100°C, and assayed with dibinding reagent bicinchroninic acid (in Chesbro et al. 1994). The dibinding reagent complexes with protein in the sample to produce a purple color the intensity of which is proportional to the mass of protein. Color is quantified from absorbance measurements at 562 nm and converted to protein based on alibrations with known standards. Possible interferences with this method include the presence of iron, glucose and ascorbic acid which may cause overestimates of protein levels.

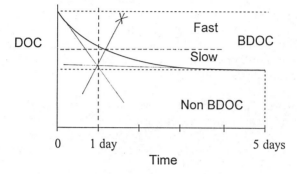

Figure 1.   Fractionation Scheme Used to Distinguish Between Fast and
              Slow Biodegradable Dissolved Organic Carbon (BDOC)
              (adapted from Klevens et al., 1996)

Table II. QA/QC Summary

Parameter	CALIBRATION		ACCURACY Readback/Spike		PRECISION Range or % RSD	
	LOD	LOQ	% Recovery	n	mg/L or %	n
DOC, mg/L	0.1	0.2-0.4	101 ± 6 L	93	<0.2	2/samp
			100 ± 3 H	49	7.2 ± 7% rsd	15%
UV$_{254}$ Abs, cm^{-1}	nd	--	--	--	<0.002	2/samp
THMs, ug/L	2-25	6-74	120-122	35	1-7 R	40
BDOC, mg/L	0.15[1]	---	.063 ± .069	---	---	18
XAD8 DOC, mg/L	0.13[1]	---	---	---	0.14 ± 0.20 R	24
					6 ± 7% rsd	24
AMW DOC, mg/L	nd	---	---		13 ± 17 (all)	17
0.5, 10k Filter, mg/L	---	---	---		0.21 ± 0.24	13
3k Filter, mg/L	---	---	---		0.03 ± 0.028	4

[1]99% cl, one-tailed t statistic. nd = not determined. R = range.
rsd = relative standard deviation.

DAPI Direct Counts. Extracted media were fixed with gluteraldehyde, stained with DAPI (4',6-diamidino-2-phenyl indole, ICN Biomedicals, Costa Mesa, CA), and filtered on 0.22 μm Nuclepore polycarbonate black membrane filters (Nuclepore, Pleasanton, CA) for examination by epifluorescent microscopy. This procedure (Standard Methods 1992) quantified all membrane-intact cells providing a count of viable and nonviable extracted organisms. Slides were stored at 4°C in the dark and counted by an experienced analyst within 1 week holding time.

HPC Plate Counts. Three dilutions of extracted bacteria were plate din triplicate on low nutrient agar (R2A, Difco Labs, Detroit MI), incubated at 15°C for counting at 5 and 10 days. HPC Method 9215 (Standard Methods 1992) was followed.

Phospholipid Analyses. Split duplicate field media samples were collected and shipped overnight to the University of Cincinnati Department of Civil and Environmental Engineering for determination of biomass by a phospholipid extraction technique (Findlay et al. 1989). The extraction was performed soaking the biofilm coated media in a coloroform/methanol/water mixture, to solubilize microbial phospholipids in the organic solvents. This procedure circumvents the inefficiencies inherent in recovering bacteria attached to the filter media, because it enables direct extraction from attached organisms. It is selective for viable bacteria and protozoa since these compounds are not stored but produced as part of microbial metabolism (Wang 1995). The procedure further includes separation of extracted phospholipids into the chloroform phase, evaporation of the solvent, oxidation and complexation of phosphate, addition of malachite green dye to form a colored solution proportional to the concentration of phosphate, quantified by the adsorption of light at 610 nm wavelength (Wang 1995). The reported detection limit was 0.5 μM phosphate or nmol phospholipid/2 mL extracted volume (Wang 1995).

Results and Discussion

The results obtained during the one year characterization efforts for Croton Reservoir pilot treatment are presented by (i) examining seasonal influences on raw water quality and NOM composition, followed by (ii) evaluation of the performance of three pilot columns with respect to various monitoring parameters. Detailed treatability evaluation of specific NOM fractions (iii) is discussed to quantify the effect of ozonation and biological treatment on such fractions. Distribution of biomass (iv) within the pilot filters are compared and related to observed removals.

Seasonal Influence on Raw Water NOM. Sampling of the Croton Reservoir was conducted regularly between Jan 1994 and Feb 1995 to observe seasonal influences on NOM. Observed raw water quality parameters are summarized in Table III. Croton raw water contained low particulate matter (1.5 $\pm$ 1.0 NTU average turbidity) and bromide levels. Apparent color levels were moderate throughout the year of study at 32 $\pm$ 13 scu, similar to 1991 characterizations when color ranged

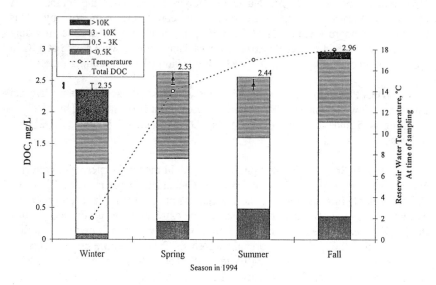

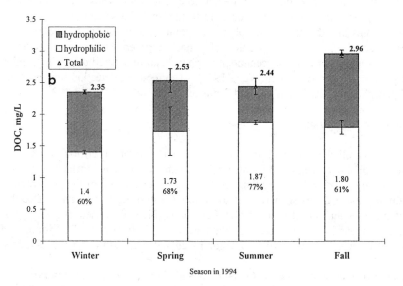

Figure 2.   Seasonal Fractionation of Croton Reservoir Water by (a) AMW Distribution and (b) Hydrophobicity (adapted in part from Klevens et al., 1996)

Table III. Croton Raw Water Quality in 1994

Parameter, units	Average ($\pm$)	Range	n
Total Organic Carbon, mg/L	3.18 $\pm$ 0.70	2.25 - 5.07	28
UV_{254} Absorbance, cm^{-1}	0.086 $\pm$ 0.01	0.069 - 0.108	29
Specific UV Absorbance, L/mg-m	2.80 $\pm$ 0.42	1.71 - 3.42	28
Chlorine Demand, mg Cl_2/L	2.68 $\pm$ 0.72	1.3 - 3.9	14
TTHMSDS[1], ug/L	79 $\pm$ 30	<20 - 199	14
HAA6SDS[1], ug/L	67 $\pm$ 10	53 - 86	5
Bromide, mg/L	<0.1	--	2
Turbidity, NTU	1.5 $\pm$ 1.0	0.7 - 6.8	33
Color, scu	32 $\pm$ 13	17 - 83	27
Water Temperature (°C)	10 $\pm$ 6.9	3.3 - 23	25
pH	7.5 $\pm$ 0.4	7 - 7.9	4

[1]Total Trihalomethane (4) and Haloacetic Acids (6) sum after simulated distribution system chlorination at 1:1 mg Cl_2/mgTOC, incubation 24h @ 20°C and pH 7-7.5.

12-35 scu (Mastronardi et al. 1993). Water temperature variations in 1994 were also typical of previous years (Mastronardi et al. 1993); from April onward, a gradual rise in water temperature was observed from the Winter levels of 2-3°C to 18-19°C in the Fall (Sep-Oct), then cooling again to 4°C by Jan 1995.

Croton's seasonal water temperature fluctuations had a marked effect on dissolved organic carbon characteristics. The difference between TOC and DOC was indistinguishable from the minimum detectable difference inherent to the variability of the TOC analyzer (0.1-0.2 mg/L) while TOC levels remained within a fairly narrow band of 2.2-3.8 mg/L and UV_{254} adsorbance was very stable (0.086 $\pm$ 0.01 1/cm average), specific fractions of NOM changed noticeably. AMW distribution in the raw water should be mostly affected by biological activity in the reservoir, in turn dependent on water temperature and nutrient availability (Thurman 1985). In Croton Reservoir, Summer and Fall samples (16 Aug and 17 Oct sample dates) were found to contain relatively higher proportions of low (<3k) molecular weight compounds, coinciding with water temperature peaks of 17-18°C as shown in Figure 2a. The error bars reflect the pooled standard deviation from analysis of each permeate and the total unfractionated sample. Of these representative samples, slightly higher levels of the smallest organics <0.5k amu were found in the Summer, while Winter samples (23 Mar 94) showed a considerable contribution (~20%) from large (>10k) molecules.

As depicted in Figure 2b, hydrophilic compounds were the predominant contributors to the DOC pool in untreated Croton water, with values ranging 60-77% in 1994 samples. The pooled variability from hydrophilic and total DOC fractions is shown by error bars. Winter and Fall samples showed identical hydrophilic/hydrophobic splits of 60/40% though they exhibited different AMW

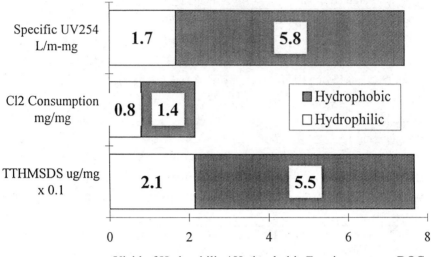

Figure 3.   Normalized Yields for Hydrophilic/Hydrophobic DOC Fractions

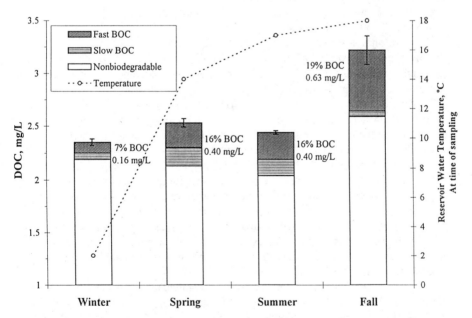

Figure 4.   Seasonal Variation in Raw Water Biodegradability (1994)

distribution as noted earlier. In general, hydrophilic DOC was highest in the summer and fall months, coinciding with a higher contribution from small MW (<3K) organics.

Mass weighted average contributions of hydrophilic/hydrophobic organics to the total raw water potentials are summarized in Figure 3. These data show the higher reactivity of the hydrophobic organics which compensated for their lower DOC contribution. The hydrophobic fraction represented 2.6 times greater TTHMSDS reactivity (ug TTHMSDS/mg DOC), but 1.5 times more chlorine consumption (mg Cl_2/mg DOC), and 3.4 times higher UV_{254} absorbance than the hydrophilic fraction. Others have documented the higher reactivity of the hydrophobic fraction of DOC (Collins et al. 1986, Owen et al. 1993), attributed principally to its higher degree of aromaticity and phenolic content (Reckhow et al. 1990).

Seasonal warming through 1994 increased raw water biodegradability, defined by BOC measurements, from 7% in the Winter to 19% in the Fall (Figure 4). Also depicted are the fractions of fast and slow BOC, characterized based on observed differences in DOC uptake during sample contact time in recirculating laboratory biofilters. The error bars show the pooled standard deviation from total DOC and BOC analyses. These results suggested that the higher BDOC concentration in the Fall sample was due to the presence of more fast biodegradable substrate, perhaps due to increased bioactivity in the reservoir during the warm months. Others have observed increased BDOC in the Fall season, associated with increased bioactivity and the presence of algae (Prévost et al. 1991) and the warmer temperatures (Owen et al. 1993). Croton Reservoir raw water biodegradability average of 0.47 mg/L (15%) was within the ranges typical for surface water resources. Kaplan et al. (1994) surveyed 53 surface water supplies across the U.S. and reported biodegradable dissolved organic carbon content ranged 0.1-0.8 mg/L (0.3 mg/L median), corresponding to 13% of raw water TOC levels.

Pilot Treatment Performance. Raw water treatment by ozone/BAC using ~0.5 mgO$_3$/mg TOC followed by 5 and 10 min EBCT BAC biofilters, was evaluated side by side with a 'control' unozonated GAC filter (10 min EBCT), to compare treatment performance with respect to various bulk monitoring parameters. Monitoring NOM surrogate parameters over the study period developed similar removal trends. For example, monitoring of UV-absorbance over the study period (Figure 5) showed that ozonation consistently produced a significant (52%) reduction in raw water UV_{254} absorbance. Subsequent BAC filtration of ozonated water reduced this parameter negligibly, only 6-7% on average. The GAC (control) filter removed 12% UV_{254} absorbance from the raw influent, which, though not statistically lower, was consistent through the monitoring period and was achieved primarily by biological reactions in the GAC, since adsorption capacity had been exhausted prior to the beginning of this research program.

A detailed summary of removals achieved through pilot treatment is provided as Table IV. These results, based on all samples collected between 5 Jan 94 and 6 Feb 1995 (excluding 30 days ripening of the 5 min BAC), suggested that the

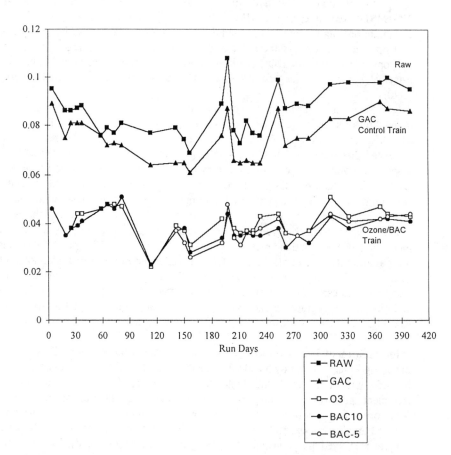

Figure 5.　　UV$_{254}$ Absorbance Removals Through Pilot Facility: 5 Jan 94 - 6 Feb 95 (adapted from Klevens et al., 1996)

combination of ozone/BAC yielded improved removals over GAC alone for all parameters monitored. Ozonation reduced principally UV-absorbing compounds, TTHM and HAA6 precursors from the raw water, while other parameters were reduced or changed less significantly. Bromated DBP species were not significant (<5%) for this source water. Biofiltration removals with respect to filter influent levels (raw water vs. GAC effluent and ozonated water vs. BAC effluents) showed similar performance by GAC and BAC filters relative to TOC and turbidity reduction. BAC treatment outperformed GAC for reduction of chlorine demand and DBP formation in their respective filter influents. BAC generally resulted in similar removals after 5 or 10 min EBCT; which suggested higher treatment efficiency was obtained using a higher flow, shorter (5 min) contact time. Not all changes quantified in Table IV were statistically significant (based on paired t-tests at 95% c.l.), but the consistency in treatment performance provided meaningful insight to the treatability of NOM.

Table IV. Average Removals for Croton Pilot Treatment

Raw Water Average mg/L Influent	n	10 min GAC	Ozonation		10 min BAC		5 min BAC	
		% of Raw	% of Raw	mg/L Effluent	% of Raw	% of O_3	% of Raw	% of O_3
TOC 3.18 ± 0.70 mg/L	17-28	14%	10%	2.85 ± 0.37	23%	14%	21%	12%
BOC 0.47 ± 0.15 mg/L	8-16	26%	123% incr.[b]	1.05 ± 0.28	22% incr.[b]	46%	37% incr.[a,b]	39%
UV_{254} Abs. 8.6 ± 1.0 m^{-1}	17-29	12%	52%	4.1 ± 0.6	56%	6%	56%	7%
SUVA (UV/TOC) 2.78 ± .41 L/mg-m	17-28	nil	48%	1.44 ± 0.26	44%	9% incr.[b]	45%	7% incr.[b]
Cl_2 DEMAND 2.68 ± 0.72 mg/L	11-18	8%	10%	2.42 ± 0.51	28%	21%	30%	22%
TTHMSDS 79 ± 21 µg/L	12-18	14% incr.[b]	38%	49 ± 23	54%	25%	44%	10%
HAA6SDS 67 ± 10 µg/L	4-5	14%	42%	39 ± 6	57%	26%	55%	23%

[a]Not significantly different from raw water BOC at 99% c.l.
[b]Incr = increased concentration across treatment.

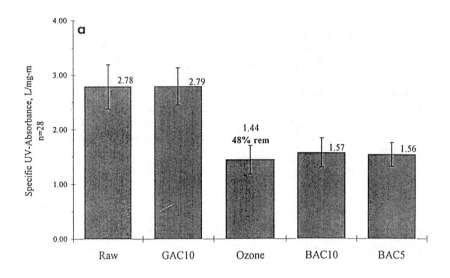

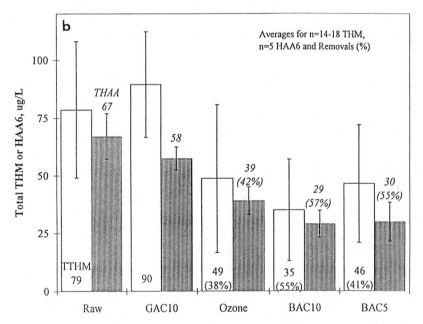

Figure 6.   Averaged Removals of (a) Specific Absorbance and (b) Disinfection By-Products Through Pilot Facility (adapted in part from Klevens et al., 1996)

Changes in TOC levels between raw and treated samples were not substantial because treatment was targeted more at altering the biodegradable character of organics rather than reducing overall concentration, though statistically significant differences (SD) resulted from all treatments applied to the raw water. Mineralization of TOC to carbon dioxide was low (10% average) because of the low doses of ozone applied. Observed biofiltration removals of 21-23% TOC from combined ozone/BAC (Table IV) were within the ranges reported for other biofiltration processes, including ozone-enhanced slow sand filtration where removals surpassed 30% (2 mgO$_3$/mg DOC, Malley et al. 1993), and ozone/BAC plants with 14-17% reported TOC reduction (Servais et al. 1992, 1994; Cipparone et al. 1993). Both 5 and 10 min BACs achieved, on average, similar removals of 12-14% from preozonated TOC. BAC effluents were significantly lower than mean ozonated influent but treated effluents were not significantly different (NSD) from each other. Subsequent (predominant) removals by biodegradation were slightly lower from the shorter EBCT and increased mass loading rate compared to the slower filters.

The control GAC filter provided TOC removals of 14% average from raw water influent, similar to the performance of BAC 10 and 5 min filters. Both this column and the 10 min BAC had been exhausted with respect to adsorption capacity prior to initiating the present monitoring program, a judgment which was based on the reduction and stabilization of TOC removals (Farrar 1994). GAC and BAC biofilters achieved removals typical of conventional slow sand filters, which reportedly reduce 10-20% of influent TOC (Collins et al. 1991).

Specific UV$_{254}$ absorbance (SUVA) proved to be a good surrogate indicator of the change in NOM character and predictor of DBP removals through treatment. Raw water and GAC effluents were not significantly different (2.78 $\pm$ 0.42 L/mg-m Raw and 2.79 $\pm$ 0.34 L/mg-m GAC, Table IV) and generally followed a parallel pattern. Ozonation again caused a significant reduction in SUVA of 48% (Figure 6a), representative of the changes in organics structure. BAC did not produce significant changes in SUVA except to provide some insignificant but noticeable TOC removal.

Limited analyses for haloacetic acids SDS potential and variable data quality for THMSDS precluded firm conclusions regarding pilot plant performance in reducing DBP formation, however, qualitative reductions can be observed in Figure 6b and Table IV. Effectiveness of ozone/BAC treatment was more clearly apparent from comparison to the control GAC filter effluent, where removals were not significantly different from raw water levels. DBPSDS as TTHM and HAA6 potential was reduced significantly by ozonation (40%) and by combined ozone/BAC (44-57%), with respect to raw water influent averages. Decreased reactivity of the organics was accomplished mostly by ozonation, since biodegradation alone failed to show a statistically significant reduction compared to ozonate influent (paired t-tests at 95% c.l.).

BOC reduction in ozonated Croton water was one of the primary purposes of this treatment process. Raw water influent average BOC of 0.47 $\pm$ 0.15 mg/L

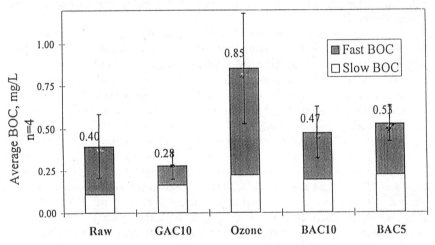

Figure 7.   Distinction Between Fast and Slow BDOC Removals

(15% of total DOC) was reduced 26% to 0.35 ± 0.18 mg/L by the control GAC, though no statistical difference could be confirmed due to the variability of the data over the study period. This reduction nevertheless supported the fact that appreciable biodegradation is fostered in GACs when operated without regeneration. Note: BOC is used interchangeably with BDOC in this study since there was no significant difference between DOC and TOC measurements.

Preozonation is known to increase the BOC content of the raw water, especially the fraction of quickly biodegradable compounds (Langlais et al. 1991), whereby bacterial degradation in subsequent BAC filters may provide improved removals for similar EBCT. Average BOC changes in Croton's pilot facility (Figure 7) showed roughly a doubling of BOC by ozonation (SD from raw water level), followed by 41 and 46% removal of ozonated BOC through 5 and 10 min BAC, respectively. BAC filtration effluents were significantly lower in the fast biodegradable fractions than ozonated influent BOC levels.

Additional analysis of average fast and slowly biodegradable organic fractions in treated samples (Table V) showed that:

- GAC removed only fast BOC in the unozonated raw water.
- Ozonation increased the fast and slow BOC fractions in the raw water.
- BAC 10 and 5 min EBCT reduced the fast BOC to the same levels initially present in the raw water.
- BAC 10 and 5 min EBCT were insufficient to produce significant removals of slow BOC present in ozonated water.
- Increased BOC substrate in treated BAC effluents was due to an increase in the contribution from slowly biodegradable organics in those samples.

Table V. Treatability of Fast/Slow Biodegradable DOC

Sample Location	Biodegradable Organic Carbon, mg/L			
	Total-All[1] Samples (n=8-11)	Total-Seasonal[1] Samples (n=4)	Fast BOC (n=4)	Slow BOC (n=4)
Raw	0.47±0.15	0.40±0.19	0.29±0.20	0.10±0.05
GAC Control	0.35±0.18	0.28±0.08	0.12±0.10	0.16±0.09
Ozonated Water	1.05±0.28	0.85±0.34	0.63±0.29	0.22±0.07
BAC 10 min EBCT	0.57±0.19	0.47±0.16	0.28±0.07	0.18±0.11
BAC 5 min EBCT	0.62±0.12	0.53±0.10	0.30±0.11	0.22±0.11

[1]Only seasonal (n=4) samples were fractionated in fast/slow BOC components.

DOC vs BDOC Removal

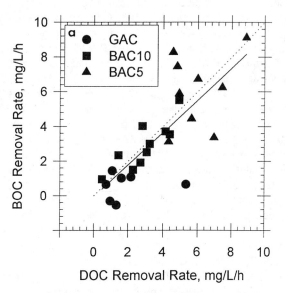

BDOC vs Fast BDOC Removal

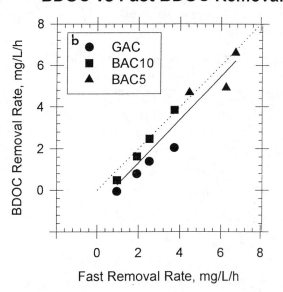

Figure 8.   BDOC Removal Rate Comparisons with (a) DOC Removal
              Rate and (b) Fast BDOC Removal Rate

Langlais et al. (1991) reviewed studies by Billen et al. (1989) where the more slowly hydrolyzable BOC was virtually unchanged through biofiltration treatment, reporting that much longer contact times (beyond 15 min) would be required to reduce this fraction. Proposed treatment by DE filtration after BAC for Croton is not likely to affect BDOC concentrations further (Spencer and Collins 1995).

Another approach to emphasize the treatability differences between the fast and slow BOC fractions can be demonstrated using the relationships explored by Huck (Mitton et al. 1993, Huck et al. 1994). Parameter removal rates can be calculated from the influent and effluent concentrations for the biofilter and the EBCT as follows:

$$\text{Removal Rate (mg/L/h)} = \frac{\text{Influent} - \text{Effluent Concentrations (mg/L)}}{\text{EBCT(h)}} \quad (1)$$

As shown in Figure 8a, the DOC removal rate calculated for the biofilters was correlated to the BDOC removal rate although there was significant scatter around the line-of-equality, especially for the BAC5 filter. A more distinct relationship developed between the fast BDOC fraction removal rate and the total BDOC removal rate as noted in Figure 8b. The excellent agreement between the regressed slope with the line-of-equality confirmed the importance of the fast BOC fraction as being the most susceptible to removal by biofiltration.

Treatability of NOM Fractions. AMW distributions based on DOC through the various treatment units, averaged for the four seasons, are summarized in Figure 9a. Data for UV absorbance are not shown. Error bars depict the pooled analytical standard deviation for seasonal samples. Ozone/BAC treatment removed principally small ($<0.5k$) and midrange (3-10k) organics to produce overall reductions of 23% TOC cited previously. Collins and Vaughan (1996) showed biodegradation reduced DOC from all AMW ranges. Goel et al. (1995) reported that *Anabaena* exudates with 60% DOC $<1k$ and SUVA <2.0 showed 37% TOC removal by biodegradation, whereas only 21% TOC reduction was achieved for a Florida groundwater (SUVA 4.0), despite its composition of 93% DOC $<1k$ amu. MW distributions of the biotreated waters were not reported by Goel et al. (1995), therefore the fractions which were preferentially biodegraded could not be compared. Changes in AMW as DOC through the control GAC filter are also depicted in Figure 9a for comparisons. DOC changes were small but showed reductions in the range of 3-10k and some $<0.5k$ amu, whereas some organics within 0.5-3k amu were leached in the GAC effluent.

Though only moderate DOC reduction was accomplished through combined ozonation and biofiltration, it is interesting to note that these removals resulted solely by changes in hydrophilic DOC as shown in Figure 9b. After conversion of ~0.2 mg/L average hydrophobic to hydrophilic DOC from ozonation, BAC 10 removed 0.5 mg/L hydrophilic DOC, slightly more than the 0.4 mg/L hydrophilic reduction achieved through BAC 5. The observation that biological removal was selective to hydrophilic DOC supported similar observations by other researchers that humic or

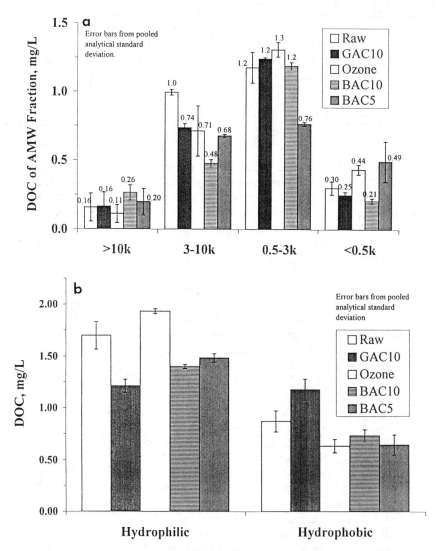

Figure 9.   Treatment Induced DOC Fractionation Changes as Quantified
by (a) AMW Distribution and (b) Hydrophilic/Hydrophobic

hydrophobic DOC is more resistant to biodegradation (Shaw 1994). However, Collins and Vaughan (1996) observed removals of both hydrophilic and hydrophobic DOC from ozone/biological treatment of raw water, and removals of only hydrophilic DOC by similar treatment of settled water, the latter exhibiting reduced hydrophobic content after conventional treatment.

Inspection of the control train (raw influent and GAC filter effluent) averages in Figure 9 showed some trends similar to BAC treatment. Since the control GAC was exhausted with respect to physical adsorption, the changes observed were principally derived from biodegradation. Bacteria again assimilated only hydrophilic DOC removals 0.5 mg/L on average, identical to BAC treatment. The GAC filter also leached organic matter in the form of hydrophobic DOC (~0.3 mg/L).

Summer and Fall samples were processed through additional resins to determine the relative contributions from organic acid basic and neutral hydrophilic/phobic groups; Table VI lists the results obtained on a representative Fall sample. The trends in total hydrophobic/hydrophilic fractions for this point estimate followed those discussed above for the averages across all seasons, in that total hydrophilic DOC was increased by ozonation then decreased by either BAC or decreased by GAC biofiltration, whereas total hydrophobic DOC was decreased from ozonation but essentially unchanged by BAC or GAC. All hydrophobic DOC was classified as 'acidic', since no bases were retained by adsorption on XAD8® at neutral pH, and no neutrals remained sorbed to the column after extraction and elution of hydrophobic acids (120% average mass recovery). This result might be expected based on the carbon media used in the pilot filters, which is specifically targeted to removal of alcohols, ketones and ethers (Calgon Carbon Corp., Pittsburgh, product literature for Type BPL carbons), compounds which are classified as basic and neutral fractions of NOM (Edzwald 1993).

Hydrophilic acids decreased very slightly through either ozone/biofiltration or direct GAC filtration (Table VI). Hydrophilic bases were isolated by one procedure on cationic resin (included with neutrals in Table VI) but showed no changes through treatment, probably due to the low levels present (~0.1 mg/L or 3-5% in all samples). However, separations based on Leenheer (1981) with sequential cation/anion resins, served to isolate hydrophilic bases and acids from neutrals. The results from this procedure suggested that hydrophilic neutrals were the principal fraction affected by ozone/BAC: 0.76 mg/L or ~26% hydrophilic neutrals in the raw water were increased 49% from ozonation, then decreased by BAC (0.6-0.65 mg/L in effluents) to levels similar to or lower than present initially. Following the faster reaction with humic substances, Bose and Reckhow (1995) reported that ozone reacted next with the hydrophilic neutrals fraction, as observed in the present characterization analysis.

GAC filtration effected very moderate reduction in hydrophilic neutrals on this date (0.7 mg/L in effluent), but instead removed about 29% hydrophilic acids from 0.93 mg/L estimated in the raw water to 0.66 mg/L (Table VI). However, this reduction was overshadowed in the total DOC by a concurrent increase in hydrophobic DOC of approximately 10%.

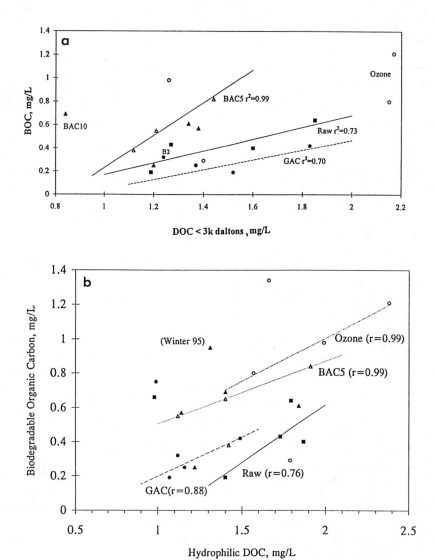

Figure 10.   Relationship Between the Biodegradability of DOC and (a)
              <3k AMW or (b) Hydrophilic Content (adapted in part from
              Klevens et al., 1996)

Table VI. Treatability of Organic Acids, Bases and Neutrals - 17 Oct 94 Sampling -

DOC Classification	Raw Water	Control GAC	Ozonated	5 or 10 min BAC
Total DOC, mg/L	2.96±0.04	2.78±0.04	3.03±0.12	2.34-2.53 ±0.25
Total Hydrophobic (≈Phobic Acids), mg/L (% of total)	1.2 (39%)	1.29 (46%)	0.65 (21%)	0.5-0.62 (21-25%)
Total Hydrophilics, mg/L (% of total)	1.8 (61%)	1.49 (54%)	2.38 (79%)	1.84-1.91 (75-79%)
Hydrophilic Acids[1], mg/L (% of total)	0.93-1.24 (31-42%)	0.66-0.96 (24-35%)	1.05-1.13 (35-37%)	0.98-1.19 (75-79%)
Hydrophilic Base/Neutrals,[2] mg/L (% of total)	0.55-0.87 (19-30%)	0.53-0.83 (19-30%)	1.25-1.33 (41-44%)	0.72-0.86 (29-37%)

[1]Extracted by XAD4® separation.
[2]Calculated by difference between XAD4 influent and eluant.

The increased biodegradability of low MW DOC is defended by the fact that smaller molecules are more readily available for transfer across the bacterial cell wall for use as substrate; whereas large molecules require the expense of additional energy by the cell, for the production of exoenzymes to oxidize the molecule into smaller compounds amenable to bioassimilation (Brock and Madigan 1988). Hydrophilic and low MW DOC should thus be more readily biodegradable because of their composition of smaller, simpler and predominantly aliphatic organics. Amy et al. (1992) and Owen et al. (1993) have shown that BDOC correlates well with low MW (<0.5k amu) DOC and with nonhumic (hydrophilic) DOC concentrations in ozonated waters.

Based on these premises, the seasonal data collected for Croton were pooled and graphed, with significant correlation coefficients obtained for BOC as a function of <3k amu DOC (Figure 10a) and with hydrophilic DOC (Figure 10b). Other correlations attempted were not significant. The positive slopes obtained for raw, GAC and BAC5 suggested that the biodegradability of DOC in these samples increased with increasing proportion of small (<3k amu) organics. These qualitative trends were based on 3-4 samples collected in different seasons and spread 2-3 months apart over the year of study. Unfortunately, due possibly to the

Table VII. Summary of Filter Biomass Distribution as Compared to the Literature

Analyses	Range in Croton Filters	Reported Values
Viable Counts, CFU/g dw media (5-10 days incubation)	1.0×10^5 - 8.9×10^9	0.001-0.1xDirect Counts (1) 10^7-10^8 (2)
DAPI Direct Counts		0.3 - 1.0×10^7 (3)
cells/g dw media	$1.5 - 7.7 \times 10^6$	0.15 - 4.0×10^9 (3a)
Protein, g protein/g dw media	5.4×10^{-6} - 8.8×10^{-2}	0.5 - 5×10^{-3} (4)
g/CFU	2.0×10^{-12} - 2.6×10^{-9}	0.15 - 2×10^{-12} (4.4a)
^{14}C-L (mixed) Aminoacids Incorporation, g/g dw media	0.83 - 4.7×10^{-2} Protein	0.53 - 8.0×10^{-6} Leucine (5)
g/CFU	0.01 - 3.3×10^{-13} g prot/CFU	2.9×10^{-15} g AA / cell (6)
Lipid Phosphate		
nmol PO_4^{3-}/g dw media	20 - 270	8 - 270 (7)
nmol PO_4^{3-}/CFU	1.7×10^{-8} - 6.3×10^{-4}	2.5×10^{-8} (8)
^{32}Phosphate Incorporation		
nmol PO_4^{3-}/g dw media	0.038 - 0.33	55 - 217 (9)
nmol PO_4^{3-}/CFU	1.6×10^{-19} - 4.0×10^{-16}	--

(1)   Balkwill and Ghiorse (1985) - Subsurface Soils; Buchanan-Mappin and Wallis (1986) - Stream Sediments; DiGiano et al. (1992) - GAC filters; Eighmy et al. 1993 - Slow Sand Filters.

(2)   Bancroft et al. (1983), Servais et al. (1991) and DiGiano et al. (1992) - GAC filters. Eighmy et al. 1993 - Slow Sand Filters.

(3)   Balkwill and Ghiorse (1985).

(3a)   DiGiano et al. (1992), Eighmy et al. 1993.

(4)   Collins et al. 1989 - Slow Sand Filters, protein from Folin Reactive Material, cell numbers from direct (acriflavine) enumeration.

(4a)   Neidhart et al. 1990 - E. coli Bacterium.

(5)   Tibbles et al. (1992) - ^{14}C-Leucine incorporation by bacteria extracted from marine sediments, cell numbers based on direct (DAPI) enumeration.

(6)   Simon (1988) - y (ugC-cells/L-h) = 0.137x (Aminoacids ng/L-h) + 0.008, r^2 = 0.74. Growth rate from incorporation of mixed aminoacids by free and attached bacteria in Lake Constance, Germany, and direct cells counts by acridine orange staining. Converted using regression slope and 4.0×10^{-13} g/cell (Balkwill et al. 1988).

(7)   Federle et al (1983, 1986) - 8-18 nmol PO_4^{3-} / g dw subsurface soils, 50-145 per g estuarine sediments; Wang (1995) - 170-270 nmol PO_4^{3-} / g dw GAC, and 12-160 per g sand.

(8)   Baldwill et al. (1988) - Subsurface Aquifer Sediments, ratio per direct (acridine orange) counts.

(9)   Bott and Kaplan (1985) - Stream Sediments.

variability of the data, ozonated samples and BAC10 effluents analyzed in this program did not behave consistently, though they would be expected to follow the same trend of increasing biodegradability with increased concentration of low MW DOC. The GAC effluent collected on 8 June 94 (Spring) was not included in the GAC correlation (0.26 r^2 with this datum point vs. 0.70 r^2 without) though it should not be rejected as an outlier based on Dixon's Q or Grubs outlier tests (Miller and Miller 1993).

Increased biodegradability of most samples tested was correlated with increased hydrophilic DOC in Figure 10b. As shown by the similar slopes of the regression lines, similar dependence on hydrophilic DOC was observed despite the different sample treatments. A sample collected during the ripening period of the BAC5 filter (8 June 94) was excluded from the correlation with hydrophilic DOC. The preferential removal of hydrophilic organics by biodegradation was also confirmed in a slow sand filtration study (Eighmy et al., 1993)

Biomass Characterization

Samples of GAC/BAC media for biomass characterization were collected as part of Summer and Fall sampling events, when the biofilters were judged to have reached steady state operation based on monitoring of TOC and UV_{254} absorbance removals. Media samples were extracted from the top (5 cm), middle (60 cm) and bottom (120 cm) of each of the three pilot filter beds. Care was taken to sample a few cm below the top of the bed in case residual ozone may have impacted the biota levels at the surface, and a few cm above the bottom layer of gravel, though the bottom of the bed was not clearly defined because of the mixing occurred during backwash. Duplicate field samples were collected from the middle sampling location from each BAC column but not from the GAC filter.

Biomass characterization as protein, direct 4',6-diaminino-2-phenyl indole (DAPI) direct counts, indirect culturing on R2A media, and activity by radioisotope labeling originated from the suspension obtained by pyrophosphate extraction of attached bacteria. Split field samples were shipped directly from the site to the University of Cincinnati for independent biomass analyses as lipid bound phosphates, which were extracted from attached bacteria. The results of biomass analyses on carbon samples obtained in Oct (Fall) 1994 will be emphasized. The general results obtained are compared to those reported in the literature as shown in Table VII. Biomass figures presented herein include error bars of one standard deviation based on analyses of field duplicates. Laboratory replicates were not depicted except for the distribution of biomass as lipid phosphate, though all biomass analyses were performed in triplicate.

Biomass levels in Croton pilot filters were generally within the ranges found by other researchers evaluating bacterial growth from natural organic matter substrate (Table VII). Two exceptions can be noted in the results from DAPI enumeration and [32]Phosphate incorporation into microbial lipids, which were 2-3 orders of magnitude lower than expected. The discrepancy in [32]Phosphate activity may be due to the limited literature found on the use of a similar procedure on

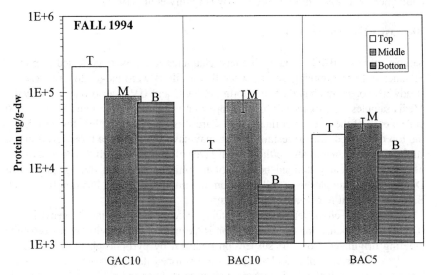

Figure 11.   Protein Levels in Pilot Plant Filters - Fall 1994

environmental samples. A suitable explanation could not be determined for the difference in DAPI enumeration therefore direct count data are not discussed further in this section.

The levels of bacterial protein quantified in Fall samples are depicted in Figure 11. The highest protein levels were found within the GAC filter where protein decreased from the top downward. The top sample contained one half log unit higher protein than corresponding middle and bottom level samples. The Fall biomass distribution in the GAC filter may have been caused by algae, which is generally higher in the reservoir in the Fall, and could have accumulated at the GAC filter surface. Similar accumulation was not observed at the top of the BACs, perhaps because ozonation was effective in partially breaking up large algal extracellular matter. Instead, protein levels in the Fall were distributed within the top half of the BAC columns, while filter bottom protein was generally lower. Mid column protein levels were greater in the higher flow column (5 min EBCT), because of the impact of flow velocity and fluid shear, but more so because of the contact time required for degradation of the organics. Biodegradation removals have been observed to occur primarily within the first few minutes EBCT (Servais et al. 1987, Wang 1995), which corresponded to the top of the 10 min EBCT BAC filter and to the middle of the 5 min EBCT BAC filter. The protein assay quantified all proteinaceous matter in dead and live biota, thus the distributions observed were not necessarily indicative of the distribution of bioactivity.

Viable bacteria recovered from media extracts were plated on R2A (low nutrient) agar and incubated at 15°C. Resulting colony forming units (CFU) were counted after 5 and 10 days (data not shown) incubation. One CFU can be assumed to result from the growth of each individual cell initially present in the sample, an assumption that modestly underestimates the number or organisms present (Standard Methods 1992). Different characteristics were observed in bacteria extracted from the GAC column compared to those extracted from BAC columns, specifically for media sampled in the Fall season. GAC bacteria types behaved like glider organisms, with growth spreading throughout the plate surface; they exhibited a distinct purple pigment. BAC (5 and 10 min EBCT) bacteria were also pigmented but of varied colors, their growth was limited to small, distinct circular and elongated colonies.

After 5 days incubation, clearer differences could be observed between viable colonies from the GAC and BAC filters sampled in Fall 1994 (Figure 12a). Viable bacteria were lowest by more than one log unit in the GAC column, but were distributed with depth, decreasing only gradually from top to bottom ($7.2 - 2 \times 10^7$ CFU/g-dw). The 10 min BAC showed the highest counts after 5 days, with viable bacteria concentrated in the top (8.9×10^8 CFU/g-dw) and middle ($2.1 \pm 0.4 \times 10^8$ CFU/g-dw), with negligible levels at the filter base (10^4 CFU/g-dw). Viable counts were more evenly distributed with bed depth in the 5 min BAC filter, likely because of the higher flow rate and shorter EBCT, with 2.6 and $3.2 \pm 1.2 \times 10^8$ CFU/g-dw top and middle samples, as well as significant levels at the filter base (4.4×10^7 CFU/g-dw). Standard deviations on middle BAC samples were derived from field duplicates.

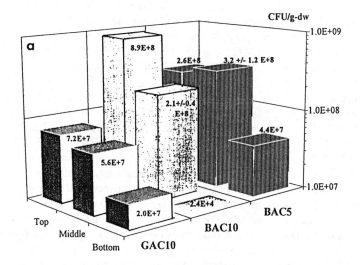

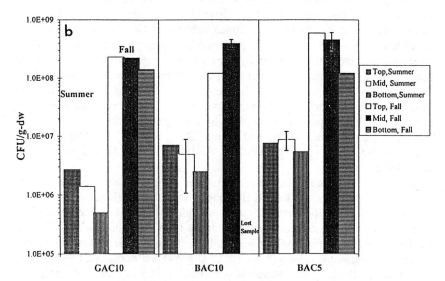

Figure 12.   Viable Bacteria as CFU/gdw After (a) 5 Day Incubation and
(b) 10 Day Incubation

Differences among the pilot filters were less evident after 10 days incubation, perhaps because the slower working GAC bacteria required a longer incubation time to match the levels of growth of bacteria from the BAC filters. Figure 12b shows the results from enumeration after 10 days for both seasons. Summer levels were still 2 orders of magnitude lower ($\sim 10^6$) compared to Fall samples, and stratification with depth was less pronounced overall compared to 5 day counts, but generally followed the trend of deceasing counts with increasing depth. The plates from the bottom 10 min BAC Fall sample were lost in the laboratory. Error bars indicated on middle BAC filter results reflect combined field and analytical variability for these samples.

The number of viable organisms quantified in the GAC/BAC pilot filters were generally one order of magnitude higher than those quantified in the surface of slow sand filters, presumably because of the additional surface area offered by GAC particles. Eighmy et al. (1992) reported viable counts after 6-10 days incubation on R2A media (20°C) ranged 10^4 - 10^8 CFU/g-dw in the top 10 cm of three full-scale, mature sand filters evaluated. In studies with pilot slow sand filters, Eighmy et al. (1993) quantified viable counts on the order of 10^7 CFU/g-dw in samples from the schmutzdecke of filters treating preozonated water. Incubation conditions in the later study were 10 days at 20°C on R2A media.

Independent characterization analyses performed on split samples from Croton biofilters were completed thanks to the collaboration of the University of Cincinnati Civil and Environmental Engineering Department. Results from extraction of bacterial lipid phosphates from attached bacteria including accompanying pore water ('total'), as received, supported the lower levels of biomass in Summer samples (Figure 13). The variability depicted by error bars in the figure corresponds to one standard deviation from triplicate analysis of each sample, except for the middle BAC locations were field duplicates are also reflected. Processing of prewashed or attached-only bacteria from media samples showed similar trends as the unwashed or total samples, but generally exhibited lower biomass concentrations.

Lipid phosphates were appreciably higher in the Fall in the GAC column, with the top sample exhibiting more than two fold greater biomass compared to the remainder of the GAC bed or the top samples from the BAC columns. Based on these observations, and if extracted lipid phosphates described the distribution of bioactivity, the GAC should provide the highest removals; however, actual BOC removals reflected the opposite. These results may have been influenced from accumulation of algae at the GAC surface, especially in the Fall, since phospholipids in algae would also be quantified in this procedure. A more plausible explanation may be that GAC bacteria although present in high numbers, accomplished less direct biooxidatin (to CO_2) because of the need to first biotransform the large, complex organics in the raw water. BAC bacteria were fed smaller, preozonated organics which could be mineralized more readily. Higher biomass at the surface of all three filters was observed in viable counts and protein levels, and is generally expected based on the relatively higher levels of substrate present. This effect was

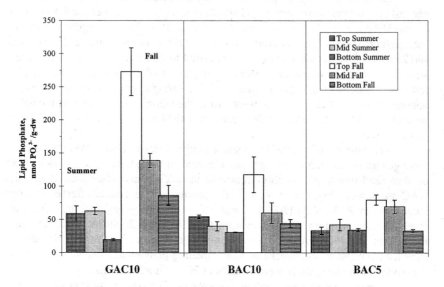

Figure 13.   Biomass Distribution as Microbial Lipid Phosphates

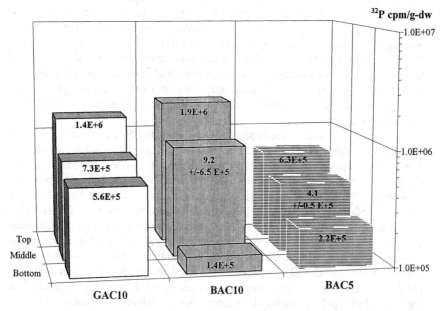

Figure 14.   Bioactivity Distribution as ^{32}Phosphate Incorporation in Microbial Lipids - Fall 1994 (adapted from Klevens et al., 1996)

more consistently observed in the 10 min EBCT GAC and BAC columns but not in the higher flow, 5 min EBCT column.

Wang (1995) also observed higher levels of lipid phosphate within the top half of pilot GAC biofilters. Levels reached 280 $\pm$ 14 nmol-PO_3^{3-}/g dw at the surface but were halved within 30 cm depth in Calgon F400® GAC/sand filters. An anthracite/sand biofilter showed this distribution but without as sharp a decrease; colonization levels averaged 65 $\pm$ 2 nmol-PO_4^{3-}/g dw. Wang developed a relationship between biomass (as lipid phosphates) and EBCT in bench scale sand biofilters, and found biomass to decrease expoentially with increasing EBCT (i.e., filter depth). Federle et al. (1983) found lipid bound phosphates in bacteria extracted from marine sediments ranged 45-145 nmol-PO_4^{3-}/g dw. Biomass levels of 20-270 nmol-PO_4^{3-}/g dw in Croton biofilters were in close agreement with the estimates reported by others.

The uptake of ^{14}C-L-aminoacids in microbial protein was used for more accurate quantification of live biota in Croton's Summer media samples. The procedure allows a short incubation period where respiring organisms incorporate the label during synthesis of new protein, then are washed to rid the sample of excess label, dead cells or debris that might interfere with radiocounting.

Because higher numbers of organisms had populated the filters by the Fall sampling event, and also by use of a more powerful, higher specific activity radiolable in ^{32}P, the quantification of bioactivity based on incorporation of ^{32}Phosphate in microbial lipids yielded clearer trends in the Fall. It should be noted that the actual distribution of activity may be different in field pilot filters, since all live biota extracted from media samples are capable of uptake of ^{32}P, but not all of these organisms contribute equally to the removals of biodegradable substrate.

As depicted in Figure 14, bioactivity was moderately higher in the BAC10 compared to the GAC, where flowrates were comparable. Possibly because of the different substrate feeds, (the BAC receiving readily biodegradable (preozonated) organics and the GAC receiving large, more refractory organics from raw water). Biological activity was concentrated in the top half in the BAC10 where most of the substrate was quickly consumed, but distributed through the full depth of the GAC bed. The full residence time in the latter was utilized to accomplish biodegradation and biotransformations. Despite showing similar total bioactivity (3.0 x 10^6 sum in BAC10 vs. 2.7 x 10^6 ^{32}P cpm/g dw sum in GAC), the BAC10 filter provided higher removals of BOC (52% on this date) compared to 38% removal of BOC through the GAC column.

Results from bioactivity distribution as ^{32}Phosphate in the 5 min BAC supported those from (5 day incubation) viable bacteria counts, suggesting bacterial colonization of this higher flow column was generally one half to one log unit lower than the 10 min BAC at the top and middle of the filter, but higher at the base of the filter. Total filter activity was one half (1.3 vs. 3.0 x 10^6 ^{32}P cpm/g dw) in the BAC5 compared to BAC10, suggesting the higher flow promoted slightly lower overall growth, and caused bacterial populations to be distributed through the full filter depth accordingly with the shorter contact time. The more even distribution in the 5 min BAC may have also been caused by greater fluid shear.

[32]Phosphate incorporation in Fall samples was selected to represent pilot filter bioactivity for comparisons to BOC removals. Averages for the top and bottom half of each column were derived from pooling the corresponding top/middle and middle/bottom results, to yield average activity levels per section of filter (Table VIII). These values were converted from mass to volume basis using a bulk density of 0.47 g/cm^3 for Calgon BPL® 4x10 carbon (Calgon Carbon Corp., Pittsburgh, PA), and multiplied by one half the filter volume (44500 cm^3) for an estimate of activity per filter section in units of ^{32}P counts per minute (cpm). Total bed activity was calculated as the sum of the filter halves.

A simulated bacterial 'efficiency' was estimated by dividing observed BOC removals on the date of sampling (17 Oct 94) by total filter activity. The results shown in Table VIII suggest that bacteria colonizing the BAC5 filter achieved the highest removal of substrate per unit bioactivity, while GAC bacteria were least efficient at mineralizing substrate. However, these ratios are only qualitative, since bioactivity levels were not significantly different (95% c.l.) among the three pilot filters. Also, on 17 Oct 94, effluent BOC levels from the GAC filter were not significantly different from influent raw water levels (95% c.l.), though both BAC effluents were significantly lower than influent ozonated water BOC.

Based on the concentration of bioactivity in the top half of the BAC10 min EBCT, observed substrate removals of 52% from ozonated influent probably occurred within the first 5 min of filter contact. Similar removals were thus expected and observed from the 5 min BAC which achieved 47% reduction on the same date. BOC removals were accomplished through assimilation of fast BOC with negligible change in slow BAC. Comparable performance by the BAC5 was achieved despite lower (though NSD) bioactivity levels and twice the organics loading (higher flowrate), possibly because the additional 5 min contact provided in the 10 min BAC contributed little to degradation of the more recalcitrant organics remaining after most of the fast BOC had been consumed.

Additional explanations for the comparable performance of the BAC5 despite lower biomass may lie in differences in the rates of bacterial metabolism between the populations colonizing either BAC, or might suggest mass transfer limitations in the diffusion of substrate and essential nutrients to the biofilm surface. The first hypothesis may be supported if BAC5 bacteria were working closer to their maximum rate of utilization (μ_{max}), where they are most efficient in converting organic substrate for cell maintenance and energy production, with little excess for biomass reproduction. External mass transfer limitations have not been found to influence biofiltration treatment based on evaluations of parallel biofilters operated at different flow rates but similar EBCT (Servais et al. 1994, Wang 1995). However, different application rates at constant contact time were not explored for Croton pilot filters.

Moll et al. (1995) showed statistically distinct differences between microbial populations colonizing biofilters of different feed sources, including ozonated humic substances and conventionally treated (unozonated) Ohio River water. Differences in community structure and metabolic capabilities were quantified from phospholipid fatty acid analysis (PLFA), DNA amplification fingerprinting (DAF), and the ability

Table VIII.   Bioactivity Levels and BOC Removals in Fall Samples

	Units	GAC10	BAC10	BAC5
Top Half Activity	^{32}P cpm/g-dw	$1.08\pm0.5\times10^6$	$1.41\pm0.7\times10^6$	$5.20\pm1.5\times10^5$
	^{32}P cpm	$2.26\pm1.0\times10^{10}$	$2.95\pm1.5\times10^{10}$	$1.09\pm0.3\times10^{10}$
Bottom Half Activity	^{32}P cpm/g-dw	$6.48\pm1.2\times10^5$	$5.27\pm5.5\times10^5$	$3.18\pm1.3\times10^5$
	^{32}P cpm	$1.35\pm0.25\times10^{10}$	$1.10\pm1.0\times10^{10}$	$6.64\pm2.8\times10^9$
Total Activity	^{32}P cpm	$3.62\pm0.25\times10^{10}$	$4.06\pm1.3\times10^{10}$	$1.75\pm0.30\times10^{10}$
BOC levels on 17 Oct 94	mg/L	0.39 ± 0.04	0.57 ± 0.10	0.63 ± 0.17
	mg/L removed[1]	0.24	0.62	0.56
	% removal[1]	38%	52%	47%
Bacterial Efficiency mg/L/cpm x 10^{12}	BOC consumed/ filter activity	6.6	15	32

[1]   Removals relative to raw water (0.63 ± 0.14 mg/L BOC) for GAC, or relative to ozonated water (1.19 ± 0.05 mg/L, BOC) for BACs.

to use different organic compounds as sole carbon sources. Their results showed that bacterial communities from the top of the filter receiving ozonated humics were significantly different metabolically and genetically, as well as more diverse, than communities developed within the biofilter treating unozonated Ohio River water. These findings support the differences in removals observed between Croton BACs (receiving ozonated NOM) and GAC (receiving unozonated raw NOM) filters.

In addition, Moll et al. (1995) showed that communities were significantly different with depth in the filter treating ozonated humics, while no differences were apparent with depth for filters receiving either raw (Ohio River) water or unozonated humics. They concluded that the homogeneity (with respect to substrate utilization, diversity and metabolic capabilities) of bacterial populations using unozonated feed was due to the relatively more homogeneous character of unozonated compared to ozonated influents. Similarly, Croton's GAC filter showed most evenly distributed biomass and bioactivity with depth through the filter bed, suggesting communities developed within this filter may have been similar throughout, and probably were active through the full 10 min bioreactor contact time.

Conclusions

This research confirmed that on a pilot scale, the combination of ozone and biologically activated (BAC) can reduce NOM and alter its characteristics in Croton Reservoir water to assist the City of New York in meeting required disinfection byproduct (DBP) limits and minimizing regrowth in the distribution system. Reduction of crucial treatment parameters including biodegradable organic carbon (BOC), chlorine demand, trihalomethanes (THM) and haloacetic acids (HAA) formation in treated water was improved by ozone enhanced biofiltration compared to biofiltration alone. However, other treatment configurations including GAC-DE-Chlorination or GAC-Ozone-DE-Chlorination may also provide efficient performances, but were not evaluated as part of this research effort. Specific conclusions derived from this program are offered below.

- Seasonal water temperature fluctuations had a marked effect on Croton raw water characteristics during 1994. The biodegradability of raw water may have been influenced by the warming trend in the reservoir, changing from a BDOC 7% (Winter) to 19% (Fall). These changes were caused principally by changes in the fast or more readily biodegradable substrate in the raw water.

- Combined Ozone/BAC (5 or 10 min EBCT) accomplished improved removals over those provided by the control (10 min) GAC filter, for all parameters monitored. Contact times of 5 and 10 min reduced only fast BDOC with negligible removals of slowly biodegradable organics. BDOC levels in BAC column effluents were somewhat higher than levels initially present in the raw unozonated water, but this increase was due to a higher contribution from slowly biodegradable compounds. Insufficient data were collected as part of this monitoring program to enable a distinction between 5 and 10 min EBCT BAC filters.

- Fractionation of NOM in treated samples provided additional insight to the effect of treatment. Observed patterns in the treatability of NOM were consistent with the literature. Increased biodegradability in samples from different seasons was correlated with the proportion of small MW ($<$3k) and the hydrophilic fraction.

- Methods for quantifying biomass distributions within Croton pilot filters included total (live and dead) biomass and (live) bioactivity measures. Bacteria extracted from samples collected in mid-Aug 1994 (Summer) were generally 2-3 log units lower than those obtained from Oct 1994 (Fall) samples. These results were supported by other biomass analyses.

- In general, biomass levels were asymmetrically distributed with the highest levels at the filter surface. The different distribution among the BAC columns may be explained by the higher hydraulic loading applied to the 5 min BAC, which accomplished greater penetration of substrate.

- Bacteria colonizing the GAC filter were different morphologically, slower working, and less efficient in providing complete biooxidation of organics compared to those colonizing the BAC filters. These characteristics might be expected based on their exposure to more complex organic substrate in the raw water.

Acknowledgements

Financial support for this project was provided by the Joint Venture - Metcalf & Eddy and Hazen & Sawyer. Special thanks are given to Austin Melly and Staff at the Croton Gatehouse (NYC-DEP), Don Brailey (Hazen & Sawyer), and Bob Mooney, S. Pepin, and Dr. Bill Chesbro (UNH Department of Microbiology) for their contribution to the study. Thanks are also extended to Dr. Scott Summers (University of Cincinnati) for his assistance and advice.

Literature Cited

Amy, G., Chowdhury, Z., Green, R., Krasner, S., Owen, D., Paode, R., Rice, E. and Summers, R.S. (1992). Biodegradability of Natural Organic Matter: A Comparison of Methods (BDOC and AOC) and Correlations with Chemical Surrogates. AWWA Proc. June 18-22, Vancouver, Canada.

Balkwill, D.L., Leach, F.R., Wilson, J.T., McNabb, J.F. and White, D.C. (1988). Equivalence of Microbial Biomass Measures Based on Membrane Lipid and Cell Wall Components, Adenosine Triphosphate, and Direct Counts in Subsurface Aquifer Sediments. Microb. Ecology 16: 73-84.

Balkwill, D.L. and Ghiorse, W.C. (1985). Characterization of Subsurface Bacteria Associated with Two Shallow Aquifers in Oklahoma. Appl. and Environm. Microbiol. 50 (3): 580-588.

Bancroft, K., Maloney, S.W., McElhaney, J. and Suffet, I.H. (1983). Assessment of Bacterial Growth and Total Organic Carbon Removal on Granular Activated Carbon Contactors. Appl. and Environm. Microbiol. 46 (3): 683-688.

Bose, P. and Reckhow, D.A. (1995). Selected Physical-Chemical Properties of NOM and Their Changes due to Ozone Treatment. Presented at NY-NEWWA Joint Conf., April 25-27, Saratoga Springs, NY.

Bott, T.L. and Kaplan, L.A. (1985). Bacterial Biomass, Metabolic State, and Activty in Stream Sediments: Relation to Environmental Variables and Multiple Assay Comparisons. Appl. and Environ. Microbial. 50 (2): 508-522.

Brock, T.D. and Madigan, M.T. (1988). Biology of Microorganisms, 5th ed., Prentice Hall, Englewood Cliffs, NJ. 835.

Bunch, W. and Kerr, K. (1995). "Now, Drink and Be Wary - Bacteria and Chlorine Plague Storied System". New York Newsday, p. A4-. 18 Jun 95.

Buchanan-Mappin, J.M. and Wallis, P.M. (1986). Enumeration and Identification of Heterotrophic Bacteria in Groundwater and in a Mountain Stream. Can. J. Microbiol. 32: 93-98.

Chesbro, W.R., Mooney, R.E., Negm, R.S., O'Neil, J., Pepin, S. (1994). Methods for the Evaluation of Attached Microbial Communities and the Efficiency of Filtration Columns, Univ. of New Hampshire Department of Microbiology. Durham, NH.

Cipparone, L.A., Diehl, A.C. and Speitel, G.E. (1993). The Effect of Biodegradable Organic Carbon Removal on the Bacterial Regrowth Potential and Disinfection By-product Formation of Ozonated Water. AWWA Proc. June, San Antonio, TX.

Collins, M.R. and Vaughan, C.W. (1996). Characterization of NOM Removal by Biofiltration: Impact of Coagulation, Ozonation and Sand Media Coating. ACS Symposium Series - Disinfection By-Products in Water Treatment: The Chemistry of Their Formation and Control, R.A. Minear and G.L. Amy (eds.). CRC, Lewis Publishers, Boca Raton, FL.

Collins, M.R., Eighmy, T.T. and Malley Jr., J.P. (1991). Evaluating Modifications to Slow Sand Filters. J. AWWA 83 (9): 62-70.

Collins, M.R., Eighmy, T.T., Fenstermacher, J.M. and Spanos, S.K. (1989). Modifications to the Slow Sand Filtration Process for Improved Removals of Trihalomethane Precursors. AWWARF, Denver, CO.

Collins, M.R., Amy, G.L. and Steelink, C. (1986). Molecular Weight Distribution Carboxylic Acidity, and Human Substances Content of Aquatic Organic Matter: Implications for Removal During Water Treatment. Environ. Sci. & Technol. 20 (10):1028-1032.

DiGiano, F.A., Mallon, K., Stringfellow, W., Cobb, N., Moorse, J. and Thompson, J.C. (1992). Microbial Activity on Filter-Adsorbers. AWWARF Final Report, Denver, CO.

Edzwald, J.K. (1993). Coagulation in Drinking Water Treatment: Particles, Organics and Coagulants. Water Sci. Technol. 27 (11): 21-35.

Eighmy, T.T., Collins, M.R., Malley Jr., J.P., Royce, J. and Morgan, D. (1993). Biologically Enhanced Slow Sand Filtration for Removal NOM. AWWARF Final Report, Denver, CO.

Eighmy, T.T., Collins, M.R., Spanos, S. and Fenstermacher, J. (1992). Microbial Populations, Activities and Carbon Metabolism in Slow Sand Filters. Water Res. 26 (10): 1319-1328.

EPA (1991). Methods for the Determination of Organic Compounds in Drinking Water. EPA 600/4/88/039.

Farrar, M. (1994). Pilot Plant Project Manager. Project Meeting Minutes and Personal Correspondence. Croton Reservoir, Yorktown Heights, NY.

Federle, T.W., Meredith, A.H., Livingston, R.J., Meeter, D.A. and White, D.C. (1983). Spatial Distribution of Biochemical Parameters Indicating Biomass and Community Composition of Microbial Assemblies in Estuarine Mud Flat Sediments. Appl. and Environ. Microbiol. 45 (1): 58-63.

Findlay, R.H., King, G.M. and Watling, L. (1989). Efficacy of Phospholipid Analysis in Determining Microbial Biomass in Sediments. Appl. and Environ. Microbiol. 55 (11): 2888-2893.

Goel, S., Hozalski, R.M. and Bouwer, E.J. (1995). Biodegradation of NOM: Effect of NOM Source and Ozone Dose. J. AWWA 87 (1): 90-105.

Huck, P.M., Zhang, S. and Price, M.L. (1994). BOM Removal During Biological Treatment: A First Order Model. J. AWWA 86 (6): 61-71.

Kaplan, L.A., Reasoner, D.J. and Rice, E.W. (1994). A Survey of BOM in U.S. Drinking Waters. J. AWWA 86 (2): 121-132.

Klevens, C.M., Collins, M.R., Negm, R. and Farrar, M.F. (1996). Characterization of NOM Removal by Biological Activated Carbon. In: Advances in Slow Sand Biological Filtration, N. Graham and R. Collins (Editors), John Wiley and Sons, Ltd., Chichester, England.

Klevens, C.M. (1995). Natural Organic Matter Characterization and Treatability by Biologically Activated Carbon Filtration of Croton Reservoir Water. M.S. Thesis, University of New Hampshire, Durham, NH.

Langlais, B., Reckhow, D.A. and Brink, D.R. (1991). Ozone in Water Treatment, Cooperative Res. Report. AWWARF. Lewis Publishers, Chelsea, MI. 569.

Leenheer, J.A. (1981). Comprehensive Approach to Preparative Isolation and Fractionation of Dissolved Organic Carbon from Natural Waters and Wastewaters. Environ. Sci. & Technol., Vol. 15 (5): 578-587.

Malley Jr., J.P. (1988). A Fundamental Study of Dissolved Air Flotation for Treatment of Low Turbidity Waters Containing Natural Organic Matter. PhD Dissertation. Univ. of Massachusetts, Dept. of Civil Engineering, Amherst, MA.

Malley Jr., J.P., Eighmy, T.T., Collins, M.R., Royce, J.A. and Morgan, D. (1993). The Performance and Microbiology of Ozone-Enhanced Biological Filtration. J. AWWA 85 (12): 47.

Mastronardi, R.A., Fulton, G.P., Farrar, M. and Collins, A.G. (1993). Prezonation to Improve and Optimize Diatomaceous Earth Filtration. Ozone Sci. & Eng. 15: 131-147.

Miller, J.C. and Miller, J.N. (1993). Statistics for Analytical Chemistry, 3rd ed. Ellis Horwood PTR Prentice Hall. NY, NY. 233.

Mitton, M.J., Huck, P.M., Krasner, S.W., Prévost, M. and Reckhow, D.A. (1993). Quantifying and Predicting the Removal of Biodegradable Organic Matter and Related Parameters in Biological Drinking Water Treatment. AWWA WQTC Proc., Nov. 7-11, Miami, FL.

Mogren, E., Scarpino, P., and Summers, R.S. (1990). Measurement of Biodegradable Dissolved Organic Carbon in Drinking Water. AWWA Proc., Cincinnati, OH.

Moll, D., Wang, J.Z. and Summers, R.S. (1995). NOM Removal by Distinct Microbial Populations in Biofiltration Processes. Presented at AWWA Conf. June 18-22, Anaheim, CA.

Neidhardt, F.C., Ingraham, J.L. and Schaechter, M. (1990). Physiology of the Bacterial Cell: A Molecular Approach. Sinauer Assoc., Inc., Sunderland, MA. 506.

Owen, D.M., Amy, G.L. Chowdhury, Z.K. (1993). Characterization of Natural Organic Matter and Its Relationship to Treatability. AWWARF, Denver, Co.

Prévost, M., Desjardins, R., Duchesne et C. Poirier, D. (1991). Chlorine Demand Removal by Biological Treatment in Cold Water. Environ. Technol. 12: 569-580.

Principe, M., Mastronardi, R., Brailey, D., Nickols, D. and Fulton, G. (1994). New York City's First Water Filtration Plant. AWWA Proc. June 19-23, NY, NY.

Reckhow, D.A., Singer, P.C. and Malcolm, R.L. (1990). Chlorination of Humic Materials: Byproduct Formation and Chemical Interpretations. Environ. Sci. & Technol. 24 (11): 1655-1664.

Servais, P., Billen, G. and Bouillot, P. (1994). Biological Colonization of Granular Activated Carbon Filters in Drinking-Water Treatment. J. Environ. Eng. 120 (4): 888-899.

Servais, P., Billen, G., Bouillot, P. and Benezet, M. (1992). A Pilot Study of Biological GAC Filtration in Drinking-Water Treatment. J. Water Supply Res. and Technol.-Aqua 41 (3): 163-168.

Servais, P., Billen, G., Ventresque, C. and Bablon, G.P. (1991). Microbial Activity in GAC Filters at the Choisy-le-Roi Treatment Plant. J. AWWA 83 (2): 62-68.

Servais, P. et al. (1987). Determination of the Biodegradable Fraction of Dissolved Organic Matter in Water. Water Res. 21 (4): 445.

Shaw, J.P. (1994). The Effects of Ultraviolet Irradiation on Organic Matter in Natural Waters. M. Thesis. Univ. of New Hampshire, Durham, NH.

Simon, M. (1988). Growth Characteristics of Small and Large Free-Living and Attached Bacteria in Lake Constance. Microbial Ecology 15: 151-163.

Spencer, C.M. and Collins, M.R. (1995). Improving Precursor Removal by Precoat Filters. J. AWWA 87 (12): 71-82.

Spencer, C.M. (1991). Modification of Precoat Filters with Crushed Granular Activated Carbon and Anionic Resin to Improve Organic Precursor Removals. Master of Science Thesis. Civil Engineering, Univ. of New Hampshire, Durham, NH.

Standard Methods for the Examination of Water and Wastewater (1992). 18th ed. Am. Public Health Assoc., Washington, DC.

Standard Methods for the Examination of Water and Wastewater (1989). 17th ed. Am. Public Health Assoc., Washington, DC.

Tibbles, B.J., Davis, C.L., Harris, J.M., Lucas, M.I. (1992). Estimates of Bacterial Productivity in Marine Sediments and Water from a Temperate Saltmarsh Lagoon. Microbial Ecology 23: 195-209.

Wang, J.Z. (1995). Assessment of Biodegradation and Biodegradation Kinetics of Natural Organic Matter in Drinking Water Biofilters. PhD Thesis, Civil and Environ. Engineering. Univ. of Cincinnati, Cincinnati, IL.

Chapter 15

The Reduction of Bromate by Granular Activated Carbon in Distilled and Natural Waters

Jennifer Miller[1,3], Vernon L. Snoeyink[1], and Joop Kruithof[2]

[1]Department of Civil Engineering, University of Illinois, 3230 Newmark Civil Engineering Laboratory, 205 North Mathews, Urbana, IL 61801
[2]KIWA N.V., Groningenhaven 7, P.O. Box 1072, 3430 BB Nieuwegein, Netherlands

An existing model describing the reduction of free chlorine by granular activated carbon (Suidan et al., 1977a and 1977b) has been applied to the reduction of bromate by granular activated carbon. Both finite batch and packed bed column studies were used to verify the model predictions. Distilled water tests were conducted under varying initial bromate concentrations, at several solution pH values, and using different carbon particle size fractions. The effect of natural organic matter was studied by spiking solutions with a fulvic acid isolate, as well as using water obtained from the Interstate Water Company in Danville, IL. Preloaded carbon was also used for batch and column tests. It was found that the model describes bromate reduction well in distilled water, but fails to account for the cumulative effect of natural organic matter in natural waters.

Recent interest in the byproducts of alternative oxidants for drinking water treatment has led to the publication of several studies of bromate reduction by granular activated carbon (GAC) in distilled water (Siddiqui et al., 1994; Miller et al., 1995). Much more research on bromate will undoubtedly be forthcoming. It would be useful to have a model to describe and extrapolate the data, both in distilled water and also in natural waters. In this paper we present batch and column test data for bromate reduction by GAC in distilled and natural waters, and describe the data with an existing model.

[3]Current address: Department of Civil Engineering, W348 Nebraska Hall, University of Nebraska, Lincoln, NE 68588

0097–6156/96/0649–0251$17.75/0
© 1996 American Chemical Society

The model which will be discussed here was developed previously by Suidan et al. (1975, 1977a, 1977b, 1978) to describe the reduction of free chlorine by GAC. Although bromate and hypochlorite are both oxyanions, bromate requires more electrons for reduction, reacts more slowly, and is found in drinking water at much lower concentrations than hypochlorous acid and hypochlorite. Never the less, the main assumptions used by Suidan (1975) in developing this model, namely that the reduction reaction is limited by both mass transfer limitations and surface reaction rate limitations, appear to hold true for bromate.

In this paper, we will discuss the applicability of the model developed by Suidan et al. (1975, 1977a, 1977b, 1978) to bromate reduction in distilled water, as well as in waters containing natural organic matter (NOM). The success and failure of the model to predict bench top GAC column performance under different laboratory conditions will be discussed.

Methods and Materials

Chemicals used in these experiments were reagent grade quality. Anion solutions were prepared using sodium and/or potassium salts of the various anions. Salts were dried for twenty-four hours in a 105°C oven, and then were cooled and stored in a desiccator. Salts were weighed using a Mettler AE2000 electronic balance (Hightstown, NJ).

Water from the Interstate Water Company in Danville, IL, was obtained from the clarifiers after lime softening. The water was stored at 4°C until use. The source water for the Interstate Water Company in Danville is Lake Vermilion. The raw water characteristics of Lake Vermillion water are given in Table 1. At the time of the experiments, the water was buffered with a phosphate buffer and the pH was adjusted.

Table 1: Raw water characteristics for Lake Vermillion

Water	pH	Total Hardness mg/L as $CaCO_3$	Total Alkalinity mg/L as $CaCO_3$	Turbidity NTU	Chloride, mg/L	TOC, mg/L
Lake Vermillion	8-9	90-120	40-60	0.3-5	25-30	2

The carbon used in these experiments was Ceca GAC 40, batch number B70921-2B. The carbon was sieved following ASTM Standard D 2862-82 (ASTM, 1988). Following sieving, the 30 x 40 mesh size fraction was rinsed with distilled water to remove dust, and then dried in a 105°C oven. Dried carbon was stored in a desiccator.

Preloaded carbon was loaded with natural organic matter at the Interstate Water Company, Danville, Illinois. The water from the filter effluent was passed through carbon which had been packed into glass columns. A one minute empty bed contact time was used for preloading, and the carbon was loaded for various lengths of time.

Finite batch tests were used to examine the effects of reaction variables such as solution pH, initial bromate concentration, and NOM concentration. Batch tests were conducted in a four liter pyrex glass beaker. The solution was stirred using a magnetic stirrer, turning at a speed sufficient to keep carbon suspended uniformly throughout the beaker. A phosphate buffer (approximately 0.1 mM) was used to maintain a constant pH during batch tests. The adjustment of solution pH was accomplished by adding either 1 M H_2SO_4 or 1 N NaOH.

Samples were taken during batch tests using a 10 mL glass syringe with a luer-lock connection, and filtered using a filter tip with 0.45 μm cellulosic filter paper. Samples were stored in high density polyethylene screw top bottles, and were kept in a cold room (temperature 3-8°C) until analysis.

Column studies were conducted in glass columns of 1.3 cm (one-half inch) or 2.5 cm (one inch) diameter, depending on carbon particle size. Carbon beds were supported on silica sand, glass beads and/or glass wool. Prior to packing, the carbon was soaked in distilled, deionized water overnight to remove air bubbles. Care was taken to insure that no air bubbles entered the carbon bed during the test. Influent solution was mixed and stored in glass carboys on a magnetic stirrer. Masterflex pumps (Cole-parmer, New Jersey) and Tygon tubing were used to pump the solution through the carbon bed. The columns were operated upflow. Samples were collected directly into sample bottles. In general, short empty bed contact times, less than one minute, were used.

Standards for anion analysis were mixed according to Standard Methods for the Examination of Water and Wastewater, procedure 4110 B (1989). A stock solution for bromate was mixed using 1.1798 g $NaBrO_3$ in distilled water; this amount of $NaBrO_3$ was calculated to give a stock solution of 1 mg BrO_3^- per 1 mL. The working standard for bromate was made as for the other anions. Working standards were mixed using distilled water, and were run at the beginning of each analysis session. The following concentrations were used for working standards: 0, 5, 10, 20, 50, and 100 μg/L of each anion. Analysis was not done if the linear correlation factor for the standards was less than 0.99.

The method used for analysis was developed by Hans van der Jagt and co-workers at KIWA, Nieuwegien, the Netherlands and is described as follows. Analysis was done on a Dionex Series 300 ion chromatograph, equipped with a gradient pump, an autosampler and an electrochemical detector. Samples were analyzed at room temperature, and the system temperature compensation factor was 1.7. The following operational parameters were used:

Eluent 1: 0.5 mM Na_2CO_3, 0.18 mM $NaHCO_3$
Eluent 2: 4 mM Na_2CO_3, 1.5 mM $NaHCO_3$
Regenerant: 25 mM H_2SO_4
Columns: AG9-SC, AS9-SC
Flowrate: 2 mL/min
Conductivity detector range = 0.01 μS
Run time=12 minutes: 4.5 minutes on Eluent 1,5.5 minutes on Eluent 2, remaining time on Eluent 1
Bromate retention time = 2.33 minutes

The detection limit was calculated to be 2 μg/L, based on the slope of the calibration curve and the reproducibility of analysis at low concentrations. Concentrations less than the detection limit are plotted in the figures as zero. Natural water samples were filtered with a silver filter (Dionex) to remove chloride prior to analysis.

Model

The details of the development of the model by Suidan et al. which describes the reduction of free chlorine by GAC can be found in previously published papers (1977a, 1977b). Here we will discuss the most important assumptions used in the development of the model which pertain to bromate reduction. The equations used in the model and a list of the variables are given in the appendix.

The reduction of bromate, in this case, is assumed to be limited by both the rate of mass transfer and the rate of the surface reaction. The surface reaction occurs in two steps: a reversible adsorption/desorption step, followed by an irreversible dissociation step. The irreversible dissociation step is considered to be rate limiting. The surface reaction (for free chlorine) can be visualized as:

$$
\begin{array}{lll}
C^* + HOCl \longrightarrow C^*\text{-}HOCl & \text{reversible adsorption/desorption} \\
\underline{C^*\text{-}HOCl \longrightarrow Cl^- + H^+ + C^*O} & \underline{\text{irreversible dissociation step}} \\
C^* + HOCl \longrightarrow Cl^- + H^+ + C^*O & \text{overall}
\end{array}
$$

In this case C^* represents an "active site" on the carbon surface. Voudrias et al. (1983) visualized the reduction of chlorite by GAC in an analogous manner:

$$ C^* + ClO_2^- \longrightarrow Cl^- + C^*O_2 $$

The above reaction also follows the pattern of reaction proposed in the model by Suidan et al. (1977a, 1977b), that one oxyanion molecule reacts at one "active site" on the carbon surface. Because bromate is chemically related to chlorite and free chlorine, a similar reaction between bromate and GAC can be visualized.

$$
\begin{array}{lll}
C^* + BrO_3^- \longrightarrow C^*\text{-}BrO_3^- & \text{reversible adsorption/desorption} \\
\underline{C^*\text{-}BrO_3^- \longrightarrow Br^- + C^*O_{(3)}} & \underline{\text{irreversible dissociation step.}} \\
C^* + BrO_3^- \longrightarrow Br^- + C^*O_{(3)} & \text{overall}
\end{array}
$$

The reduction of bromate by a reaction similar to the free chlorine reaction would be pseudo first order with respect to bromate concentration. The reaction would not be first order over all, since bromate reduction requires the transfer of six electrons. If only one step is rate limiting, however, the reaction rate could be approximated as first order. Over time, the reduction of bromate would result in a build-up of oxides on the carbon surface, which would decrease the rate of bromate reduction (i.e., the carbon behaves as a poisoned catalyst).

Suidan et al. (1977a, 1977b) coupled the above assumptions with ideal reactor analyses for a finite batch reactor and a packed bed column. To obtain the kinetic parameters for the reduction reaction, the data from a bench scale batch reactor is analyzed. The constants are then used to predict bromate reduction in a packed bed column over time. The information needed in the model for a finite batch reactor includes: the initial bulk concentration, the pore volume of the carbon, the diffusivity, the "pore length" which is one sixth of the carbon particle diameter (Levenspiel, 1962), the carbon concentration, and four kinetic constants which are found by trial and error. For the packed bed column model, the information needed includes: the influent concentration, the axial dispersion coefficient, the bed length, the cross sectional area of the packed bed, the bed porosity, the carbon density in the bed, the flow rate, and nine empirical constants from an algebraic rate expression developed by Suidan (1975). The determination of the kinetic parameters for the model is discussed below.

Parameter Estimation

A large number of parameters are used in the model. For the batch test description there are four variables which were derived in the original model: SIT, K8, K9, and K10. The parameter SIT describes the number of "active sites" on the carbon surface. There is no independant method for determining this parameter, since we do not know exactly how bromate is reacting with the carbon surface. In general, however, SIT can be thought of as a relative measure of carbon activity. A carbon with a larger SIT value will be more effective for bromate reduction.

The parameters K8, K9, and K10, are kinetic constants which were derived in the original model. Again, there is no independent method of evaluating these constants, and the constants vary for different carbons and solutions (reflecting effects on the bromate reduction reaction.) The parameter K8 is the equilibrium constant for the reversible adsorption/desorption step. The value of this constant was determined by trial and error using batch test data. Once determined, the value of 0.00015 was used for all batch tests conducted in a solution of pH 7, since the partitioning of bromate is a constant for that species.

The parameter K9 is a rate constant for the formation of oxidized sites, ie. for the irreversible dissociation step. The value of K9 varied during batch tests, but was, with one exception, between 10 and 100 during this study. The remaining rate constant, K10, describes the spontaneous degradation of oxidized sites (or the regeneration of sites on the carbon surface.) There is no direct evidence for the regeneration of carbon sites after bromate reduction, however, other researchers (Kim, 1977) have noted this phenomenon when dealing with other oxidants. In the case of bromate, spontaneous regeneration of the carbon surface does not appear to be important, and K10 was held constant at 380.

The variables SIT, K8, and K9 were determined for the virgin 60x80 carbon by comparing the model predictions to the batch test results and column test results. The parameters determined for the 60x80 carbon data were used to calculate values for SIT and K9 for virgin carbon of other mesh sizes. According to Suidan (1975), SIT and K9 are proportional to the square of the pore length. The pore length is equal to one sixth of the particle diameter (Levenspiel, 1962), so SIT and K9 were easily calculated. The value for K8 was held constant, since K8 does not depend on carbon particle size (Suidan, 1975).

For batch tests conducted at a solution pH other than 7, the value for K8 was changed. At low pH, bromate is more reactive (Siddiqui et al., 1994). Thus, there should be less bromate adsorbed to the surface, since the bromate will decompose rapidly to form bromide and the oxidized carbon surface. In batch tests conducted in low pH solutions, the value for K8 was lower compared to the value for neutral solutions, reflecting the lower amount of adsorbed bromate. Conversely, in high pH solutions, the value of K8 was higher, reflecting the build-up of unreacting bromate on the carbon surface. In batch tests with solution pH other than 7, the values for SIT and K9 were held constant since these parameters are characteristic of the carbon and are not affected by the solution pH.

For batch tests conducted using preloaded carbon and in natural water solutions, SIT and K9 were found by trial and error to match the batch test data. A summary of values for the four kinetic parameters is given in Table 2. A simple optimization, such as least squares fit, was not used to determine the values of the parameters because the four constants are considered to have a physical significance. Arbitrarily selecting the parameters would diminish this aspect of the model.

Table 2: Kinetic parameters for batch test model

Water	Carbon Size	pH	SIT	K8	K9	K10
Distilled	0.51mm (30x40)	7	0.0077	0.00015	41	380
Distilled	0.2 mm (60x80)	7	0.0013	0.00015	19	380
Distilled	0.16 mm (80x100)	7	0.00076	0.00015	10	380
Distilled	1.44 mm (12x16)	7	0.61	0.00015	110	380
Distilled	0.51 mm	4	0.0077	0.000015	41	380
Distilled	0.51 mm	10	0.0077	0.0015	41	380
Distilled + fulvic acid	0.51 mm	7	0.003	0.00015	25	380
Danville	0.51 mm	7	0.0007	0.00015	36	380
Preloaded	0.51 mm	7	0.0006	0.00015	35	380

The model developed by Suidan (1975) also contains an approximate rate expression which contains nine variables. The only reason for the approximate rate expression was to simplify the computer solution of the non linear differential equations. The nine rate expression constants fit the equation:

$$R = (J1 \times C)/\{(J2 + C)[1 + J3\exp(-J4 \times C) \times X]^{J5} [1 + (J6\exp(J7 \times C) \times X)^{J8}]^{J9}\} \quad (1)$$

The rate expression constants were determined using the batch test model, after the values for SIT, K8, K9, and K10 were chosen. The values for the nine rate expression constants were not chosen arbitrarily, but were based on correlations developed by Suidan et al. (1977a). According to Suidan et al. (1977a), the values for J4 and J7 are constant, 10700 and 26600, respectively. The values for J5, J8 and J9 are related by the equation: J5 + J8*J9 = 1. In addition, J8 is equal to two. Suidan et al. (1977a) was able to correlate the remaining variables, J1, J2, J3 and J6, to the pore length of the carbon. In this study, the correlations were found not to be true, however, it was found that J3 and J6 could be held constant at the values that Suidan (1975) used in his initial work, 5900 and 100, respectively. The remaining variables, J1, J2, J5 and J9, were found by trial and error fit of the rate curve. The value for J5 (an exponent) ranged from 0.01 to 2, while the value for J9 (also an exponent) ranged from 0.445 to -0.5.

Results and Discussion
The effect of Initial Concentration

The reduction of bromate to bromide requires the transfer of six electrons. Generally, the transfer of electrons occurs individually (Larson and Weber, 1994). Thus, bromate reduction by GAC would not be expected to be an elementary reaction. The model developed by Suidan (1975) assumes that the reduction reaction occurs in one step, after a surface complex is formed. This assumption is adequate for bromate reduction if the overall reaction is limited by the rate of one step.

The rate of bromate reduction is affected by the initial concentration of bromate, as shown in figure 1. The rate of bromate reduction cannot be described as having a simple first or second order dependence on the initial bromate concentration, since the reaction is not elementary. Moreover, as will be discussed later, the reduction of bromate by GAC is limited by pore diffusion at low concentrations.

The initial concentrations of bromate for the batch test data shown in figure 1 differ more than one order of magnitude. The very high initial concentration of 460 ppb is not realistic of bromate concentrations encountered in water treatment, but is presented for discussion purposes. All of the data sets shown in figure 1 have been described using the model developed by Suidan. The kinetic parameters (discussed above) were the same for all three cases, and only the initial concentration was varied in the model calculation.

The batch data shown in figure 1 were used to predict the performance of a packed bed reactor with a high initial bromate concentration. The packed bed column experiment was conducted using 30x40 mesh virgin Ceca carbon, a one minute empty bed contact time (EBCT), an average initial bromate concentration of 470 ppb in distilled water, and a solution pH of 7. The predicted breakthrough curve and the column effluent data are shown in figure 2. There is good general agreement between the predicted curve and the data. The data show some scatter caused by pump and influent concentration variations over the course of the experiment, which lasted for 43 days. The carbon was not exhausted at the end of the study, but at that time 27 mg bromate per gram carbon had been reduced.

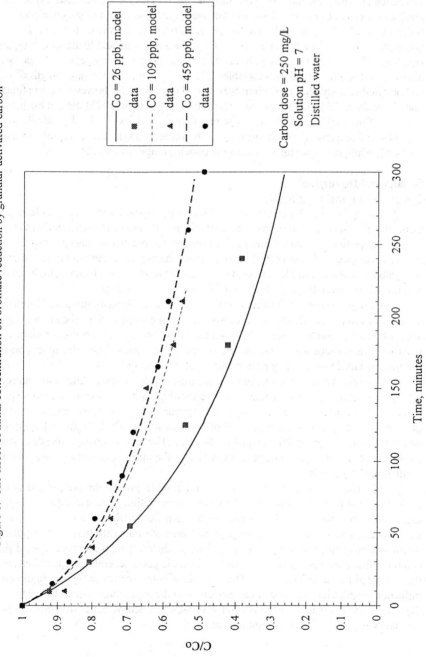

Figure 1: The effect of initial concentration on bromate reduction by granular activated carbon

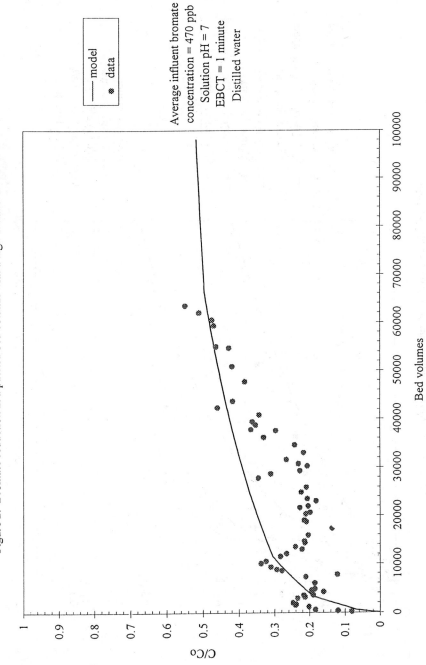

Figure 2: Bromate reduction in a packed bed column with a high initial bromate concentration

model
data

Average influent bromate
concentration = 470 ppb
Solution pH = 7
EBCT = 1 minute
Distilled water

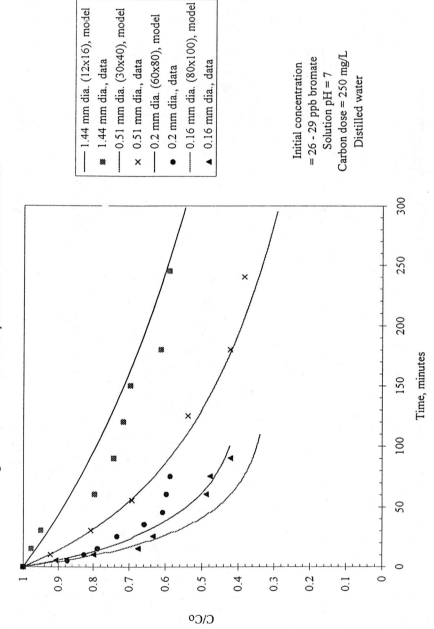

Figure 3: The effect of carbon particle size on bromate reduction

The effect of pore diffusion
    The reduction of bromate is limited by pore diffusion, which has been shown in this work. The effect of pore diffusion can be nicely shown by comparing bromate reduction by GAC of various particle sizes. Figure 3 is a comparison of batch test data obtained using four different size fractions of virgin Ceca carbon. With smaller GAC particles, the average pore length, defined for porous spherical particles as one sixth of the particle diameter (Levenspiel, 1962), is shorter. Thus, bromate reaches "active sites" on the carbon surface more rapidly and the overall rate of bromate reduction is more rapid. Figure 4 shows batch test data for batch tests conducted using slightly higher initial bromate concentrations. The same model parameters were used to obtain the curves in both figures 3 and 4, with only the initial concentration being changed. Note that the kinetic parameters for the various carbon particle sizes are mathematically related (SIT and K9 are related to particle size, per Suidan et al., 1975). Once data have been obtained for one carbon size fraction, it is simple to calculate the effect of using a different particle size for bromate reduction. The model curves for the different size fraction particles represent predictions, based on the 30x40 data.
    At very high initial bromate concentrations, in the mg/L range for example, bromate reduction is not limited by pore diffusion, but by the rate of the surface reaction itself. Figure 5 shows batch test data collected at very high initial bromate concentrations, using various size fractions of the same Ceca carbon.
    It is possible to estimate when diffusion limits a reaction by using the Weisz-Prater criterion (Cwp). (For more information on this criterion, see Fogler, 1992). The Weisz-Prater criterion is calculated as:

$$Cwp = -r_A'(obs) * \rho_p * R^2/(D_e * C_o) \qquad (2)$$

where $-r_A'(obs)$ is the observed rate of reaction, ρ_p is the particle density, R is the particle radius, D_e is the effective diffusivity (assumed, based on Fogler, 1992), and C_o is the initial batch test concentration. For a single batch test, an observed rate of reaction was calculated which was then used to calculate Cwp. In general, if Cwp >> 1, the reaction is severely limited by internal diffusion. If, on the other hand, Cwp << 1, the reaction is not limited by diffusion. For some of the batch tests shown in figures 3, 4, and 5, Cwp has been calculated. A summary of these values is given in table 3. To avoid the limitation of pore diffusion, either very small carbon particles may be used, or the initial bromate concentration must be very high.

Table 3: Weisz-Prater criterion for several batch test experiments

Carbon	Radius, mm	Rate, observed $m^3/sec\text{-}g_{cat}$	Initial concentration	Cwp
30x40 mesh	0.255	$2.5 \times 10^{-7} * Co$	26 ppb	1.1
30x40 mesh	0.255	$1.5 \times 10^{-7} * Co$	459 ppb	0.65
30x40 mesh	0.255	$2 \times 10^{-8} * Co$	10.9 ppm	0.08
80x100 mesh	0.08	$5.8 \times 10^{-7} * Co$	27 ppb	0.25
PAC (p325)	<0.000225	$4 \times 10^{-8} * Co$	10.9 ppm	1.4×10^{-7}

Figure 4: The effect of carbon particle size on bromate reduction at higher initial concentration

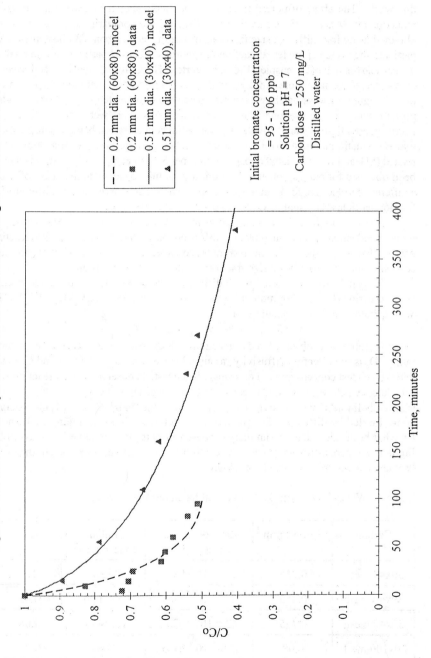

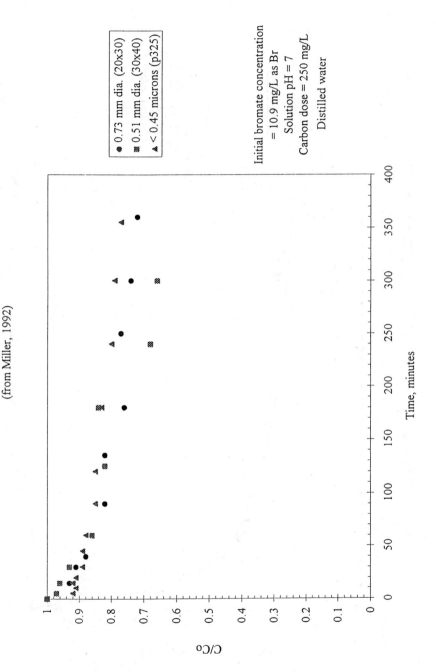

Figure 5: The effect of carbon particle size on bromate reduction at high initial bromate concentrations (from Miller, 1992)

◆ 0.73 mm dia. (20x30)
▩ 0.51 mm dia. (30x40)
▲ < 0.45 microns (p325)

Initial bromate concentration
= 10.9 mg/L as Br
Solution pH = 7
Carbon dose = 250 mg/L
Distilled water

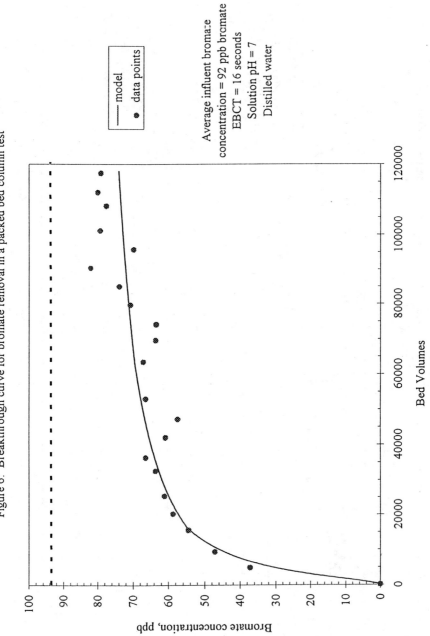

Figure 6: Breakthrough curve for bromate removal in a packed bed column test

The limitation of pore diffusion on bromate reduction can be seen in column test analysis, also. The data for the 60x80 mesh virgin Ceca carbon were used to predict packed bed reactor performance. The column test conducted with 60x80 mesh carbon had an average influent bromate concentration of 92 ppb in distilled water, a solution pH of 7, and an EBCT of 16 seconds. These experimental conditions were not meant to be realistic of water treatment conditions. The high influent concentration and the very short EBCT were used to exhaust the carbon in a short time period. The predicted bromate breakthrough curve and the column effluent data are shown in figure 6. There is very good agreement between the prediction and the data. The general shape of the breakthough curve is the same as the breakthrough seen using 30x40 mesh carbon and a higher influent bromate concentration.

The immediate breakthrough of bromate in both the 60x80 mesh carbon and the 30x40 mesh carbon column tests shows the limit of diffusion. If a longer EBCT is used, bromate has more time to diffuse into the carbon particles and react, and the initial breakthrough time is delayed. The effect of using longer EBCTs for the 60x80 mesh carbon test was simulated using the model. The curves are shown in figure 7. The only parameter changed for these model predictions was the flow rate. The initial breakthrough of bromate is controlled by diffusion, while the gently sloping portion of the curve indicates the region where the rate of the surface reaction is limiting bromate removal. Similar findings were made by Gonce and Voudrias (1994) for the removal of chlorite by GAC.

The Effect of pH

Although the model developed by Suidan (1975) is not a rigorous chemical model and does not directly address variables such as the effect of solution pH, the model can still be used to describe data obtained under various conditions. When working with free chlorine, Suidan et al. (1977b) found that pH affected the rate of free chlorine reduction only as far as it affected speciation between HOCl and OCl$^-$. They found that only considering the fraction of free chlorine as HOCl, they could describe the rate of reduction (ie. OCl$^-$ did not react appreciably with carbon). Unlike chlorine species, however, bromate does not exist in the acid form; the pKa of bromate is around zero (Christen, 1968). The model can still describe the effect of solution pH on bromate reduction in a qualitative manner.

Bromate reduction in batch tests as different solution pH values is shown in figure 8. In this modeling, it was found that the effect of solution pH could be described in the model by changing the kinetic parameter K8 only. In this case, other kinetic parameters were held constant. The initial rate of bromate reduction can then be calculated by the model. The relationship between the initial rate of reaction and the hydrogen ion concentration is shown in figure 9.

Based on the batch test data collected at different solution pH values, the performance of a packed bed reactor can be predicted. In this case, the packed bed experiments were assumed to have an influent bromate concentration of 30 ppb, and a one minute EBCT. These curves are shown in figure 10. Note that except for the influent concentration, the curve for pH 7 is the same as shown in figure 2 for a much higher influent bromate concentration. There is clearly a benefit to having a lower solution pH when reducing bromate. Other researchers (Siddiqui et al., 1994) have also noted this effect.

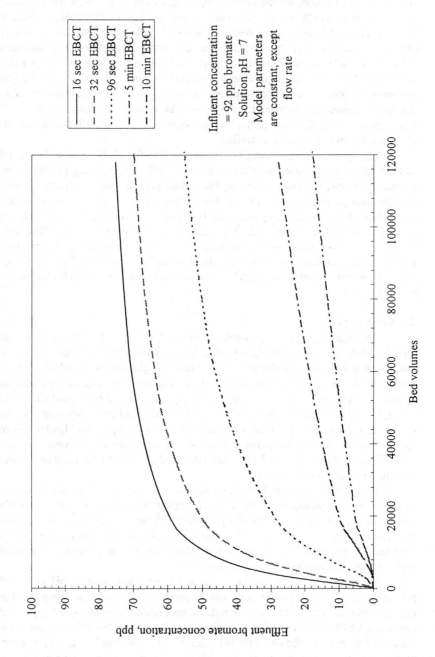

Figure 7: The effect of empty bed contact time on bromate reduction by 60x80 mesh Ceca carbon

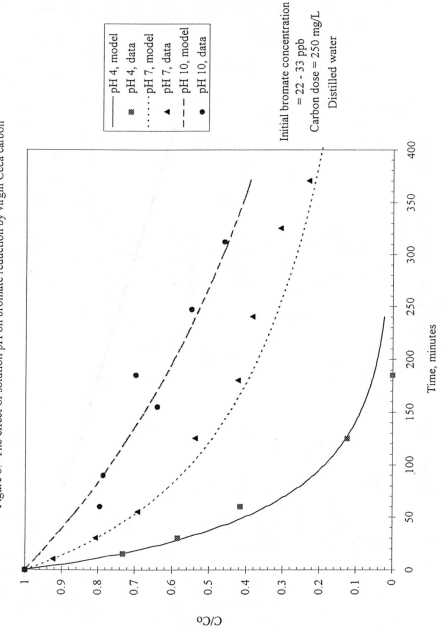

Figure 8: The effect of solution pH on bromate reduction by virgin Ceca carbon

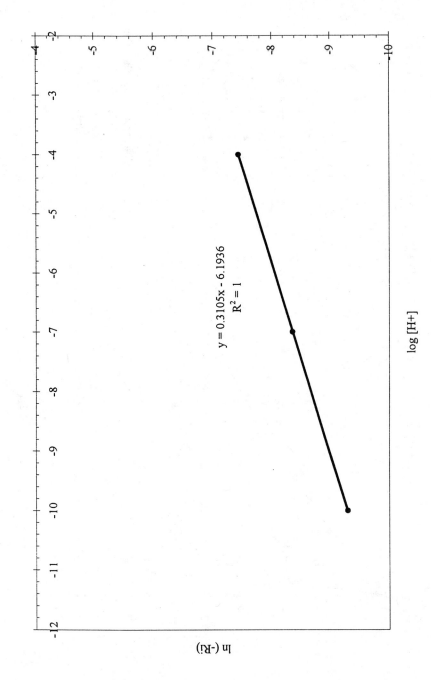

Figure 9: The effect of pH on the initial rate of bromate reduction by GAC

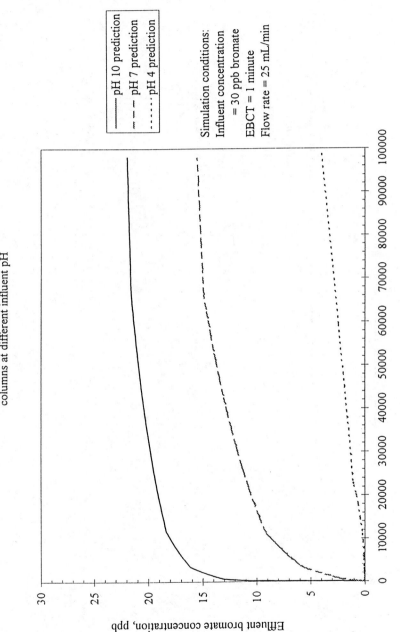

Figure 10: Predicted breakthrough curves for bromate reduction by virgin Ceca carbon in packed bed columns at different influent pH

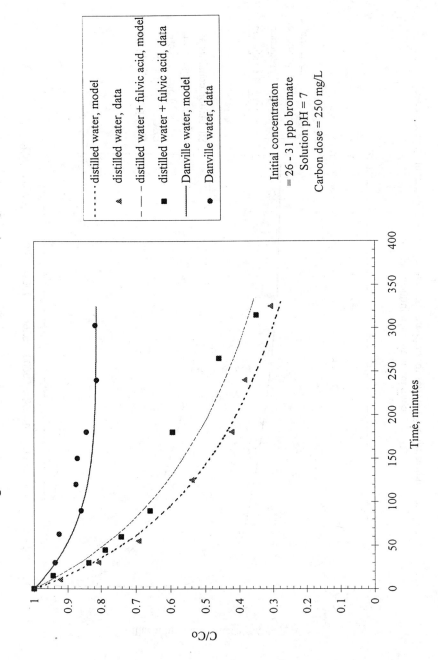

Figure 11: The effect of NOM on bromate reduction by virgin Ceca carbon

The effect of Natural Organic Matter

    The previous batch and column data have all been obtained in distilled water. The model used here was developed and tested by Suidan et al. (1975, 1977a, 1977b, 1978) using distilled water solutions spiked with free chlorine. It is clear that the model is adequate for describing the reduction of free chlorine and of bromate by GAC in distilled water. However, in drinking water treatment, bromate reduction will be necessary in natural water. Typically, drinking water contains natural organic matter (NOM) in concentrations of several mg/L, two or three orders of magnitude greater than the concentrations of bromate expected to be encountered in drinking water. Gonce and Voudrias (1994) found that NOM blocked the carbon surface and affected the capacity of the carbon for the reduction of chlorite. Since bromate reduction is also a surface reaction, it is expected that NOM will affect the capacity of the carbon for bromate reduction.

    The reduction of bromate by fresh GAC was compared in batch tests using three different waters: distilled water, distilled water spiked with a fulvic acid isolate, and water obtained from the clarifiers of the Interstate Water Company in Danville, IL. The batch test data are shown in figure 11. There is a large difference in the effect of the fulvic acid verses the effect of the NOM from the Danville water. The NOM in the Danville water contains more than just fulvic acid, although all of the organic compounds cannot be characterized. An isolated fulvic acid does not appear to behave the same as an uncharacterized mixture. Moreover, the fulvic acid was stored in a cold room, and some chemical changes may have occurred over time.

    It is apparent that the presence of the natural water NOM has an immediate effect on the rate of bromate reduction, even in the short term batch tests. In the model, the effect of water type can be described by decreasing the number of "active sites" on the carbon (the parameter SIT) for bromate reduction in water containing NOM, as compared to bromate reduction in distilled water.

    The fulvic acid isolate solution was used in a packed bed column test. The effluent bromate data and the predicted breakthrough curve based on the batch test data are shown in figure 12. The average influent bromate concentration for this test was 16 ppb, the solution pH was 7, the solution TOC was 2 mg/L, and the EBCT was 1 minute. The effluent bromate concentration for the column is always underestimated by the model. In other words, the carbon in this test reduces less bromate than would be expected, based on the batch test. The model fails to account for the cumulative effect of NOM on the bromate reduction ability of the carbon over time. This is due to the fact that the model depends on a constant value of SIT, and a constant rate expression.

    It is possible to describe the column data by the model, as shown in figure 13, by changing the kinetic parameters used to describe the batch test. To obtain the prediction shown in figure 13, the model parameter describing the number of "active sites" on the carbon surface was decreased. In figure 14, the actual model curve and the batch test data are compared to the model curve that is necessary to describe the column data well. The carbon in the column behaves as though there are fewer "active sites", which is consistent with the idea of NOM building up on the carbon surface over time and blocking the surface from bromate.

    Based on the experience with fulvic acid isolate, it was expected that the general effect of NOM would be to gradually build up on the carbon surface and to cause a loss

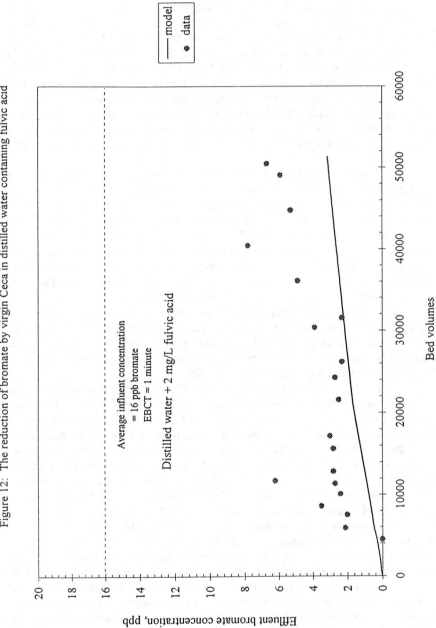

Figure 12: The reduction of bromate by virgin Ceca in distilled water containing fulvic acid

Average influent concentration
= 16 ppb bromate
EBCT = 1 minute

Distilled water + 2 mg/L fulvic acid

model
data

Effluent bromate concentration, ppb

Bed volumes

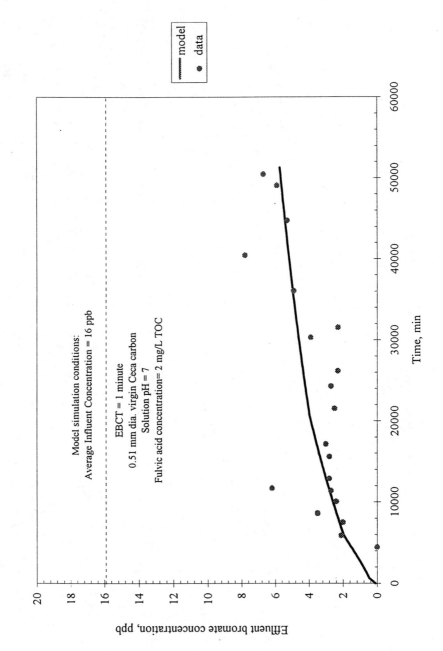

Figure 13: Refit of virgin Ceca carbon, fulvic acid packed bed column test

Model simulation conditions:
Average Influent Concentration = 16 ppb

EBCT = 1 minute
0.51 mm dia. virgin Ceca carbon
Solution pH = 7
Fulvic acid concentration= 2 mg/L TOC

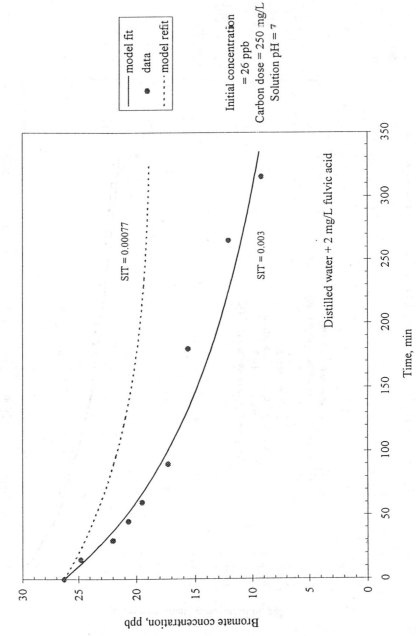

Figure 14: The refit of virgin Ceca carbon, fulvic acid batch test data

in bromate reduction ability over time. To test this idea, fresh GAC was preloaded at the Interstate Water Company facility in Danville, IL. Carbon was preloaded using a one minute EBCT for times of 2 days, 10 days, 15 days, and 55 days. Aliquots of the preloaded carbons were used for batch tests, and the results are shown in figure 15. The preloading caused an immediate loss of capacity, and even after two days of preloading the carbon showed almost no ability to reduce bromate.

It should be noted that the Interstate Water Company facility in Danville, IL, does not ozonate the water or do any additional NOM removal after the softening process. The water used for preloading was obtained after filtration, and had a TOC of about 2 mg/L. The effect of NOM on bromate reduction may be less for treatment plants which use additional processes to lower the amount of NOM in the water, such as enhanced coagulation or ozonation followed by a biologically active filter. Also, the concentration of NOM may not be as important as the chemical characteristics of the NOM. Ozonation, in particular, may affect the chemistry of the NOM enough to make a difference for bromate reduction.

The batch test data obtained using the carbon which had been preloaded for 55 days was used to predict bromate removal by preloaded carbon in a packed bed reactor. The average influent bromate concentration for this column test was 25 ppb, the solution pH was 7, the TOC was 2 mg/L, and the EBCT was one minute. The predicted curve and the column effluent data are shown in figure 16. The preloaded carbon did not reduce bromate in a packed bed column. This finding suggests that GAC which has been in use at a water treatment plant for months or years will not be effective for reducing bromate.

Conclusions

1. The reduction of bromate can be described as having a pseudo first order dependance on bromate concentration. The reaction is also limited by pore diffusion. Both of these assumptions are inherent in the model being applied here.

2. It is possible to describe batch test results for bromate reduction by GAC under a variety of conditions using the model. Trends in the kinetic parameters can be identified with respect to solution pH and NOM content.

3. In distilled water it is possible to predict column results based on batch tests which were conducted under similar conditions.

4. The model cannot describe the effects of NOM build-up on fresh carbon in a packed bed column test. It may be possible to modify the model to describe this effect. The fulvic acid isolate used to spike distilled water does not appear to approximate the effect of NOM in natural water particularly well.

5. The applicability of the model is not a confirmation of mechanism. Some assumptions used in the development of the model have been verified, such as the pseudo first order dependance on bromate concentration and the limitations of mass transfer.

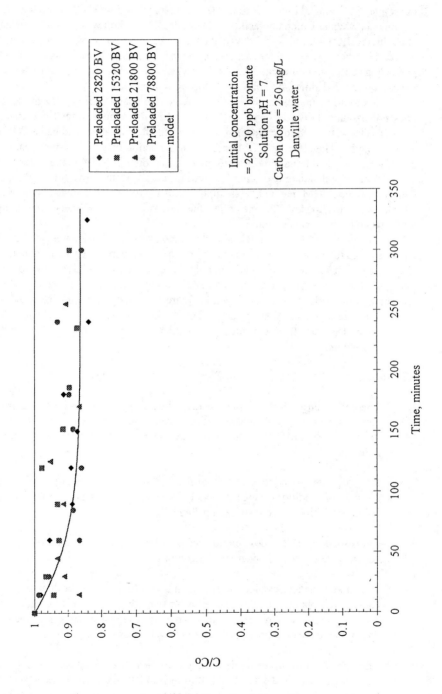

Figure 15: The effect of preloading virgin Ceca carbon on bromate reduction

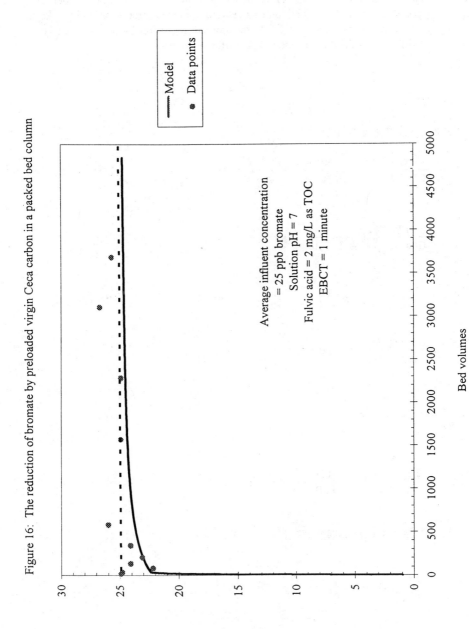

Figure 16: The reduction of bromate by preloaded virgin Ceca carbon in a packed bed column

6.   GAC was not effective for reducing bromate in natural water under the conditions tested. Ozonation or other water treatment processes could make a difference in the effect of NOM on bromate reduction by activated carbon. Different results may be obtained in different waters.

Appendix

Finite Batch Reactor Equations

$$\frac{\partial G}{\partial \theta} = \frac{G}{GPo + K8}\left(SIT - \frac{K9Q}{1 + K10Q}\right) + \frac{\partial^2 G}{\partial \xi^2} \tag{1}$$

$$\frac{\partial Q}{\partial \theta} = \frac{GPo}{GPo + K8}\left(SIT - \frac{K9Q}{1 + K10Q}\right) \tag{2}$$

$$Q = 0; \quad \text{for } \theta = 0, 0 \le \xi \le 1 \tag{3}$$

$$\frac{\partial^2 G}{\partial \xi^2} - \frac{GSIT}{GPo + K8} = 0; \quad \text{for } \theta = 0, 0 \le \xi \le 1 \tag{4}$$

$$\frac{\partial G}{\partial \xi} = 0; \quad \text{for } \theta \ge 0, \xi = 1 \tag{5}$$

$$G = 1 + \int_0^\theta \left(mVp\left.\frac{\partial G}{\partial \xi}\right|_{\xi=0} - \frac{VpLp^2}{PoDc}Rd\right)d\theta; \quad \text{for } \theta \ge 0, \xi = 0 \tag{6}$$

Packed Bed Reactor Equations

$$\frac{\partial C}{\partial t} = \frac{-m}{\varepsilon}R(X,C) + D_A\frac{\partial^2 C}{\partial z^2} - V\frac{\partial C}{\partial Z} \tag{7}$$

$$\frac{\partial X}{\partial t} = R(X,C) \tag{8}$$

$$X = 0; \quad \text{for } t = 0, 0 \le z \le L_b \tag{9}$$

$$D_A \frac{\partial^2 C}{\partial z^2} - V \frac{\partial C}{\partial z} - \frac{m}{\varepsilon} R(0,C) = 0; \quad \text{for } t = 0, 0 \le z \le L_b \tag{10}$$

$$V C_0 = V C - D_A \frac{\partial C}{\partial z}; \quad \text{for } z = 0, t \ge 0 \tag{11}$$

$$\frac{\partial C}{\partial z} = 0; \quad \text{for } z = L_b, t \ge 0 \tag{12}$$

$$R(X,C) = \frac{J_1 * C}{\left(J_2 + C\right)\left[1 + J_3 \exp(-J_4 C) * X\right]^{J_5}\left[1 + \left(J_6 \exp(J_7 C) * X\right)^{J_8}\right]^{J_9}} \tag{13}$$

Variables

G	dimensionless free chlorine concentration in pore model
θ	dimensionless time variable in pore model
Po	initial or bulk concentration per unit mass carbon in finite batch carbon model
SIT	reactive sites per unit mass of carbon
K8	adsorption/desorption equilibrium constant x Vp
K9, K10	constants describing formation and degradation of oxidized sites
Q	mass reacted per unit mass of carbon
ξ	dimensionless position variable in pore model
m	mass of carbon per unit reactor volume
Vp	pore volume per unit mass of carbon
Lp	pore half length
Dc	pore diffusion coefficient
Rd	rate of decay in blank reactor; zero for bromate modeling
C	free chlorine concentration per unit volume of solution
t	cumulative time for batch or column run
ε	packed bed porosity
R(X,C)	mass removal rate per unit mass of carbon
DA	packed bed axial dispersion coefficient
Z	distance variable in packed bed column model
Lb	packed bed length
X	total free chlorine reacted per unit mass of carbon
J1 - J9	constants in expression for R(X,C)
V	axial velocity thru packed bed reactor

References

ASTM, Annual Book of ASTM Standards, Vol. 15.01, American Society for Testing and Materials, Philadelphia, 1988.

Christen, H. R., Grundlagen der Allgemeinen und Anorganischen Chemie, Sall und Sauerlander, Frankfurt am Main, 1977.

Fogler, H. Scott, Elements of Chemical Reaction Engineering, Prentice Hall, Englewood Cliffs, New Jersey, 1992.

Gonce, Nancy, and Evangelos A. Voudrias, "Removal of Chlorite and Chlorate Ions from Water using Granular Activated Carbon," Water Research, Vol. 28 (5), 1994, pp. 1059-1069.

Kim, Byung Ro, "Analysis of Batch and Packed Bed Reactor Models for the Carbon-Chloramine Reactions," PhD. Thesis, University of Illinois-Urbana, 1977.

Larson, Richard A., and Eric J. Weber, Reaction Mechanisms in Environmental Organic Chemistry, Lewis Publishers, Boca Raton, 1994.

Levenspiel, Octave, Chemical Reactions Engineering, John Wiley and Sons, Inc., New York, 1962.

Miller, Jennifer, "The Reduction of Aqueous Bromate by Granular Activated Carbon," Master's Thesis, Unversity of Illinois - Urbana, 1992.

Miller, J., V. L. Snoeyink, and S. Harrell, "The Effect of Granular Activated Carbon Surface Chemistry on Bromate Reduction," in Disinfection By-Products in Water Treatment: The Chemistry of their Formation and Control, Lewis Brothers Publishers, to be published in 1995.

Siddiqui, Mohamed, Gary Amy, Kenan Ozeki, Wenyi Zhai, and Paul Westerhoff, "Alternative Strategies for Removing Bromate", Journal of the American Water Works Association, Vol. 86(10), Pp. 81-96, 1994.

Standard Methods for the Examination of Water and Wastewater, 17th Edition, edited by Lenore S. Clesceri, Arnold E. Greenberg, and R. Rhodes Trussell, American Public Health Association, Washington, D. C., 1989.

Suidan, Makram T., Reduction of Aqueous Free Chlorine with Granular Activated Carbon, University of Illinois, PhD Thesis, 1975.

Suidan, Makram T., Vernon L. Snoeyink, and Roger Schmitz, "Reduction of Aqueous HOCl with Activated Carbon", Journal of the Environmental Engineering Division, ASCE, Vol. 103, No. EE4, Proc. Paper 13138, Pp. 677-691, August 1977a.

Suidan, Makram T., Vernon L. Snoeyink, and Roger A. Schmitz, "Reduction of Aqueous Free Chlorine with Granular Activated Carbon - pH and Temperature Effects", Environmental Science and Technology, Vol. 11, No. 8, Pp. 785-789, August 1977b.

Suidan, M. T., V. L. Snoeyink, W. E. Thacker, and D. W. Dreher, "Influence of Pore Size Distribution on the HOCl-Activated Carbon Reaction," in Chemistry of Wastewater Technology, edited by J. F. Malina, Jr., Ann Arbor Science Publishers, Ann Arbor, Michigan, 1978.

Voudrias, E. A., L. M. J. Dielmann, V. L. Snoeyink, R. A. Larson, J. J. McCreary, and A. S. C. Chen, "Reactions of Chlorite with Activated Carbon and with Vanillic Acid and Indan Adsorbed on Activated Carbon," Water Research, Vol. 17, 1983, p. 1107.

Chapter 16

Impact of Bromide Ion Concentration and Molecular Weight Cutoff on Haloacetonitrile, Haloketone, and Trihalomethane Formation Potentials

Steve H. Via[1] and Andrea M. Dietrich

Department of Civil Engineering, Environmental Division, Virginia Polytechnic Institute and State University, 330 Norris Hall, Blacksburg, VA 24061–0246

This research examined HANFP and HKFP formation and specific formation potentials with time (1 - 168 hours) and speciation in the context of THM formation and speciation. Controlling DOC concentration, DOC AMW, bromide concentration and Cl_2:C in raw and treated waters, coagulation resulted in a decrease in HANFP while THMFP and specific THMFP increased, shifting toward more brominated THMs. DCANFP was the most reduced HAN species. Lower MWCO NOM fractions generated the greatest DCAN formation and maintained a higher formation potential with time. TCPFP appeared unaffected by coagulation. Changes in the Br⁻:Cl ratio: (1) increased the magnitude of specific molar THANFP more than TTHMFP; (2) created a higher percentage of brominated HAN species compared to formation of brominated THM species; and (3) decreased DCANFP more than $CHCl_3$FP. THANFP increased with increasing Br⁻:Cl ratio, and the increase in Br-THANFP occurred to a greater extent on a relative basis than Br-THMFP.

In 1979, health risks associated with halogenated DBPs resulted in the promulgation of a National Primary Drinking Water Standard for total trihalomethanes. In July 1994, the U.S. Environmental Protection Agency (EPA) proposed the Disinfectant / Disinfection Byproduct (D/DBP) Rule (*1*) to set the stage for future inclusion of haloacetonitriles (HANs), haloketones (HKs), and other DBPs into the nation's drinking water standards. A parallel regulation, the Information Collection Rule (ICR) (*2*), establishes monitoring requirements for HANs and HKs, to assess the extent of their formation in U.S. drinking water.

Conventional treatment (coagulation-flocculation-filtration-disinfection) dominates water treatment plant (WTP) design in the U.S.. The proposed D/DBP Rule establishes enhanced coagulation as a best available technology for WTPs that: (1) are of conventional design with sedimentation basins, (2) utilize a water source that is influenced by surface water, and (3) experience total organic carbon (TOC) levels greater than 2 mg/L prior to disinfection. Enhanced coagulation is defined as addition of a coagulant at a dose beyond which an additional 10 mg/L will remove less than 0.3 mg-TOC/L. TOC abundance is a primary driver behind DBP formation; in Alberta, Canada, average 168 hour TTHMFP yields (pH 7, 5 mg-Cl_2/L

[1]Current address: 3631 Ranchero Drive, #102, Ann Arbor, MI 48108

residual) of 11 ug/mg-TOC were observed in a survey of water treatment systems employing conventional treatment (*3*).

Although exact structural characterization of dissolved organic carbon (DOC) is critical to determining which DBPs will be generated upon chlorination, general trends can be attributable to characteristics like apparent molecular weight (AMW) (*4,5,6,7*). Venstra and Schnoor (*4*) found 88% of the TOC and 87% of TTHMFP to occur in the <3,000 AMW fraction. Joyce et al. (*8*) attributed greater $CHCl_3FP$ by weight to the greater number of active sites in the <10,000 AMW fraction and particularly the 1,000-5,000 AMW range.

Peak THAN and TTHM formation were observed in higher AMW fractions (>30,000 AMW) (*9*). Positive correlation between nitrogen content and increasing humic substance size was suggested as a reason for increased HANFP. However, Mulford et al. (*10*) demonstrated that achieving 90% removal of 26 evaluated DBPs required removing all NOM>500 molecular weight cutoff (MWCO). Pilot scale membrane filtration experiments and simulated distribution system (SDS) tests (72 hour, 0.1-0.55 Cl_2/L residual, 20°C, ambient pH) showed ultrafiltration (100,000 MWCO) did not change SDSFPs for THMs, but HANSDS response decreased 25% relative to raw water trials (*11*). Nanofiltration (400-600 MWCO) reduced both THMSDS and HANSDS greater than 70-80% respectively. DBPSDS was reduced further when the MWCO was reduced to 200 to 300 AMW.

Coagulation preferentially removes larger (i.e., >30,000), typically more humic acid type DOC. The following hierarchy of coagulation effect on DBPFP removal was proposed by Reckhow and Singer (*9*) after a study of DBPFP associated with 10 southeastern U.S. raw water sources: "DCANFP > TCAAFP > DCAAFP > THMFP (~TOXFP) > TOC > TCPFP" (bench scale coagulation; formation potential tests at 72 hours, 20°C, pH7, 20 mg-Cl_2/L). DCANFP responsivity was attributed to removal of the hydrophobic humic fraction. Normalized THM, trichloroacetic acid (TCAA), TCP, and total organic halide (TOX) formation were found to vary less than 23% among the eight colored, raw waters; HAN and dichloroacetic acid (DCAA) formation were more variable, particularly DCANFP which had a coefficient of variance of 40% (*9*). More recently researchers observed a similar hierarchy of DBPFP response (THMFP > THAAFP > TOX > UV absorbance) to enhanced coagulation (*12,13*). Coagulation reduced CHFP and THANFP when TOC was greater (test conditions of 4 and 2 mg-TOC/L) (*13*). Previous research has found THKFP insensitive to coagulation (*13*).

Differences in THM speciation have been attributed to "chlorine acting preferentially as an oxidant, whereas bromine is a more effective halogen-substituting agent" (*14*). Previous authors have found that speciation of brominated DBPs is not just the product of stoichiometry; addition of bromide appears to enhance chlorine oxidation reaction rates and result in increased "total halogen consumption" (*15*). Increased THM yield and shifts in THM speciation occur with increasing bromide concentration (*16*). DBP speciation effects have been observed at 0.5-1.5 mg-Br⁻/L (0.4-1 mg-residual Cl_2/L, pH 6.4-9.4, time 2-40 hours) (*17*). Specific effects on THMFP (168 hour, pH 7, 25°C, 11.2 mg-Cl_2/L dose, 2.8 mg-NVTOC/L) include decreasing $CHCl_3FP$, increasing $CHBr_3FP$, and passage of $CHCl_2Br$ and $CHClBr_2$ through a formation peak with increasing bromide (*17*). TTHM formation (24 hour, 20°C, pH 7, 25 mg-Cl_2/L, 2.5 mg-DOC/L) has also been observed to increase and shift to more brominated THMs with increasing bromide concentration. Bromine incorporation as a percentage of TTHMFP at a range of temperatures (10°, 20°, and 30°C), chlorine doses (2.5, 5, 10 mg-Cl_2/L) and pH ranges (pH 6-9) varied with bromide (*4,18*). Symons et al. (*19*) suggested that the initial Br⁻:average Cl^+ molar ratio controls bromide substitution reactions in THM formation. The rate of THM-Br formation is faster than THM-Cl formation (*19,20*). Kinetics drive the substitution reaction to completion even in the presence of HOCl which will only form $CHCl_3$ if it is present in excess when the bromide substitution reaction is complete. NOM precursors limit THM formation due to TTHM-Br consumption of available sites prior to TTHM-Cl formation and thus bromide concentration increases total DBP yield (19).

Laine et al. (11) employed Br⁻:Cl weight ratios to illustrate the effect of bromide ion on THM and HAA speciation (1.5-1.7 Cl_2:TOC weight:weight ratios, 0.06-0.64 Br⁻:Cl_2 dose weight:weight ratios, 3 day, 20°C, ambient pH SDS tests). The research suggested that total DBP production was constrained by DOC concentration even when bromide ion was present in abundance. Also, while water quality and ultrafiltration membrane type affected total THMSDS and HAASDS formation, the Br:Cl ratio was more significant than the type of precursor material in determining the distribution of THMSDS and HAASDS species.

Peters et al. (21) investigated DCAN, BCAN, DBAN, $CHCl_3$, $CHClBr_2$, $CHClBr_2$, and $CHBr_3$ formation, at six Dutch WTPs applying chlorine (1.7-5.6 mg-DOC/L, 0.34-186.7 DOC:Cl_2 dose weight:weight ratio). HAN formation was 5% of THM formation on a weight basis. Brominated HANs were more abundant than chlorinated HANs, totaling 60% of the observed HAN concentration. Brominated HAN species concentrations were greatest when the DOC:Cl_2 dose ratio was between 5.6 and 11.8 but insufficient data were developed to explain HAN speciation. A shift to brominated THM species and higher brominated THM formation was observed under low DOC concentrations (e.g., high Br⁻:DOC ratio) (21).

Several researchers have investigated the role of time in DBP formation. A trend of increasing DBPFP with time is true of HAAs and THMs generally with THM formation essentially complete after four days (22). Two DBPs, chloral hydrate (CH) and DCAN, have been observed to achieve peak formation potentials and decline with time. DCAN was observed to decreased as a function of reaction time decaying completely over a seven day period at a pH of 7 (7,23).

Water Characteristics and Treatment Train

The water studied in this research has low alkalinity, moderate color, and low ammonia; it was obtained from a reservoir located in eastern Virginia (24,25). Storage of water in the reservoir is believed to alter raw water characteristics prior to the WTP treatment train, by promoting settling and algal growth. Comparison of DOC concentration in water samples prior to 0.45 um filter and after filtration indicates that NOM > 0.45 um accounts for less than 15% of raw water TOC. Seasonal effects on water sample character were observed; such effects on TTHM and THAN are known to be small relative to other waters (25).

WTP staff have optimized the treatment train, and at the time of this research, plant records indicated 50% reductions in THMFP and TOC and near complete color and turbidity removal. The WTP employs 30-45 mg-alum/L and 0.3 mg-cationic polymer/L for coagulation. Pretreatment includes addition of potassium permanganate at 0.3 mg/L. Benchscale comparisons of the current treatment train parameters with enhanced coagulation as specified in EPA guidance reflected more positively on the current treatment train (13,25).

NOM concentrations observed in this research varied with sample and MWCO. Full-scale treatment caused an apparent increase in <1,000 AMW NOM and a decrease in higher AMW NOM after coagulation. General trends observed indicated that NOM < 0.45 um was dominated by <10,000 AMW and that coagulation-flocculation demonstrated little removal of this fraction. Removal of NOM by coagulation-flocculation varied little with increasing AMW from 10,000 to 4,500,000 (i.e., material passing through a 0.45 micron filter).

Approach

Samples were taken at the raw water intake and immediately after the coagulation basin. Subsequent analyses of coagulated water formation potentials reflected effects of full scale coagulation treatment on: (1) AMW distribution, (2) NOM composition, and (3) chemical reactivity. All sample waters were filtered using a 0.45 um filter to limit biological activity. Nominal MWCO fractions were prepared using ultrafiltration (Amicon, Beverly, MA). After ultrafiltration but prior to chlorination, the DOC concentration of each MWCO fraction was equalized to a common concentration.

Formation potential tests were performed using Standard Method 5710 B. The method was modified to (1) a consistent 3:1 Cl_2:DOC ratio and (2) multiple time points to monitor trends in formation potential over time. HAN, HK, and THM formation potential tests were conducted on 1,000, 10,000 and 30,000 MWCO fractions; filtered raw water; and filtered coagulated water. Conditions were controlled such that initial bromide concentration, DOC concentration, pH (pH 7), temperature (20°C), and Cl_2 dose:DOC concentration ratio (Cl_2:DOC) were constant. Formation potential tests were performed at <1, 24, 48, 96 and 168 hour time points.

Standard EPA methods were used to determine analyte concentrations resulting from the formation potential tests. EPA Method 501.1, Analysis of Trihalomethanes in Finished Waters by Purge and Trap Method, was employed to quantify $CHCl_3$, $CHCl_2Br$, $CHClBr_2$, and $CHBr_3$. EPA Method 551, Determination of Chlorination Disinfection Byproducts and Chlorinated Solvents in Drinking Water by Liquid-liquid Extraction and GC with ECD was used to quantify HAN and HK species. Method detection levels were evaluated and found to be similar to those specified in method documentation and to be adequate for this research (Table I).

Table I. Analytical Method Detection Limits

Analyte	Detection Limit (ug/L)
BCAN	0.05
$CHCl_3$	1.68
$CHCl_2Br$	0.04
$CHClBr_2$	0.11
$CHBr_3$	1.77
DBAN	0.15
DCAN	0.03
TCAN	0.03
TCP	0.03

DBAN was the most difficult analyte to quantify, while TCAN was present only on rare occasions. Bromide concentrations were determined using ion chromatography as described in Standards Method 4110 B. Chlorine dosage and residuals were measured using amperometric titration, Standard Method 4500 Cl - D. DOC concentration was determined using persulfate-ultraviolet oxidation method, Standard Method 5310 C.

Results

In this research, unless otherwise specified, formation potentials were conducted under the following conditions: (1) 7 days, 20°C and pH 7; (2) 2 mg-DOC/L after passage through MWCO filters and adjustment by dilution; and (3) 6 mg-Cl_2/L dose

(Cl_2:C = 3:1); and (4) 0.07 mg-Br⁻/L. DOC concentration, chlorine dose, and initial bromide concentration were held constant in order to generate directly comparable formation potentials and specific formation potentials (ug-DBP/mg-DOC). Error bars on graphs in subsequent discussion represent +/- 1 standard deviation.

Effect of Full Scale Treatment on DBPFP.

Raw Water. Data for raw water DBPFP are shown in Figures 1 (THMs) and 2 (HANs and TCP). The raw water THMFP experiment yielded values of about 95 ug-$CHCl_3$/L, 30 ug-$CHCl_2Br$/L, 3 ug-$CHClBr_2$/L, and no measurable $CHBr_3$. The TTHMFP for this water sample was about 120 ug/L (64 ug-TTHMP/mg-DOC), which is approximately 40% of the amount of raw water THMFP recorded at the full scale treatment plant for unfiltered water. Concentrations of individual THMs occurred in the anticipated hierarchy of: [$CHCl_3$] > [$CHCl_2Br$] > [$CHClBr_2$] > [$CHBr_3$], which is the same as reported by other researchers performing experiments on this water (13,25). As is typical for THMFP experiments, there was a distinct increase in TTHMFP and individual THMFPs over time (18).

Raw water HANFP and TCPFP data (i.e., TCP was the only HK detected) demonstrated a different response pattern from that observed for the THM species (Figure 2). THAN and TCP concentrations peaked at about 24-48 hours in the formation potential test and then declined in concentration for the remainder of the seven day period. This pattern is consistent with that previously reported (7,23,25). The dominant HAN species was DCAN; its maximum concentration was about 5 ug/L at 24-48 hours, approximately 10% of the measured $CHCl_3$ concentration for this same water sample at the same time point. BCAN followed a formation potential pattern with time similar to DCAN but at reduced concentrations; DBAN was seldom observed. TCP was observed to form in the concentration range of 1-3 ug/L. The TCP concentrations and formation pattern with respect to time were comparable to that observed for DCAN and BCAN.

Coagulated Water. Treatment of the raw water by full scale coagulation and upflow clarification resulted in similar formation potential patterns at slightly increased THM concentrations when compared with the THMFP of the raw water (Figure 1). The TTHMFP value at 7 days was about 160 ug/L (79 ug-TTHMP/mg-DOC) for treated water. Formation of $CHClBr_2$ was greater in the coagulated water than the raw water for all time points. Increased TTHMFP with treatment has also been reported by other authors (26). Increased THMFP after treatment may result from pretreatment with potassium permanganate or increased reaction of chlorine and bromine with THM-precursor material after coagulation.

After treatment, the DCANFP remained the dominant HAN but its concentration was reduced to 65-80% of the raw water DCANFP values at all time points (Figure 2). Its maximum concentration was 3.8 ug/L at 48 hours, which was about 5% of the measured $CHCl_3$ concentration at the same time point. Treated water BCANFP values were similar in concentration to those observed for the raw water but at continuously declining concentrations with time. DBAN was more frequently detected in formation potential experiments for treated water samples compared to raw water.

TCPFP results for the treated and raw water were very similar in concentration and pattern, with peak concentrations occurring at about 24 hours. In this water, haloketone formation appeared unaffected by coagulation; this trend was previously reported (13).

A Matched Pair Wilcoxon Ranked Test was performed to compare specific formation potentials in raw and coagulated water across all time periods; statistical results are presented in Table II. Differences ($\alpha = 0.2$) were determined for $CHCl_3$,

Figure 1. Observed Raw and Coagulated Water THM Formation Potentials

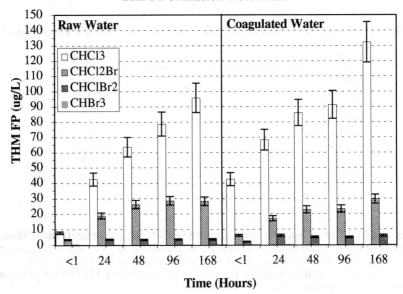

Figure 2. Observed Raw and Coagulated Water HAN and TCP Formation Potentials

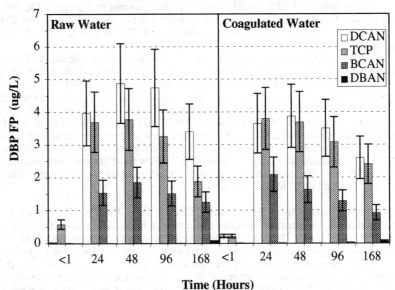

DBAN, and TCP. The differences in formation of $CHClBr_2$ and BCAN were indicated to be minimal by this test. Overall, there were more similarities than differences in DBP profiles and concentrations for raw and coagulated water.

Table II. Coagulation Effect, Matched Pair Wilcoxon Ranked Sum Analysis

Analyte	Probability
BCAN	0.959
$CHCl_3$	0.136
$CHCl_2Br$	0.534
$CHClBr_2$	0.859
$CHBr_3$	0.317
DBAN	0.173
DCAN	0.683
TCP	0.173

Note: Significant difference indicated at p<0.2.

Effect of MWCO on Specific Formation Potentials for Raw and Treated Water.
Comparison was made of specific formation potential data for fractionated raw and treated waters; DOC concentration in all fractions was adjusted to 2 mg/L. Comparisons of specific formation potentials as a function of MWCO are presented in Figure 3 for 96 hour formation potential data.

In general, TDBP formation increased with time across all MWCO fractions in both raw and coagulated waters. This increase was due to THM formation as HANs and HKs remained constant or decreased with time. $CHCl_3$ was the dominant DBP in all fractions at all times; $CHCl_2Br$ typically occurred at about 20% of the chloroform value. THANFP varied from the detection limit to 8.3 ug/L. DCAN was the dominant HAN species and its concentration did not exceed 10% of the $CHCl_3$ concentration. Peak concentrations of HANs and TCP occurred between 48-96 hours for all fractions.

The species with the greatest 96 hour specific formation potentials (Figure 3) were $CHCl_3$ among the THMs, and DCAN among the HANs. Consistently higher specific formation potentials were associated with the smallest MWCO fraction, i.e., <1,000 MWCO. Values for raw water THM and HAN specific formation potentials in the <1,000 MWCO fraction were consistently higher than those for the treated water. In most other MWCO fractions (10,000, 30,000, 4,500K) the values of the specific formation potentials for most DBP species were very similar in both raw and treated water samples. The most dramatic effects were observed for $CHCl_3$ and DCAN. The pattern and magnitude of $CHCl_3FP$ were different between the two waters with the treated water yielding a specific formation potential that was relatively constant at all MWCO values.

Abundance of Brominated Species and Effect of Br^-:Cl Ratio. Experiments were conducted to determine the effect of varying the Br^-:Cl ratio on the formation of THMs and HANs. Three Br^-:Cl ratios (mM:mM) were investigated: 0.01, 0.03, and 0.074. Coagulated and fractionated water were utilized for these experiments. The chlorine dose was constant at 6 mg/L and the Cl_2:C ratio was constant at 3:1. Figures 4 and 5 summarize the effect of Br^-:Cl ratio on specific molar TTHMFP and THANFP for the <30,000 MWCO fraction as a function of time. Similar plots were generated for <1,000, <10,000, and <450,000 MWCO fractions; these data are not shown.

**Figure 3. MWCO and Coagulation Trends in Formation
Potential**

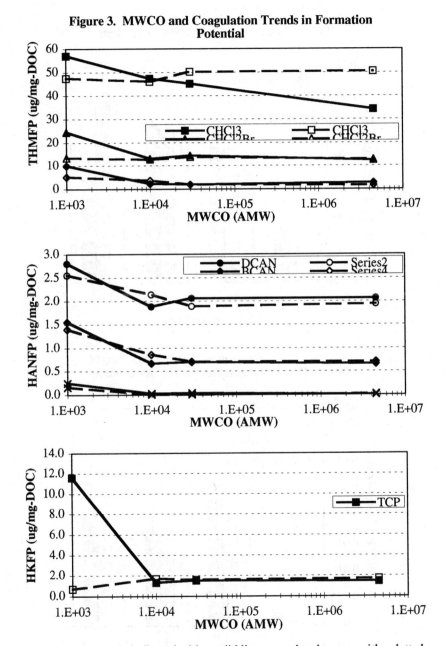

Note: Raw waters are indicated with a solid line, coagulated waters with a dotted
line.

Figure 4. DOC Normalized TTHMFP in <30,000 MWCO Fraction as a Function of Br-:Cl Ratio

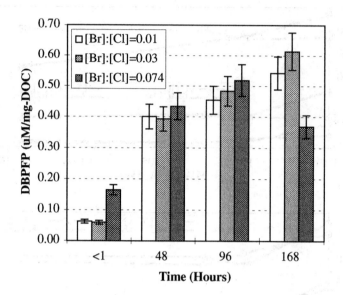

Figure 5. DOC Normalized THANFP in <30,000 MWCO Fraction as a Function of Br-:Cl Ratio

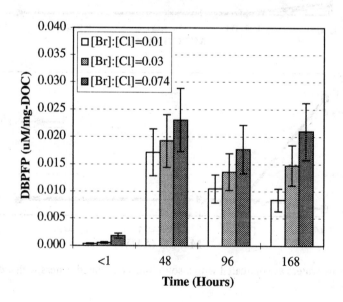

The time trends observed were the same as previously reported: TTHMFP (individual THMFPs and THMFP/mg-DOC) generally increased with reaction time at most Br⁻:Cl ratios; the exception was the 168 hour time point for the highest Br⁻:Cl ratio. THM species demonstrated different trends with changes in Br⁻:Cl ratio and MWCO. DOC normalized specific $CHCl_3FP$ increased with decreasing Br⁻:Cl ratio but varied relatively little with MWCO. DOC normalized specific $CHCl_2BrFP$ and $CHClBr_2FP$ increased with increasing Br⁻:Cl ratio in all MWCO fractions. $CHBr_3$ also increased with increasing Br⁻:Cl. The increase in $CHBr_3$ was most rapid in larger MWCO fractions but reached the greatest final concentration (168 hours) in the <1,000 AMW fraction.

Figure 5 presents data for THANFP. THAN specific molar formation potentials continually increased with increased Br⁻:Cl ratio; peak concentrations of THANs (and also individual HANs) occurred at about 48 hours. The percent increase in THANs due to increased bromide was greater than that for TTHMs. For example, at 48 hours in the <30,000 MWCO, the TTHMFP increased 7.5% from the lowest to the highest Br⁻:Cl ratio; at the same time point and Br⁻:Cl ratios, THANFP increased 35%.

The individual HANs responded differently to changing Br⁻:Cl ratios. DCANFP decreased with time after peaking at 48 hours; DBANFP was stable with respect to time, and BCANFP peaked at 48 hours. DCANFP decreased with increasing initial Br⁻:Cl ratio in all AMW fractions, thus its contribution to THANFP declined at higher Br⁻:Cl ratios. DBANFP and BCANFP increased with increasing Br⁻:Cl ratios for all MWCO fractions, but specific formation of these two species was greatest in <1,000 MWCO. As THANFP increased with increasing Br:Cl, the decrease in DCANFP was more than compensated by the increase in DBANFP and BCANFP. Although data for TCPFP are not shown, the TCPFP decreased with increasing Br⁻:Cl ratio; the <1,000 AMW fraction demonstrated the greatest effect and in some samples TCP was not detected.

Figure 6 shows specific DBPFP as a function of time for the three Br⁻:Cl ratios. Trends for THMs demonstrated the expected shift from $CHCl_3$ to more brominated species as the Br⁻:Cl ratio increased. The HANs demonstrate a similar trend toward incorporation of more bromide as the Br⁻:Cl ratio increased. Bromide was incorporated to a greater extent in the HANs compared to the THMs at all time points and MWCO values. For example, THAN incorporated twice as much as TTHM in the <10,000 MWCO at 96 hours, 2 mg-DOC/L, 6 mg-Cl_2/L, and 0.07 Br⁻-mg/L. Overall observations indicated that: (1) Br-THMs were less than 15% of the TTHMs; (2) Br-HANs were greater than 30% of the THANs; and (3) DCANFP decreased with increasing Br⁻:Cl ratio; (4) TCPFP decreased with increasing Br⁻:Cl ratio although no brominated haloketones were detected and identified; (5) increased and earlier formation of DBAN occurred with increasing Br⁻:Cl ratio. The Spearman Correlation Coefficient, a non-parametric correlation statistic which does not assume a linear trend, was employed to evaluate relationships between specific molar formation potentials (mM DBP/mg-DOC) and initial Br⁻:Cl ratio as a function of MWCO. Data are presented in Table III.

Certain DBPFPs; BCAN, DBAN, $CHCl_2Br$, and $CHClBr_2$; showed positive correlation (>0.60) with initial Br⁻:Cl ratio. Other DBPFPs; DCAN, TCP, and $CHCl_3$; were negatively correlated (< -0.60) with initial Br⁻:Cl ratio. Interestingly, TTHMs and THANs were generally not correlated with initial Br⁻:Cl ratio; this was because although individual species were correlated with Br⁻:Cl, some were negatively correlated while others were positively correlated. The correlation analysis also demonstrated that Br⁻:Cl influence on DBP speciation was more constant and stronger when the DBP contained only one type of halogen, Cl or Br, rather than a combination of halogens.

Figure 6. Effect of Br:Cl Ratio on DBP Speciation in <30,000 MWCO Coagulated Water Fraction

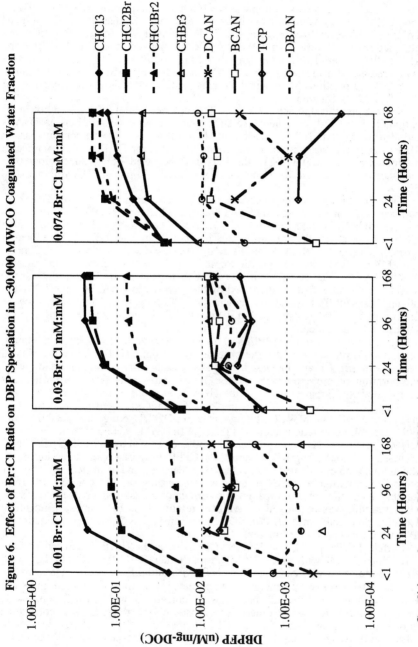

Conditions: 2 mg-DOC/L, 6 mg-Cl2/L

Table III. Correlation Between DOC Normalized DBPFP and Initial Br⁻:Cl Ratio by MWCO Fraction, Spearman Correlation Coefficients

Analyte	<1,000	<10,000	<30,000	<4,500 K
BCAN	0.196	0.196	0.353	0.321
CHCl$_3$	-0.624[1]	-0.686[1]	-0.686[1]	-0.748[1]
CHCl$_2$Br	0.089	0.321	0.089	0.045
CHClBr$_2$	0.383	0.722[1]	0.508	0.722[1]
CHBr$_3$	0.811[1]	0.935[1]	0.811[1]	0.935[1]
DBAN	0.686[1]	0.766[1]	0.766[1]	0.722[1]
DCAN	-0.641[1]	-0.659[1]	-0.613[1]	-0.631[1]
TCP	-0.761[1]	-0.720[1]	-0.671[1]	-0.671[1]
TTHM	-0.890[1]	0.080	-0.365	-0.134
THAN	0.196	0.321	0.445	0.385
TDBP	-0.089	0.080	-0.365	-0.027

Note: (1) Spearman Coefficient indicates correlation between independent parameter and dependent variable, values >|0.6| are significant.

The Paired Wilcoxon Signed Rank Sum Test was used to further investigate relationships between MWCO, normalized DOC concentration, and bromide concentration. This analysis found that high chlorine content DBPFPs (e.g., CHCl$_3$ and TCP) did not significantly change with MWCO fraction (α=0.1) The formation potential for brominated species in the <1,000 MWCO fraction was significantly different, than for >1,000 MWCO fractions, reflecting differences in reactivity with bromide ion between larger and smaller AMW NOM.

Discussion

This research examined in detail HAN formation and speciation in the context of THM formation and speciation. The research focused on controlling DOC, Cl$_2$:C ratio, and bromide concentration in raw water samples and treated water samples from a full scale coagulation-clarification WTP. The dominant form of NOM in this raw water supply is DOC. In evaluating the data collected, it is important to recognize that the DOC concentration was equalized in all fractions to eliminate any concentration related effects. Equalization emphasizes the reactivity of the < 1,000 AMW NOM with respect to larger MWCO fractions. Controlling bromide concentration was essential as bromide is a conservative substance both with respect to the water treatment process and ultrafiltration. Controlling DOC and Cl$_2$:C ratio was required to evaluate effect of NOM size on DBPFP independent of concentration effects from varying DOC and chlorine dose.

The effect of full scale coagulation of this water showed that when DOC concentration, bromide concentration and Cl$_2$:C ratio were held constant in the raw and treated water samples, coagulation caused a slight increase in THMFP and THM specific formation potential with a slight shift toward formation of more brominated THMs after coagulation. After coagulation a slight decrease was observed in THANFP; DCANFP was the most reduced HAN species. Similar effects of coagulation have been previously reported, notably a decrease in DCANFP concomitant with an increase in TTHM when comparing raw and coagulated water (20 mg-Cl$_2$/L, 4 days) and HANSDS decreases while THMSDS increased in SDSFP tests of raw and ultrafiltered waters (9,11). In all three of these examples the treated waters should have shifted toward a greater percentage of smaller AWM implying that total TOC effects (e.g., equalization) on specific formation potential are less important than AMW effects.

This is further confirmed by the experiments with fractionated water which indicated that the <1,000 AMW fraction had the greatest DBP formation potentials (Figure 3). Smaller AMW NOM is poorly removed by coagulation and lower AWM fractions are more reactive to chlorine and bromine (5,7,11). Because DOC concentration and bromide concentration were equalized in all chlorinated fractions for both raw and coagulated waters, it seems reasonable that coagulation would increase the percent of <1,000 AMW in DOC equalized fractions causing the shift to increased THMFPs after coagulation.

DCAN formation appeared to drive the HAN formation in the smaller AMW NOM precursor material. While the smaller AMW NOM generated the greatest DCAN formation it also maintained a higher formation potential with time suggesting that the quantity of specific precursor substrates, as well as the chemical character of the < 1,000 AMW MWCO, accounted for the observed formation potential. These results are consistent with coagulation removing DCANFP (9), but contradictory to research that indicated peak HAN formation occurs in higher AMW fractions such as >30,000 MWCO (27).

TCPFP appeared unaffected by coagulation in this research. This is similar to results reported by Smith et al. (13), who performed bench scale coagulation and chlorination experiments on eight waters from throughout North America and found that TCPFP was relatively unaffected by coagulation.

For this water, coagulation caused minor shifts in the distributions and formation of TTHMs and THANs, but changes in the Br$^-$:Cl ratio: (1) increased the magnitude of specific molar THANFP more than TTHMFP; (2) created a higher percentage of brominated HAN species compared to formation of brominated THM species; and (3) decreased DCANFP more than CHCl$_3$FP. The <1,000 MWCO fraction exhibited the greatest effect. For THMs, this research corroborated previous findings concerning the effect of bromide concentration of THMFP and distribution (15,19,28,29). This research also demonstrated that THANFP increased with increasing Br$^-$:Cl ratio, and that the increase in Br-THANFP occurred to a greater extent on a relative basis than Br-THMFP. This suggests that bromine interacts with more reactive sites in NOM, and to a greater extent than chlorine, to produce HANs.

Change in Br$^-$:Cl ratio within typical WTP operations were sufficient to change HAN and TCP formation. While the percentage results observed were quite dramatic it is important to remember that in this research HAN formation was typically less than 5% of the specific TTHM formation, a value similar to that reported by Peters et al. (21). This research demonstrates that HAN concentrations resulting from conventional treatment of high bromide waters could be comparable to concentrations of bromoform. With increased specificity in DBP regulation anticipated, awareness of HAN speciation's responsiveness to bromide concentration may become more important to water plant operators.

Unlike THMs, HAN and HK formation potentials were greatest at 24-48 hours. This suggests that the standard formation test employing 7 days is inappropriate for HANs and HKs. Similarly, field monitoring for HANs should focus on travel times of 24-48 hours for compliance with regulations, rather than the longer travel times for THM compliance monitoring.

References

(1) U.S. EPA. *Federal Register*. 1994, 59:38668.
(2) U.S. EPA. *Federal Register*. 1994, 59:6332.
(3) Peterson, H. G.; Milos, J. P.; Spink, D. R.; Hurdey, S. E.; Sketchell, J. *ES&T*. **1993**, 14:877-884.

(4) Venstra , J. N.; Schnoor, J. L. *Jour. AWWA.* **1980**, 72(10):583-590.
(5) Oliver, B. G.; Thurman, E. M. In *Water Chlorination: Environmental Impact and Health Effects*; Editor, R. L. Jolley, Ann Arbor Sci. Publ.: Ann Arbor, MI, 1990; Vol. 3; pp 141-147.
(6) El-Rehaili, A. M.; Weber Jr., W. J. *Water Res.* **1987**, 21(5):575-582.
(7) Reckhow, D. A.; Bose, P.; Bezbarua, B.; Hesse, E.; MacNeill, A. *Transformations of Natural Organic Material During Preozonation*; U.S. EPA: Cincinnati, OH, 1992.
(8) Joyce, W. S.; DiGiano, F. A.; Uden, P.C. *Jour. AWWA.* **1984**, 76(6):102-106.
(9) Reckhow, D. A.; Singer, P. C. *Jour. AWWA.* **1990**, 82(4):173-180.
(10) Mulford, L. A. et al. *DBP Precursor Removal by Reverse Osmosis*; AWWA Membrane Technol. in the Water Industry Conf.; AWWA, Denver, CO, 1991.
(11) Laine, J.; Jacangelo, J. G.; Cummings, E. W.; Carns, K. E.; Mallevialle, J. *Jour. AWWA.* **1993**, 85(6):87-99.
(12) Randtke, S. J.; Hoehn, R. C.; Long, B. W. *A Comprehensive Assessment of DBP Precursor Removal by Enhanced Coagulation and Softening*; The Universities Forum; AWWA: Denver, CO, 1994.
(13) Smith, L. A.; Dietrich, A. M.; Mann, P. D.; Hargette, P. H.; Knocke, W. R.; Hoehn, W. R. *Effects of Enhanced Coagulation on Halogenated Disinfection Byproduct Formation Potentials*; The Universities Forum; AWWA: Denver, CO, 1994.
(14) Cooper, W. J.; Zika, R. G.; Steinhauer, M. S. *Jour. AWWA.* **1985**, 77(4):116-121.
(15) Hutton, P. H.; Chung, F. I. *Jour. Water Res. Plan. & Man.* **1994**, 120(1):1-15.
(16) Bunn, W. W. et al. *Envir. Letters.* **1975**, 10:205.
(17) Siddiqui, M. S.; Amy, G. L. *Jour. AWWA.* **1993**, 85(1):63-72.
(18) Minear, R. A. In *Water Chlorination: Environmental Impact and Health Effects*; Editor, R. L. Jolley, Ann Arbor Sci. Publ.: Ann Arbor, MI, 1990; Vol. 3; pp 151-160.

(19) Symons, J. M.; Krasner, S. W.; Simms, L. A.; Sclimenti, M. *Jour. AWWA.* **1993**, 85(1):51-62.
(20) Summers, R. S.; Benz, M. A.; Shukairy, H. M.; Cummings, L.; *Jour. AWWA.* **1993**, 85(1):88-95.
(21) Peters, R. J. B.; de Leer, E. W. B.; de Galan, L. *Water Res.* **1990**, 24(6):797-800.
(22) Symons, J. M.; Speitel Jr., G. E.; Diehl, A. C.; Sorensen Jr., H. W. *Jour. AWWA.* **1994**, 86(6):48-60.
(23) Stevens, A. A.; Moore, L. A.; Miltner, R. J. *Jour. AWWA.* **1989**, 81(8):54-60.
(24) Knocke, W. R.; West, S.; Hoehn, R. C. *Jour. AWWA.* **1986**, 78(4):189-195.
(25) Owen, D. M.; Amy, G. A.; Chowdhury, Z. K. *Characterization of Natural Organic Matter and Its Relationship to Treatability.* AWWARF: Denver, CO. 1993;
(26) Gray, K. A.; McAuliffe, K. S.; Simpson, A. H. *Use of PY-GC/MS for Characterization of DBP Precursors*; ACS Disinfection By-Products and NOM Precursors: Chemistry, Characterization, Control Symp.; ACS, Chicago, IL, 1995.
(27) Reckhow, D. A.; Singer, P. C.; Malcolm, R. L. *ES&T.* **1990**, 24:1655-1664.
(28) Harrington, G. W.; Chowdhury, Z. K.; Owen, D. M. *Jour. AWWA.* **1992**, 84(11):78-87.
(29) Pourmoghaddas, H.; Stevens, A. A.; Kinman, R. N.; Dressman, R. C.; Moore, L. A.; Ireland, J. C. *Jour. AWWA.* **1993**, 85 (1):82-87.

OZONE AND OTHER PROCESSES

Chapter 17

Interactions Between Bromine and Natural Organic Matter

Rengao Song[1,4], Paul Westerhoff[2], Roger A. Minear[1], and Gary L. Amy[3]

[1]Institute for Environmental Studies, Department of Civil Engineering,
University of Illinois, 1101 West Peabody Drive, Urbana, IL 61801–4723
[2]Department of Civil and Environmental Engineering,
Arizona State University, Tempe, AZ 85287–5306
[3]Department of Civil, Environmental, and Architectural Engineering,
University of Colorado, Boulder, CO 80309

Bromine reacts quickly with natural organic matter (NOM) to form bromide and organo-bromine species. Bromine consumption by NOM occurs rapidly and the pseudo first order rate constants for bromine decomposition are on the order of 0.12 to 1.3×10^{-3} s^{-1} (DOC = 3 mg/L, pH = 7.5, and bromine dose = 5 μm); values decrease for preozonated NOM (0.11 to 0.24×10^{-3} s^{-1}). Based on multiple pH experiments, both hypobromous acid (HOBr) and hypobromite (OBr-) appear to react with equally with NOM under experimental conditions in this study. Bromine reduction by NOM appears to contribute to 80% of the reaction, while bromine incorporation into NOM accounts for the remaining 20%. The formation of organo-bromine species was noted to be extremely rapid (< 5 minutes) for DOC = 3 mg/L, pH = 7.5, and bromine dose = 5 μm. Overall, reactions between bromine and NOM may significantly influence the formation of disinfection by-products (DBPs) such as bromate during ozonation of bromide containing waters.

As an alterative disinfectant/oxidant, ozone has many benefits over chlorine. However, concern over potential ozonation by-products may limit its use. While many organic by-products, such as aldehydes, have been well documented, the potential for bromide-containing inorganic by-products has not been thoroughly investigated. Ozonation of bromide (Br-) containing waters has been shown to cause the formation of bromate, a suspected carcinogen with a proposed USEPA MCL of 10 μg/L. The mechanism of bromate (BrO_3^-) formation appears to rely upon the formation of an important transient intermediate species, namely aqueous

[4]Current address: Illinois State Water Survey, 22 Griffith Drive,
Champaign, IL 61820

0097–6156/96/0649–0298$16.00/0
© 1996 American Chemical Society

free bromine, a combination of hypobromous acid (HOBr) and hypobromite ion (OBr⁻). Reactions between bromine and NOM may play an important role in affecting bromate formation, possibly producing organo-bromine compounds as a result.

Ozone can rapidly oxidize bromide to form aqueous bromine (Taube, 1942). The distribution between the two forms is determined by the pH of the water; low pH favors HOBr and visa versa. At pH 8.86 and 20 ºC, the two forms are equally distributed (Westerhoff, 1995).

Aqueous bromine is a highly reactive substance, and as a result has a transient existence in water (Amy et al., 1995). In the absence of natural organic matter (NOM) (e.g. Milli-Q water, MQW), aqueous bromine oxidation by molecular ozone (O_3) and/or hydroxyl radical (HO•) results in bromate (BrO_3^-) formation (Richardson et al., 1981, Siddiqui and Amy, 1993; von Gunten and Hoigne, 1993). In the presence of NOM, bromine species also react to form organo-bromine species, such as bromoform and dibromoacetic acid, measured as total organic bromine (TOBr) (Siddiqui, 1992; Glaze et al., 1993), in addition to BrO_3^- formation. At the same time aqueous bromine can be reduced by NOM to regenerate bromide ions. Therefore, NOM-bromine interaction plays a very important role in controlling the formation of brominated disinfection byproducts in water treatment. While bromate and organo-bromine are terminal products and sinks for bromine species, HOBr and OBr⁻ are transient reactive intermediates. Thus, an examination of the transient behavior of aqueous bromine, especially NOM-bromine interactions, is of interest and importance in order to get a clear understanding of bromate and TOBr formation.

The goal of this paper is to provide insight into NOM-bromine reactions and the potential impact of these reactions on bromide containing by-products (BrO_3^- and TOBr). The experiments were conducted in batch reactors to which bromine was added, in the absence of ozone. Changes in bromine concentrations over time were monitored in NOM free water (MQW) and in the presence of NOM isolates (both unaltered and preozonated forms).

EXPERIMENTAL SECTION

Experimental Approach

A batch reactor was used to study NOM-bromide interactions (Westerhoff et al., 1993). The reactor (modified 500 ml graduated cylinder) was filled with synthesized model solutions containing NOM and phosphate buffer (2 mM) in NOM-free Milli-Q (Millipore) water. Then a bromine solution was injected below the water surface of the reactor with a syringe (Song et al., 1993). Following the addition of bromine solution, the reactor was immediately covered with an adjustable Teflon cover and the solution completely mixed with a magnetic stirrer. Samples for bromine and total organo-bromine species were collected over time from a sample port at the base of the reactor. All experiments were conducted at 20 ºC.

NOM was isolated and concentrated by an ultrafiltration/reverse osmosis

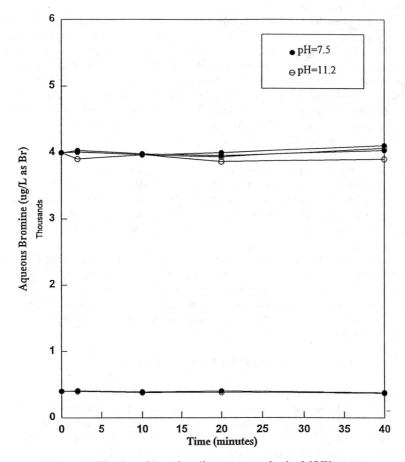

Figure 1. Kinetics of bromine disappearance in the MQW system

(UF/RO) technique which permitted isolation of three size fractions. The isolate with a molecular weight cutoff (MWC) of greater than 30,000 dalton was termed UF 30K, MWC of 1K to 30K was termed UF 1K, and MWC of less than 1K was termed RO isolate (Nanny, 1995). The results presented herein only consider the UF 1K isolate concentrated from four water sources: Teays Aquifer (TYS), IL, Lake Michigan (LMW), IL, California State Project Water (SPW), CA, and the Mississippi River (MRW). NOM isolation and concentration were conducted at 4 °C.

Measured Parameters

Each batch kinetic experiment produced a time series of samples which were analyzed for bromide (Br-), bromate (BrO_3^-), total bromine (HOBr+OBr-), and brominated organic by-products (TOBr). A Dionex ion chromatograph (DX300) coupled with an IonPac™ AS9-SC anion column and an AG9-SC guard column was used to determine bromate and bromide concentrations; a borate eluent (20 mM NaOH/40.3 mM H_3BO_3) and a 250 μL injection loop were used. The minimum reporting level for bromate is 3 μg/L and 5 μg/L for bromide. A N,N-diethyl-p-phenyldiamine (DPD) method was used to determine total bromine (HOBr +OBr-) (APHA, 1989). The method provides a lower detection limit of 0.02 mg/L as Br_2 (0.01 mg/L as Br-). The individual HOBr and OBr- concentrations can be calculated from the total bromine concentration, using the pKa value (8.86 at 20°C), and measured pH. TOBr was measured with a Dohrmann DX-20A TOX Analyzer equipped with an AD-3 Adsorption Module; this method is comparable to TOX measurements, but in the absence of chlorine only organo-bromine compounds are expected to form. A Shimadzu TOC-500 instrument was employed to measured the dissolved organic carbon (DOC).

RESULTS AND DISCUSSION

Kinetics of Aqueous Bromine Disappearance

For forty minutes following bromine addition to NOM-free (MQW) water, both HOBr and OBr- concentrations decrease only slightly (Figure 1). At pH 11.2, approximately 2.3 pH units above its pKa (8.86), hypobromite is the major bromine species; the converse is true for hypobromous acid at pH 7.5. The rate of bromine decomposition is independent of initial concentration within the time period of interest (Figure 1).

Whereas bromine decomposition did not occur significantly in MQW, the presence of NOM in solution upon bromine addition leads to a rapid consumption of both OBr- and HOBr. Data in Figure 2 illustrate the effect of NOM (DOC = 3 mg/L) on bromine consumption; bromine decomposes much faster than in NOM-free water (Figure 1). Furthermore, the reactions between NOM and bromine appear to be very rapid, as illustrated by significant bromine consumption during the first few minutes of reaction. After this initial reaction period, bromine consumption continues. In a broad sense, bromine consumption occurs as bromine addition to NOM and bromine oxidization of and/or incorporation into NOM occur.

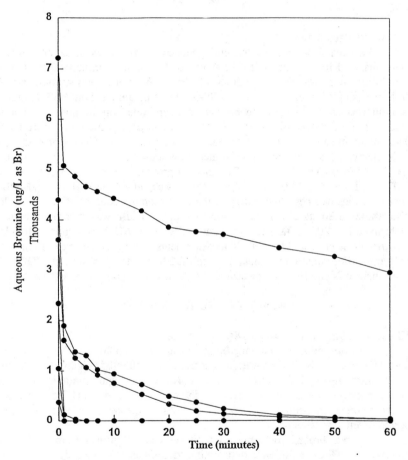

Figure 2. Kinetics of bromine disappearance
(SRFA isolate, DOC = 3 mg/L, pH = 7.5)

Bromine addition to NOM results in organo-bromine compounds, represented by the following reactions:

$$HOBr + NOM \rightarrow TOBr \tag{1}$$
$$OBr^- + NOM \rightarrow TOBr \tag{2}$$

where total organo-bromine (TOBr) is a measure of carbon-assimable substitution by-products. It has been suggested that HOBr is more effective than OBr^- in substituting with NOM.

On the other hand, oxidation of NOM by bromine results in reduction of bromine to bromide, and is represented by the following reactions:

$$HOBr + NOM \rightarrow Br^- + NOM_{oxid} \tag{3}$$
$$OBr^- + NOM \rightarrow Br^- + NOM'_{oxid} \tag{4}$$

where NOM_{oxid} and NOM'_{oxid} represent oxidized forms of NOM, but do not include organo-bromine compounds. These reactions do occur in the presence of NOM. In addition to bromine substitution to NOM to form TOBr, bromine can undergo oxidation-reduction reactions with NOM.

NOM isolation method and source influence bromine-NOM interaction. Under identical conditions (pH = 7.5, DOC = 3 mg/L, bromine dose = 5 μM), similar experiments were performed on the NOM isolated from different sources and different rates of bromine consumption were observed (Figure 3). The fulvic acid isolate from the Suwannee River (SRFA) reacts more rapidly with bromine than NOM isolates from several sources in the size range of 1K-10K daltons. Lake Michigan, State Project Water, Mississippi River Water and Teayes aquifer isolates exhibit varying degrees of bromine reactivity, respectively, in decreasing order.

Bromine-NOM interaction is also a function of pH. Increasing the pH from 7.5 to 11.2 shifts the dominant bromine species from HOBr to OBr^-. This also influences the structure of NOM (e.g. functional group deprotonation). The later effect can influence the reduction potential of NOM. Parallel experiments to those conducted at pH 7.5 (Figure 3) were also conducted at pH 11.2 (Figure 4). While SRFA was still the most reactive and the rates of bromine disappearance increase slightly except for Teays Aquifer 1K isolate.

While bromide oxidation by ozone forms bromine, ozone will also oxidize NOM and potentially change its structure. These transformations in NOM structure were examined by preozonating NOM isolates (all ozone decomposed) before bromine addition. Two forms of NOM isolates were examined, unaltered NOM isolate and pre-ozonated NOM isolates. Solutions of the unaltered NOM and pre-ozonated NOM isolates were brominated with aqueous bromine solutions. Information obtained from such experimental design could be hopefully used to explain NOM-bromine interactions during ozonation. Preozonation at an ozone dose to dissolve organic carbon (DOC) of 2 to 1 imparts certain physicochemical

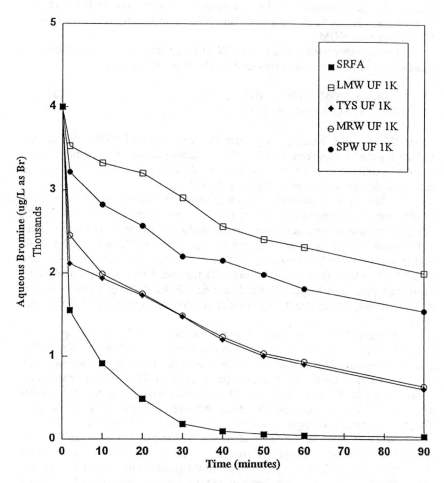

Figure 3. Kinetics of HOBr/OBr⁻ disappearance for different NOM isolates at pH 7.5
(DOC = 3 mg/L, Br$_2$ dose = 4000 μg/L)

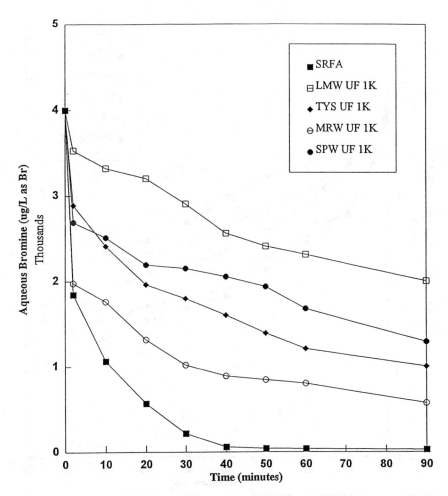

Figure 4. Kinetics of HOBr/OBr⁻ disappearance for different NOM isolates at pH 11.2
(DOC = 3 mg/L, Br₂ dose = 4000 μg/L)

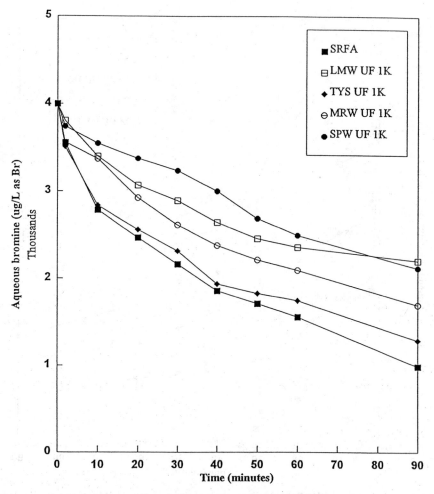

Figure 5. Kinetics of HOBr/OBr⁻ disappearance for preozonated NOM isolates at pH 7.5
(O_3 dose = 6 mg/L, DOC = 3 mg/L, Br_2 dose = 4000 μg/L)

changes in the NOM isolates.   For example, preozonation decreases NOM extinction coefficient at 254 nm.

Two sets of experiments at pH 7.5 and 11.2 were performed (Figure 5 and 6) with preozonated NOM; the results can be compared with unaltered NOM isolates (Figure 3 and 4).   Preozonation decreases the reactivity of NOM with bromine, but does not completely its demand.

The observed trends in NOM-bromine interactions suggest a two-stage behavior.   In the first-stage a rapid aqueous bromide concentration drop occurs during the first two minutes of reaction time; Afterwards a second-stage occurs in which aqueous bromine disappears relatively slowly.   However, overall bromine decomposition exhibits an exponential consumption, represented by $d\{[HOBr]_T\}/dt = -k[HOBr]_T$. $[HOBr_T]$ is total concentration of HOBr plus OBr^-.

Rapid reactions between bromine and NOM may play an important role in bromate formation.   The first-stage fast aqueous bromine disappearance can be mathematically described as follows:

$$\Delta_{02} = \{[HOBr_T]_0 - [HOBr_T]_2\}/120 \qquad \text{for } t \leq 2 \text{ minutes} \qquad (3)$$

where Δ_{02} is the parameter representing the loss of aqueous bromine during the first two-minute of reaction.   This approximation only represents a pseudo zero-order reaction based just two data points, since the first aqueous bromine concentration experimentally measured is at two-minute ($t = 120$ seconds).

The second-stage bromine disappearance exhibits an exponential decomposition behavior.   A first-order rate expression for bromine disappearance achieved a high statistical fit.   Therefore, the second-stage is assumed to be pseudo first-order with respect to aqueous bromine concentration:

$$d[HOBr_T]/dt = -k_2[HOBr_T] \qquad \text{for } t > 2 \text{ minutes} \qquad (4)$$
$$\text{or } [HOBr_T]_t = ([HOBr_T]_0 - \Delta_{02})\exp(-k_2 t) \qquad \text{for } t > 2 \text{ minutes} \qquad (5)$$

Where k_2 is the pseudo first-order rate constant for second-stage bromine disappearance.   Values of aqueous bromine consumption rate parameters for unaltered and preozonated NOM isolates from different water sources at pH 7.5 and 11.2 are presented in Tables 1 through 4.   It was observed that except for the SRFA isolate, the k values for all UF 1K isolates fall into a relatively narrow range with k values of approximately 1.0 to 3.0×10^{-4} s^{-1}.   Preozonation does have an obvious "leveling effect" on aqueous bromine disappearance rates, the average k value was reduced from 4.3×10^{-4} to 1.6×10^{-4} s^{-1} at pH 7.5 and from 4.7×10^{-4} to 1.8×10^{-4} s^{-1} at pH 11.2, respectively.   However, the order of bromine disappearance rates remains unchanged.   This suggests that preozonation does not significantly affect certain sites in the NOM isolates which react with bromine.

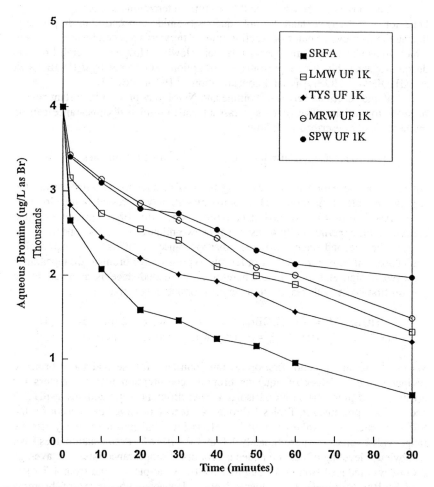

Figure 6. Kinetics of HOBr/OBr⁻ disappearance for preozonated NOM isolates at pH 11.2
(O$_3$ dose = 6 mg/L, DOC = 3 mg/L, Br$_2$ dose = 4000 μg/L)

Table 1

Aqueous bromine disappearance parameters for unaltered NOM isolates at pH 7.5

(DOC = 3 mg/L, Br_2 dose = 50 μM)

NOM source/isolate	$\Delta_{02} \geq$ ($\times 10^7$ Ms^{-1})	k_2 ($\times 10^3$ s^{-1})	r^2 for k_2	k ($\times 10^3$ s^{-1})	r^2 for k
TYS 1K	1.96	0.22	0.975	0.27	0.870
LMW 1K	0.488	0.11	0.927	0.12	0.925
SPW 1K	0.817	0.13	0.958	0.16	0.897
MRW 1K	1.61	0.26	0.973	0.30	0.900
SRFA	2.55	1.18	0.993	1.31	0.967
Average	1.48	0.38	0.965	0.43	0.912

Table 2

Aqueous bromine disappearance parameters for unaltered NOM isolates at pH 11.2

(DOC = 3 mg/L, Br_2 dose = 50 μM)

NOM source/isolate	$\Delta_{02} \geq$ ($\times 10^7$ Ms^{-1})	k_2 ($\times 10^3$ s^{-1})	r^2 for k_2	k ($\times 10^3$ s^{-1})	r^2 for k
TYS 1K	1.16	0.20	0.959	0.23	0.896
LMW 1K	0.882	0.14	0.986	0.17	0.935
SPW 1K	1.36	0.15	0.977	0.17	0.862
MRW 1K	2.11	0.23	0.926	0.31	0.802
SRFA	2.24	1.41	0.985	1.49	0.982
Average	1.55	0.43	0.967	0.47	0.895

Table 3
Aqueous bromine disappearance parameters for preozonated NOM isolates
at pH 7.5
(DOC = 3 mg/L, Br_2 dose = 50 μM)

NOM source/isolate	$\Delta_{02} \geq$ ($\times 10^7$ Ms^{-1})	k_2 ($\times 10^3$ s^{-1})	r^2 for k_2	k ($\times 10^3$ s^{-1})	r^2 for k
TYS 1K	0.510	0.18	0.967	0.20	0.948
LMW 1K	0.201	0.10	0.904	0.11	0.898
SPW 1K	0.267	0.11	0.979	0.11	0.929
MRW 1K	0.458	0.14	0.978	0.15	0.965
SRFA	0.461	0.23	0.985	0.24	0.973
Average	0.379	0.15	0.963	0.16	0.943

Table 4
Aqueous bromine disappearance parameters for preozonated NOM isolates
at pH 11.2
(DOC = 3 mg/L, Br_2 dose = 50 μM)

NOM source/isolate	$\Delta_{02} \geq$ ($\times 10^7$ Ms^{-1})	k_2 ($\times 10^3$ s^{-1})	r^2 for k_2	k ($\times 10^3$ s^{-1})	r^2 for k
TYS 1K	1.22	0.15	0.987	0.18	0.884
LMW 1K	0.473	0.12	0.948	0.13	0.929
SPW 1K	0.620	0.11	0.948	0.12	0.898
MRW 1K	0.593	0.16	0.995	0.17	0.977
SRFA	1.40	0.27	0.980	0.31	0.922
Average	0.861	0.16	0.972	0.18	0.922

A comparison between the rate of bromine formation during ozonation and the rate of bromine consumption due to reactions with NOM yield insight into the relative kinetic controls for bromate formation. The values of Δ_{02} range from 0.201 to 2.11×10^{-7} Ms^{-1}, excluding SRFA isolate. Notably, the rates of bromine formation (0.3×10^{-7} Ms^{-1}) during ozonation are similar; the rates of bromine formation can be estimated from Haag and Hoigne model (1983) with initial conditions of bromide concentration of 100 μg/L, DOC = 3 mg/L, and O_3/DOC ratio of approximately 2 mg/mg: $d[OBr^-]/dt = 160[O_3][Br^-]$. Therefore, during the first two-minute reaction time period the rates of bromine formation and disappearance through NOM-bromine interactions are similar. This suggests that bromate formation pathways involving aqueous bromine as an intermediate such as the direct ozonation pathway contribute little to bromate formation during the same time period. Consequentially, conceptual models involving bromine as an important intermediate in bromate formation is very sensitive to NOM-bromine reactions.

Table 5
Kinetic analyses for OBr⁻ disappearance pathways
(pH = 7.5, O_3 dose = 3 mg/L, DOC = 3 mg/L)

Reactants [S]	Products	k ($M^{-1}s^{-1}$)	Concentration (M)	k [S] (s^{-1})	% pathway
O_3 + OBr⁻	BrO_2^-	100	6.25×10^{-5}	6.25×10^{-3}	20.1 ~ 22.6
O_3 + OBr⁻	Br⁻	330	6.25×10^{-5}	2.06×10^{-2}	66.1 ~ 74.6
HO• + OBr⁻	BrO•	4×10^9	$10^{-12 ~ -13}$	$4.0 \times 10^{-3 ~ -4}$	12.8 ~ 1.45
NOM + OBr⁻	Br⁻ /TOBr	NA	NA	*3.2×10^{-4}	1.00 ~ 1.16

NA = not available; * = average k value of unaltered and preozonated NOM isolates

Table 6
Kinetic analyses for HOBr disappearance pathways
(pH = 7.5, O_3 dose = 3 mg/L, DOC = 3 mg/L)

Reactants [S]	Products	k ($M^{-1}s^{-1}$)	Concentration (M)	k [S] (s^{-1})	% pathway
O_3	BrO_2^-	0.013	6.25×10^{-5}	8.12×10^{-7}	0.04 ~ 0.16
HO•	BrO•	2×10^9	$10^{-12 ~ -13}$	$2.0 \times 10^{-3 ~ -4}$	40 ~ 87
NOM	Br⁻ /TOBr	NA	NA	*3.0×10^{-4}	13 ~ 60

NA = not available; * = average k value of unaltered and preozonated NOM isolates

During ozonation, bromine can react with NOM and O_3 or HO•. The relative rates of bromine disappearance, to several products, are summarized in Table 5 and 6. Considering an overall k value for bromine-NOM reactions, they

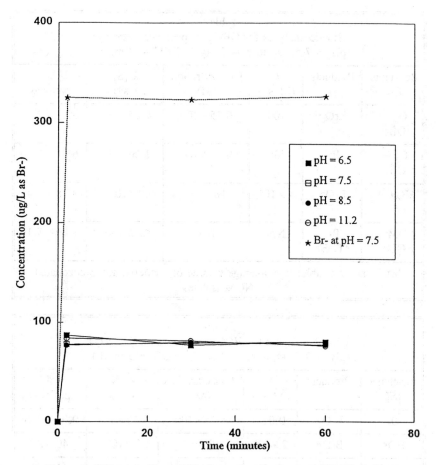

Figure 7. Effect of pH on TOBr formation for unaltered SRFA isolate
(DOC = 3 mg/L, Br$_2$ dose = 400 μg/L as Br$^-$)

do not appear to be as strong as that of the first-stage. However, at relatively low O_3/DOC ratio (approximately 1 mg/mg), the overall NOM-bromine interactions, especially NOM-HOBr interactions are still important.

Kinetics of TOBr Formation

Rapid reactions between NOM and bromine can lead to both Br regeneration or TOBr organic-bromine formation. Since also important to attempt to differentiate between oxidation alone and substitution into NOM, a series of experiments were conducted to investigate TOBr formation. Data in Figures 7 and 8 illustrate the effect of pH on TOBr formation for bromine addition to unaltered and preozonated NOM, respectively. The most dramatic observation from these experiments is that TOBr forms rapidly after bromine addition (400 μg/L as Br-) to either unaltered or preozonated NOM isolates. Secondly, pH has little effect on TOBr formation, suggesting that both OBr- and HOBr can substitute into NOM and form TOBr; however, HOBr may react with NOM followed by rapid equilibrium OBr- protonation to regenerate HOBr levels. Finally, approximately 20% (80 μg/L as Br-) of the initial bromine becomes substituted into the NOM isolates as TOBr, which is about two and half times as much as TOBr formation observed by Glaze et al. (1993) and Amy et al. (1995) who reported a 8% organo-bromine yield at typical ozonation conditions. This reflects the competitive aspects between bromate and TOBr formation during ozonation of bromide-containing waters.

TOBr formation in SRFA at different initial bromine concentrations were also investigated. Figures 9 and 10 demonstrate the effect of initial bromide level on TOBr formation for unaltered and preozonated SRFA isolate solutions, respectively. Once again, a rapid formation of TOBr was observed. However, a proportional increase in TOBr yield was not observed when initial bromine levels increased. Even though preozonation does not have a significant effect on TOBr formation at low initial bromine doses (lower than 400 μg/L), a relative 20% increase in organo-bromine production occurs as initial bromine concentration increases from 400 μg/L to 1000 μg/L as Br-. Preozonation did not seem to affect the extent or rate of TOBr formation, implying ozone does not totally destroy organic-bromine precursors.

The effects of preozonation and NOM isolate source on TOBr formation are presented in Figures 11 and 12. At initial bromine dose of 400 μg/L as Br-, while similar TOBr production is associated with both unaltered and preozonated SRFA isolates, preozonation does decrease organo-bromine yields significantly for all three UF 1K NOM isolates examined, a 37, 46, and 50% reduction in TOBr formation for MRW, SPW, and TYS NOM isolates, respectively. However, at bromine dose of 1000 μg/L as Br-, a positive effect of preozonation on TOBr formation was observed for SRFA isolate. This suggests that 1) the reactive sites may be impacted by preozonation for these membrane NOM isolates, and 2) the SRFA isolate exhibits the fastest bromine disappearance mainly through reduction-oxidation process, thus upon preozonation, its bromine demand through the reduction-oxidation reactions is expected to decrease. This may in turn increase

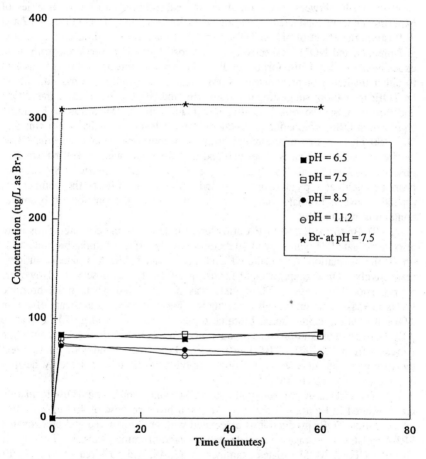

Figure 8. Effect of pH on TOBr formation for preozonated SRFA isolate
(O_3 dose = 6 mg/L, DOC = 3 mg/L, Br_2 dose = 400 μg/L as Br-)

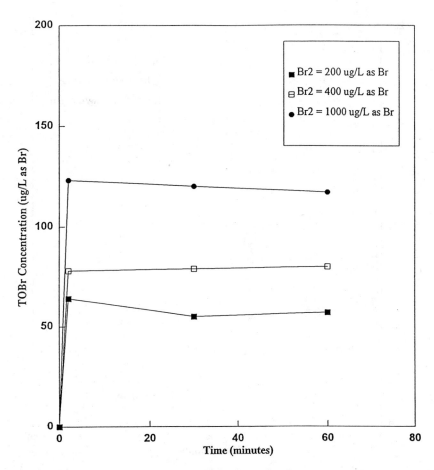

Figure 9. Effect of bromine dose on TOBr formation for unaltered SRFA isolate
(DOC = 3 mg/L, pH = 7.5)

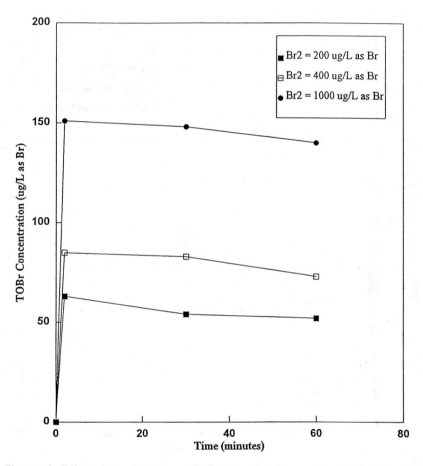

Figure 10. Effect of bromine dose on TOBr formation for preozonated SRFA isolate (O_3 dose = 6 mg/L, DOC = 3 mg/L, pH = 7.5)

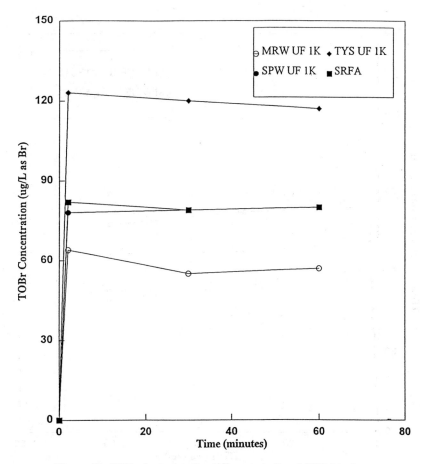

Figure 11. TOBr formation for different unaltered NOM isolates
(pH = 7.5, DOC = 3 mg/L, Br_2 dose = 400 μg/L as Br⁻)

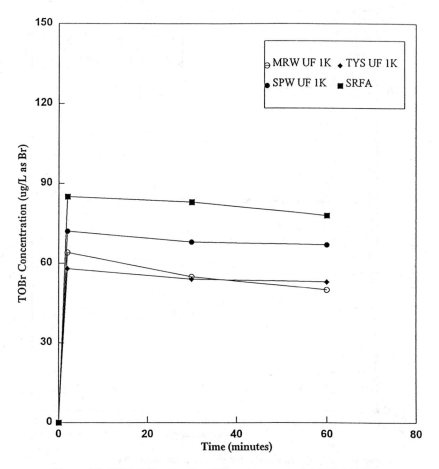

Figure 12. TOBr formation for different preozonated NOM isolates
(O_3 dose = 6 mg/L, pH = 7.5, DOC = 3 mg/L, Br_2 dose = 400 μg/L as Br^-)

TOBr production even though some reactive sites for TOBr formation may also be destroyed.

SUMMARY AND CONCLUSIONS

Unlike the slow decomposition of bromine in the MQW system, aqueous bromine rapidly decomposes in the presence of NOM isolates as bromine oxidizes and/or substitutes into NOM. This suggests that bromine reduction by NOM and organo-bromine formation may effectively compete for bromine with molecular O_3 and HO• radicals. Kinetic analysis indicates that during the first two minutes of reaction time, bromide oxidation reactions involving aqueous bromine as an intermediate may contribute little to bromate formation. Even during the second-stage NOM-bromine interactions, NOM related bromine disappearance at low O_3/DOC ratio such as 1 mg/mg is still important in order to better understand bromate formation. Therefore, it appears that NOM isolates decrease the efficiency of bromine conversion to bromate in comparison to MQW.

Although only 20% of the direct aqueous bromine becomes incorporated as TOBr, with the total magnitude of TOBr formation expected to be even lower (8%) during ozone-bromide-NOM interactions (Glaze et al., 1993; Song, 1996), TOBr formation is still important since TOBr formation appears to be as quick as direct oxidation of bromine. Also, organo-bromine is a sink for aqueous bromine.

ACKNOWLEDGEMENT

The authors gratefully acknowledge the American Water Works Association Research Foundation (AWWARF) for its financial support, and the valuable advice of the project panel advisory committee members: Dr. Werner Haag, Mr. Robert Powell, Dr. Philip Singer, and Dr. James Symons, as well as project officers: Robert Allen and Jeff Oxenford. Stuart Krasner, Brady Coffey, and Richard Yates of Metropolitan Water District of Southern California as well as Drs. Vernon Snoeyink, Richard Larson, and Gary Peyton of University of Illinois are also appreciated for their professional help.

REFERENCES

Amy, G., Minear, R., Westerhoff, P., and Song, R. (1995). Bromide Ozone Interactions in Water Treatment: Bromate versus Organo-Bromine Disinfection By-Product Formation, *American Water Works Association Research Foundation*.

Glaze, W.H., Weinberg, H.S., and Cavanagh, J.E. (1993). Evaluating the Formation of Brominated DBPs During Ozonation, *Journal of American Water Works Association*, 85:1:96-103.

Gorden, G. (1993). The Chemical Aspects of Bromate Ion Control in Ozonated Drinking Water Containing Bromide Ion, *Proceedings of the International Water Supply Association International Conference*: Bromate and Water Treatment, Paris, November 22-24, pp.41-50.

Haag, W.R. and Hoigne, J. (1983). Ozonation of Bromide-Containing Waters: Kinetics of Formation of Hypobromous Acid and Bromate, *Environmental Science and Technology*, 17:5:261-267.

Haag, W.R., Hoigne, J., and Bader, H. (1984). Improved Ammonia Oxidation by Ozone in the Presence of Bromide Ion During Water Treatment, *Water Research*, 18:9:1125-1128.

Haruta, K. and Takeyama, T.(1981). Kinetics of Oxidation of Aqueous Bromide Ion by Ozone, *Journal of Physical Chemistry*, 85:2383-2388.

Krasner, W., Glaze, W., Weinberg, W., Daniel, P., and Najm, I.N. (1993). Formation and Control of Bromate During Ozonation of Waters Containing Bromide, *Journal of American Water Works Association*, 85:1:73-81.

Nanny, M.A. (1995). Ph.D. Dissertation, University of Illinois, Urbana-Champaign.

Richardson, L.B., Burton, D.T., Helz, G.R., and Rhoderick, J.C. (1981). Residual Oxidant Decay and Bromate Formation in Chlorinated and Ozonated Sea-Water, *Water Research*, 15:1067-1074.

Siddiqui, M.S. (1992). Ph.D. Dissertation, University of Arizona, Tucson.

Siddiqui, M.S. and Amy, G.L. (1993). Factors Affecting DBP Formation During Ozone-Bromide Reactions, *Journal of American Water Works Association*, 85:1:63-72.

Song, R., Minear, R., Westerhoff, P., and Amy, G. (1993). Comparison of Bromide-Ozone Reactions with NOM Separated by XAD-8 Resin and UF/RO Membrane Methods, *206th American Chemical Society - Division of Environmental Chemistry Conference Proceedings*, Chicago, IL, August, 22-27, 33:2:225-228.

Song, R., Westerhoff, P., Minear, R., and Amy, G. (1994). Effects of Carbonate Alkalinity on Bromate Formation, *208th American Chemical Society - Division of Environmental Chemistry Conference Proceedings*, Washington, D.C., August 21-26.

Song, R., (1996). Ph.D. Dissertation, University of Illinois at Urbana-Champaign.

Standard Methods for the Examination of Water and Wastewater (1989), APHA, AWWA, and WPCF, Washington, D.C.

Taube, H. (1942). Reactions in Solutions Containing O_3, H_2O_2, H^+, and Br^-. The Specific Rate of reaction $O_3 + Br^- ->$, *Journal of American Chemical Society*, 64:2468-2474.

Von Gunten, U. and Hoigne, J. (1994). Bromate Formation During Ozonation of Bromide-Containing Waters: Interaction of Ozone and Hydroxyl Radical Reactions, *Environmental Science and Technology*, 28:7:1234-1242.

Von Gunten, U. and Hoigne, J. (1993). Bromate Formation During Ozonation of Bromide-Containing Waters, *Proceedings of the International Water Supply Association International Conference*: Bromate and Water Treatment, Paris, November 22-24, pp.51-56.

Westerhoff, p., Siddiqui, M., Amy, G., Song, R., and Minear, R. (1993). Evaluation of Rate Constants for Dissolved Ozone Decay and Bromate Formation, *206th American Chemical Society - Division of Environmental Chemistry Conference Proceedings,* Chicago, IL, August 22-27, 33:2:229-232.

Westerhoff, P (1995). Ph.D. Dissertation, University of Colorado, Boulder.

Xiong, F., Croue, J-P, and Legube, B. (1992). Long-Term Ozone Consumption by Aquatic Fulvic Acids Acting as Precursors of Radical Chain Reactions, *Environmental Science and Technology*, 26:1059-1064.

Yates, R.S. and Stenstrom, M.K. (1993). Bromate Production in Ozone Contactors, *AWWA Annual Conference Proceedings,* San Antonio, TX, June.

Chapter 18

Simplifying Bromate Formation Kinetic Analysis with a Linear Bromate Yield Concept

Paul Westerhoff[1,2], Gary L. Amy[2], Rengao Song[3], and Roger A. Minear[3]

[1]Department of Civil and Environmental Engineering,
Arizona State University, Tempe, AZ 85287–5306
[2]Department of Civil, Environmental, and Architectural Engineering,
University of Colorado, Boulder, CO 80309
[3]Institute for Environmental Studies, Department of Civil Engineering,
University of Illinois, 1101 West Peabody Drive, Urbana, IL 61801–4723

Under completely-mixed batch-reactor conditions, bromate formation exhibits an exponential kinetic relationship that inversely mirrors the exponential loss of ozone. Bromate formation correlates linearly with ozone consumption; this relationship is defined as the *bromate yield* of a water. The magnitude of the *bromate yields* varies depending upon water quality characteristics and water treatment processes. Example applications of the *bromate yield* concept demonstrate techniques for understanding bromate formation mechanisms, evaluation of chemical bromate control options, and evaluation of bromate control strategies in larger-scale, continuous-flow, ozone contactors.

Formation of new inorganic and organic by-products may offset the potential benefits of ozone over the traditional disinfectant or oxidant, namely chlorine, in drinking water treatment. Chlorine reactions with natural organic matter (NOM) precursors and bromide form detectable disinfection by-products (DBPs). Proposed and regulated maximum contaminant levels (MCLs) for these chlorinated, brominated, or mixed halogen species (e.g., trihalomethanes, haloacetic acids) may limit chlorine applications (1-3). With increasing concern for microbial disinfection, ozone emerges as an alternative

0097–6156/96/0649–0322$17.00/0
© 1996 American Chemical Society

disinfectant and oxidant for chlorine. However, alternative oxidants, including ozone, can form new classes of by-products (4). Proposed regulatory constraints on ozonation by-products may preclude use of ozone as an alternative to chlorine under some treatment conditions. Research into the mechanisms and control techniques for these by-products may provide strategies to use ozone and still comply with current and future regulations.

As ozone decomposes in water, both molecular ozone (O_3) and hydroxyl (HO) radicals, a daughter product of ozone decomposition, typically exist at levels capable of oxidizing particulate and dissolved phases (5). Both molecular ozone and HO radicals are highly powerful oxidants, although molecular ozone tends to react selectively with solutes while HO radicals react rather unselectively (6-9). In addition to a high disinfection potential, numerous other benefits of ozonation have been reported, among which are the following: degradation of organic taste and odor compounds (e.g., MIB and geosmin) and pesticides (e.g., atrazine), improved particle destabilization, oxidation of NOM, and oxidation of reduced metals (e.g., ferrous iron) (5, 10, 11). However, ozone can produce organic by-products (e.g., aldehydes) from ozone oxidation of NOM and inorganic by-products (e.g., bromate) during ozonation of bromide containing waters (13-16). Several ozonation by-products, including bromate, may be carcinogenic and regulatory levels have been proposed by the USEPA, European Union, and World Health Organization. By-product formation may preclude the use of ozone in some applications, or at least require additional treatments to control these by-products (15, 17).

Bromate exhibits carcinogenic effects in two species of laboratory animals and evidence supports its ability to cause chromosome damage in humans (18-21). The USEPA proposes an interim MCL of 10 μg/L for bromate until new epidemiological data and analytical methods become available; the World Health Organization proposes a 25 μg/L maximum bromate concentration in drinking waters. Bromate carcinogenity and regulatory limits have served as the impetus for understanding bromate formation.

Potentially high carcinogenic risk from bromate has spurred research into the mechanisms and levels of bromate formation. Laboratory and field measurements clearly show that bromate forms to levels exceeding regulatory levels during ozonation of some bromide containing waters (15). Current and proposed ozonation facilities now find it necessary to consider and evaluate bromate formation in each water supply. To this end, this chapter examines the applicability of a simple technique capable of predicting the kinetics and ultimate level of bromate formation.

Background

Since the mechanisms and kinetics of bromate formation depend largely upon the decomposition of ozone and a daughter by-product, HO radicals, a brief summary of ozone decomposition processes is presented prior to discussions on bromide occurrence and bromate formation kinetics.

Ozone Decomposition Processes. In pure water, hydroxide initiates ozone decomposition to several highly reactive intermediates, including HO radicals, through a series of electron transfer steps (5). Specific ozone decomposition pathways and rate constants differ in the literature, but a generalized series of reactions tend to be representative of the numerous reported models, given as follows:

$$O_3 + OH^- \Rightarrow HO_2 + O_2^- \tag{1}$$
$$HO_2 \Leftrightarrow O_2^- + H^+ \tag{2}$$
$$2O_2^- + 2O_3 \Rightarrow 2O_3^- + 2O_2 \tag{3}$$
$$2H^+ + 2O_3^- \Rightarrow 2HO + 2O_2 \tag{4}$$
$$H_2O \Leftrightarrow H^+ + OH^- \tag{5}$$

Cumulative addition of Equations 1 through 6 results in the following simplified equation:

$$3O_3 + H_2O \Rightarrow 2HO + 4O_2 \tag{6}$$

Equation 6 represents the HO radical yield from ozone decomposition; the theoretical HO radical yield (η) corresponds to a stoichiometric ratio of HO/O_3 which equals 0.67 mole per mole. Literature reports on η values calculated from laboratory experiments range from 0.5 to 1.0 mole per mole (22, 23). Despite the formation of significant HO radical quantities, their rapid reactivity with ozone itself, inorganic solutes, and organic solutes generally lead to HO radical concentrations on the order of 10^7 times lower than measured ozone residuals (6, 24).

Inorganic and organic constituents in water complicate the mechanism of ozone consumption and consequentially the mechanisms of oxidation (6-9, 25). For example, carbonate scavenges HO radicals forming a secondary oxidant and resulting in stabilization of ozone consumption (26, 27). Model organic compounds, such as tertiary butanol, likewise scavenge HO radicals, although tertiary butanol reactions with HO radicals do not form secondary oxidants (28). Other model organic compounds influence ozone decomposition through initiation or promotion mechanisms that result from direct reactions with molecular ozone or HO radicals (28). The heterogeneity of structure and functionality which characterizes NOM suggests that NOM can serve as an initiator, promoter, and/or inhibitor of ozone decomposition through reactions with molecular ozone and HO radicals (29-31).

Competing reactions among solutes influence the rate of ozone consumption. Overall ozone consumption in "pure" laboratory water typically approaches first- or second-order kinetics with respect to ozone (29, 31, 32). Inorganic and organic by-product formation can occur through both direct (molecular ozone) and indirect (HO radical) pathways. Therefore, compounds affecting molecular ozone or HO radical concentrations can influence the rate, and potentially the mechanisms, of by-product formation (33).

Bromide Occurrence. Bromide (Br^-) serves as the precursor for bromate (BrO_3^-) formation. Even drinking waters containing very low bromide levels may form bromate upon ozonation. Amy et al. (34) surveyed 101 drinking water sources and found an average national bromide concentration of almost 100 $\mu g/L$ (range: <10 to 2,500 $\mu g/L$).

Dutch drinking water sources exhibited bromide concentrations ranging from 100 to 400 µg/L (35). Finnish waters contain bromide concentrations into the hundreds of micrograms per liter, with 30 to 100 µg/L being common (36). Legube et al. (37) surveyed 25 European water utilities and observed bromide concentrations in the range of <10 to 1,040 µg/L (average = 130 µg/L). The results of Amy et al. (34) suggest that connate sea-water, salt water intrusion, aerosol sea-water precipitation, mineral salt deposits, and anthroprogenic activities are the major sources of bromide in drinking water supplies.

Bromate Formation. Bromate formation mechanisms rely on a series of molecular ozone and HO radical oxidation steps (14, 31). Current theories postulate that initial bromide (Br⁻) oxidation to aqueous bromine (HOBr/OBr⁻) occurs predominantly through a molecular ozone pathway (13, 14, 31), although HO radicals may play an important role in this oxidation step at higher pH levels, during advanced oxidation treatments, or in the presence of NOM (29, 31, 38, 39). Measurable quantities of aqueous bromine accumulate during ozonation, serving as evidence that aqueous bromine is an important intermediate in bromate formation. Oxidation of aqueous bromine to bromite (BrO_2^-) occurs through a series of HO radical reactions (14, 31), although only about 30% of the aqueous bromine forms bromite while disproportionation reactions convert the remainder back to bromide (40). The final oxidation step of bromite to bromate occurs rapidly in the presence of molecular ozone (13, 14, 31). Overall bromate formation depends upon both molecular ozone and its daughter by-product, HO radicals, to oxidize bromide to bromate.

The majority of lab-, pilot-, and full-scale bromate formation studies have observed a positive correlation between bromate formation and ozone dose (14, 15, 21, 41, 42). Many of these same studies demonstrate a lower bound limit of ozone dose, below which detectable levels of bromate are not observed; thus, there is an ozone dose threshold for bromate formation. The magnitude of this ozone threshold for bromate was found to vary between 0.2 and 0.75 mg ozone per mg DOC, based on a reported bromate detection limit of 2 µg/L (43). This variation presumably arises from differences among experimental conditions, inorganic water matrices, and the source of NOM.

In contrast to chlorination by-product formation potentials or simulated distribution system tests, limited guidance exists for bromate studies and for scale-up procedures. von Gunten and Hoigne (14, 39) applied a simplifying technique to bromate formation by linking ozone decomposition kinetics to bromate formation kinetics. Integrating the area beneath a kinetic ozone decomposition profile, describing the time history of ozone, was termed *ozone exposure*. Although this approach relies on time-consuming analyses of ozone consumption kinetics, the underlying principle links ozone consumption to bromate formation.

Research Objectives

This chapter presents examples, applications, and a recommended protocol for a technique to simplify the analysis of bromate formation kinetics. This technique incorporates the stoichiometric conversion of bromide to bromate as a function of the quantity of ozone consumed, and is termed *bromate yield*. The following objectives will be met:

- Demonstrate that the efficiency of bromide conversion to bromate depends upon pH and the presence of inorganic and organic solutes.
- Introduce and support the *bromate yield* concept with examples from experimental data.
- Address the influence of initial bromide concentrations on *bromate yields*.
- Evaluate bromate control strategies through the use the *bromate yields*.
- Demonstrate the application of *bromate yield* in ozone contactor design and evaluation.
- Propose recommendations for bench-scale *bromate yield* determinations

Experimental Protocols and Analytical Methods

Bench-scale ozonation was conducted in batch reactors with model and bulk waters containing bromide. NOM was extracted and purified from several bulk waters for use in model waters. NOM isolates were obtained from the following source waters:
- Teays Aquifer, IL (TYS)
- California State Project Water, CA (SPW)
- Lake Michigan, IL (LMW)
- Silver Lake, CO (SLW)
- Mississippi River, IL (MSR)

An orthogonal experimental matrix was employed by independently varying one parameter at a time from a set of baseline conditions (pH 7.5, Br$^-$= 400 µg/L, ozone dose = 3 to 6 mg/L, DOC = 3 mg/L except in Milli-Q water (DOC < 0.2 mg/L));. Pilot- and demonstration-scale ozonation studies were performed on bulk waters by the Contra Costa Water District (CCWD) and the Metropolitan Water District of Southern California (MWD), respectively. Kinetic time-series of ozone decomposition and bromate formation were monitored.

Batch Ozonation.   Aliquots of an ozone stock solution (~40 mg/L or ~0.8 mM), generated by bubbling gaseous ozone through ice-cooled NOM-free (Milli-Q) water, were transferred to a continuously-mixed reactor via a glass syringe (Figure 1). Impurities in the gas phase were removed in a gas washing bottle containing a pH 6 phosphate buffer prior to introduction of the ice-cooled NOM-free water via glass syringe. The transferred ozone dose equaled the applied ozone dose and was reported based upon final dilution of the sample with the ozone stock water added. The bench-scale reactor, a true-batch reactor, was constructed from a 500-mL graduated cylinder. An adjustable Teflon cover in the reactor rested upon the water surface, reducing volatile loss of ozone from solution, while a magnetic stir bar provided mixing. Time series samples were collected from a port near the base of the reactor for ozone and bromate analysis.

Ozone concentrations in NOM-free solutions were measured spectrophotometrically (Shimadzu UV-160U) at 254 nm using an extinction coefficient of 2950 $M^{-1}cm^{-1}$ that is experimentally equal to 3100 $M^{-1}cm^{-1}$ at 258 nm. In the presence of NOM, ozone concentrations were determined via the indigo method since NOM absorbs light between 250 and 260 nm and interferes with direct spectroscopic measurements (44).

Bromate concentrations were determined by ion chromatography (Dionex DX300). A borohydrate eluent (20 mM NaOH / 40.3 mM H_3BO_3) with an Ionpac guard (Dionex AG9) and analytical column (Dionex AS9-SC) provided low bromate detection (reportable as 2 µg/L (0.016 µM); the molecular weight of bromate is 127.9 µg/µmole). Silver filters (Dionex OnGuard Ag) were used to precipitate and remove chloride from bulk water samples prior to bromate analysis; our experience demonstrates that chloride can interfere with bromate detection. Bromide was also measured by ion chromatography, but with a carbonate eluent (2 mM Na_2CO_3 / 0.75 mM $NaHCO_3$) with a reportable detection limit of 10 µg/L ((0.125 µM); the molecular weight of bromide is 79.9 µg/µmole).

NOM Isolation and Purification. Varying chemical and size fractions of NOM were used in model solutions to investigate the effects of NOM properties on bromate formation. Hydrophobic organic acids (i.e., humic and fulvic acids) were fractionated using XAD-8 resin adsorption following established protocols; hydrophilic organic acids were fractionated with XAD-4 resin (45). Desalting using hydrogen saturated resin chromatography was also performed prior to lypholization. DOC recoveries ranged from 35% to 60% of the initial DOC. Lyophilized Suwannee River humic (SRH) and fulvic (SRF) acids were obtained from the International Humic Substances Society (Golden, CO).

Three different size fractions of NOM were obtained from tangential flow ultrafiltration-reverse osmosis (UF-RO) membrane separations after passing a sample through a 0.22 µm tangential flow filter (46). A 30K isolate represents NOM retained by a 30,000 dalton polysulfonate membrane. A 1K isolate represents NOM which passed the 30,000 dalton membrane but was retained by a 1,000 dalton cellulose acetate membrane. A RO isolate represents NOM that passed through both the 30,000 and 1,000 dalton membranes but was retained by a reverse-osmosis membrane. DOC recoveries were typically greater than 90%.

Dissolved organic carbon (DOC) was measured by low-temperature combustion at $680°$ C (Shimadzu TOC-5000). Model solutions containing NOM isolates were formulated from stock NOM solutions, Milli-Q water, and inorganic salts. All solutions were pH buffered with 1 mM phosphate.

Pilot- and Demonstration-Scale Ozone Contactors. The pilot-scale system, as shown in Figure 2, was a serial-operated multi-column (six-inch diameter) system with ozone introduced through diffusers. This facility is located at a CCWD utility; the source water

Ozone Generation System

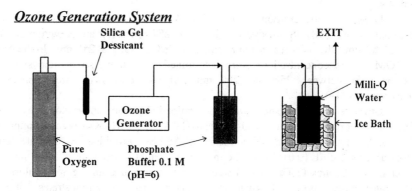

Batch Ozone Reactor

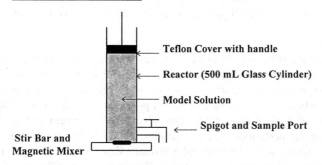

Figure 1 - Schematics of ozone stock solution generation and batch ozonation reactor

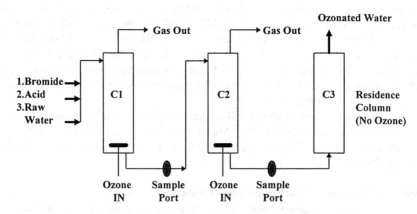

Figure 2 - CCWD Pilot-Scale ozone contactor

is from the Sacramento River Delta. Water was introduced countercurrent to ozone in the first contactor (C1) at 2.5 gallons per minute; the second column (C2) was likewise operated in countercurrent mode; the third column (C3) was simply a residence column without ozone application. Ozone could be applied solely to the first reactor or tapered to each of the first two reactors. The hydraulic residence time for each column, at the indicated flow rate, was 3 minutes. Samples for tracer, dissolved ozone, and bromate were taken at the outlet from each of the three columns, representing 3, 6, and 9 minutes of reaction. A pulse tracer test and hydrodynamic modeling indicated that this nonideal reactor could be simulated as N CSTRs in series, based upon its mode of ozone application.

The demonstration-scale system, considered to be representative of a full scale system since tests were conducted at 2.75 MGD (total hydraulic residence time was 12 minutes), is located at MWD. The contactor is shown in Figure 3; the baffled over/under contactor consists of six alternating counter-current and co-current chambers, with ozone introduction possible in each through a diffuser. Three sampling locations were selected with cumulative residence times of 3.7, 8.0, and 12 minutes which corresponded to the outlets of the three co-current chambers, identified as 1B, 2B, and 3B, respectively (Figure 3). Transferred ozone doses in both units were based on influent and effluent gas phase concentrations.

Results and Discussion

This section presents a discussion of experimental results and introduces the *bromate yield* concept. Additional applications of the *bromate yield* concept include evaluating bromate formation mechanisms and comparing bromate control strategies.

Conversion of Bromide to Bromate. Experimental results demonstrate an effect of inorganic and organic solutes on bromide conversion to bromate. Therefore, these results support a rationale for why *bromate yield* values may differ among various water supplies.

Effect of NOM Fraction. The physical and chemical characteristics of NOM influence the rate of ozone consumption and rate of bromide conversion to bromate. The kinetics of bromate formation under (true) batch ozonation, with all the ozone applied at time zero, varies in each of five NOM isolates from SPW water, with all other conditions constant (Figure 4). Bromate formation in NOM-free water (same inorganic matrix and temperature) exhibits an exponential increase (Figure 4). Bromate formation approaches an asymmetric level at approximately 120 minutes (not shown), corresponding to near complete ozone consumption. In comparison, the presence of NOM results in much more rapid bromate formation, and an asymmetric level is reached within 15 to 60 minutes. Again, the asymmetric level corresponds to complete ozone consumption. Since bromate formation ceases following dissipation of the ozone residual, all discussions focus on the period when oxidation by ozone occurs.

The hydrophobic and hydrophilic organic acid fractions (XAD-8 and XAD-4 isolates) form similar levels of bromate, and result in approximately 12% bromide

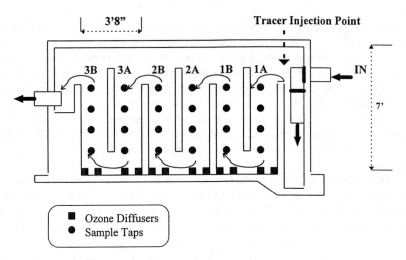

Figure 3 - MWD Demonstration-Scale ozone contactor

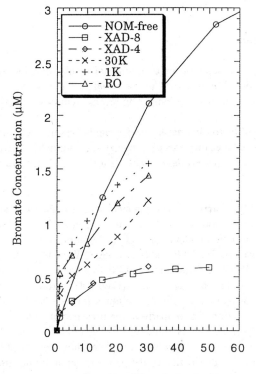

Figure 4 - Influence of NOM fraction from SPW on bromate formation (pH = 7.5, Ozone dose = 6 mg/L, [Br⁻] = 400 µg/L, DOC = 3 mg/L)

conversion to bromate upon complete ozone consumption (i.e., within 60 minutes) (from Figure 4). For a comparable NOM-free solution, approximately 75% bromide conversion to bromate occurs after complete ozone consumption. The UF-RO isolates exhibit bromide conversions to bromate of 28%, 36%, and 37% after complete ozone consumption in the 30K, 1K, and RO isolates, respectively. The presence of NOM and its physical-chemical *nature* can strongly influence the rate and extent of bromate formation.

Further evidence for the influence of NOM *nature* on bromate formation emerges through evaluation of similar NOM isolates from different sources. Three of the 1K isolates from SPW, MSR, and LMW exhibit bromide conversions to bromate of 36%, 13%, and 25%, respectively, under constant ozone doses (6 mg/L), DOC (3 mg/L), and inorganic matrix (pH = 7.5, $[Br]_0$= 400 µg/L) conditions.

Physical-chemical characteristics of NOM relate to the rate of reaction with molecular ozone and HO radicals (30, 31). Increasing NOM concentrations will increase the rate of ozone consumption (5, 29). Therefore, NOM *competes* with bromide specie (Br^-, $HOBr$, OBr^-, BrO_2^-, BrO_3^-, etc.) for oxidation by molecular ozone and HO radicals. This competition results in lower conversions of bromide to bromate than in NOM-free solutions. Furthermore, direct oxidation or substitution reactions between intermediate species of bromate formation (e.g., HOBr) with NOM may further reduce the overall conversion of bromide to bromate (46).

Influence of pH. Decreasing pH affects the rate and overall conversion of bromide to bromate. Bromate formation kinetics in NOM-free water exhibit an increase in both the rate and extent of bromide conversion to bromate with increasing pH; presented as bromate formation in Figure 5. Companion ozone consumption kinetic curves (Figure 6) indicate that increasing pH results in a more rapid loss of ozone. Overall the rate of bromate formation appears to be related to the rate of ozone consumption. In addition to varying the rate of bromate formation, the extent of bromide conversion to bromate varies with pH (Figure 5). At pH 6.5, only 10% of the bromide becomes oxidized to bromate in NOM-free waters (Figure 5). Significantly higher conversions (>50%) occur at both pH 7.5 and 8.5. Although the rate of bromate formation proceeds more rapidly at pH 8.5, similar extents of bromide conversion to bromate occur at pH 8.5 and 7.5 upon *complete* ozone consumption; due to varying rates of ozone consumption.

Experimental results under a wide range of DOC levels, NOM sources, and NOM fractions consistently exhibit a decrease in the extent of bromate formation from pH 7.5 to pH 6.5. Similar to NOM-free water, the presence of NOM can result in similar extents of bromate formation at pH 7.5 and 8.5, although the rate of bromate formation always increases with pH. This increased rate corresponds to increased rates of ozone consumption. Faster ozone consumption leads to higher rates of bromate formation.

A number of factors attribute to the observed pH effects. First, oxidation of aqueous bromine species occurs primarily with the dissociated form, hypobromite (OBr^-), rather than hypobromous acid (HOBr). Hypobromite concentrations decrease with lower pH levels. Consequentially, lower OBr^- concentrations result in lower extents of bromate

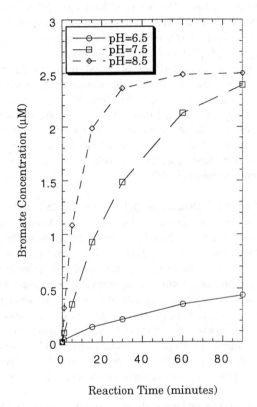

Figure 5 - Effect of pH on bromate formation in NOM-free water (Ozone dose = 3 mg/L, [Br⁻] = 400 µg/L)

formation. Second, HO radical concentrations decrease with pH depression. Theory (Equations 1-6) predicts equal *quantities* of HO radical formation at any pH, although the higher HO radical concentrations occur during more rapid ozone consumption (i.e., at higher pH levels). Since aqueous bromine oxidation depends upon both OBr^- and HO radical concentrations, factors (e.g., pH) affecting their concentrations will influence bromate formation.

Influence of Some Inorganic Constituents. The presence of some inorganic constituents or additives in water will influence the conversion of bromide to bromate by scavenging HO radicals, increasing HO radical concentrations, or interacting with intermediate aqueous bromine species. In addition to bromide concentration, we considered the effect of three major inorganic constituents: inorganic carbon (i.e., carbonate alkalinity), ammonia, and hydrogen peroxide. The percentage of bromide converted to bromate, with and without these three inorganic constituents, separately, vary in both rate and extent.

Carbonate and bicarbonate ions scavenge HO radicals, forming secondary radical specie (e.g., HCO_3 and CO_3^-) and stabilizing ozone consumption. The secondary radicals can oxidize bromide species, such as hypobromite. This selective oxidation can increase the efficiency of bromide conversion to bromate, resulting in an increased rate and extent of bromate formation. Furthermore, through inhibition of ozone consumption, more molecular ozone becomes available for oxidation of bromide to aqueous bromine. Higher aqueous bromine concentrations result in increased bromate formation (i.e., increase conversion of bromide to bromate) (Table I). In NOM-free water, bicarbonate addition (2 mM) only slightly increases the overall conversion of bromide to bromate; conversions increased from approximately 50% to 60% ($[Br]_0 = 400$ µg/L, pH=7.5, ozone dose= 3 mg/L). However, in waters containing NOM these differences are greatly magnified. For example, the XAD-8 isolate from SPW exhibits a bromide to bromate conversion of 6% (DOC= 3 mg/L, $[Br]_0 = 400$ µg/L, pH=7.5, ozone dose= 4.5 mg/L) as compared to 25% and 45% upon 2 mM and 10 mM bicarbonate addition, respectively, with all other conditions constant.

Ammonia addition can decrease the conversion of bromide to bromate during ozonation. In NOM-free water, bromide conversions to bromate decrease from 50% to 3% upon ammonia addition (Table I). Ammonia rapidly reacts with aqueous bromine to form bromamines. Molecular ozone and HO radicals then convert bromamines to bromide, resulting in a cyclical processes that consumes oxidant without forming bromate. This result in a *lag* in bromate formation (39). Once all the ammonia becomes depleted, bromine can be oxidized to bromate. Ammonia addition is less efficient in reducing bromide conversions to bromate in the presence of NOM (31).

Hydrogen peroxide addition to water can increase, have no effect, or decrease bromate formation depending largely upon pH conditions and the mass ratio (0.35 mg/mg) of hydrogen peroxide added with respect to the ozone dose. At pH 7.5 in NOM-free water, a stoichiometric hydrogen peroxide addition of 0.35 mg H_2O_2 / mg O_3 reduces the overall extent for conversion of bromide to bromate (Table I). A rapid rate of conversion upon hydrogen peroxide addition was noted in contrast to the experiment

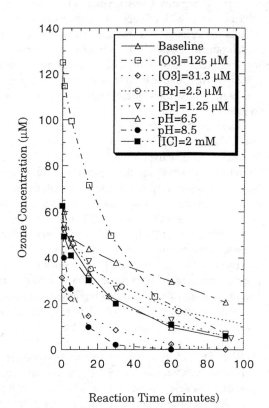

Figure 6 - Experimental results for ozone consumption in NOM-free water from an orthogonal matrix design (Baseline condition: pH = 7.5, Ozone dose = 62.5 μM (3 mg/L) , [Br⁻] = 5 μM (400 μg/L))

without hydrogen peroxide (Table I). Ozone decomposes very quickly in the presence of hydrogen peroxide. Increased rates of bromate formation result from higher HO radical concentrations during advanced oxidation applications (e.g., hydrogen peroxide and ozone addition simultaneously), but still relate inversely to the rate of ozone consumption.

Overall these three inorganic constituents, and numerous other solutes, can influence the extent of bromide conversion to bromate, for the same ozone doses. This arises from their role on influencing the oxidative driving force (i.e., molecular ozone and HO radical concentrations) and important intermediate specie (e.g., OBr⁻) concentrations.

Applying the Bromate Yield Concept. During batch ozonation, where ozone dosing occurs instantaneously at time zero, the amount of bromate formed at any time (t) directly relates to the amount of ozone consumed between time zero and time t. Plotting corresponding data pairs for the amount of ozone consumed (x-axis) and the amount of bromate formed (y-axis) during a given experiment results in a linear relationship. The slope of this line represents the *bromate yield*. A non-zero, positive, x-axis intercept of this line represents the initial consumption of ozone prior to any bromate being formed, and is termed the *ozone threshold*. The rate of bromate formation inversely tracks the rate of ozone consumption.

Analysis of bromate formation through the *bromate yield* concept facilitates comparison of water quality and water treatment variables in a simplified form. Increasing magnitudes in *bromate yields* (μmole BrO_3^-/μmole O_3) indicate higher conversions of bromide to bromate. Thus, bromate control strategies should decrease *bromate yields* in order to decrease the concentration of bromate after ozonation.

Demonstrative Examples of *Bromate Yields*. Upon batch ozonation of bromide containing waters, bromate forms rapidly as ozone decomposes. Figure 7 illustrates bromate formation in NOM-free water under a wide range of initial water quality and water treatment conditions. Ozone application occurs at time zero and bromate begins to form immediately. Companion ozone decomposition profiles for the same experiments illustrated in Figure 7 are shown in Figure 6. Non-quantitatively the shapes of both the bromate formation and ozone decomposition kinetic profiles are characterized by similar (inverse) exponential relationships.

Linear relationships can be developed for time dependent pairs of experimental bromate formed and ozone consumed data (Figure 8). Linear regressions through the data shown in Figures 6 and 7, result in high statistical correlations ($r^2 > 0.98$) for nearly all conditions. The slope of these lines represent individual *bromate yields* (μM BrO_3^- / μM O_3). Note that nearly all the linear regression lines in Figure 8 intercept the positive x-axis near zero. This indicates that bromate formation begins almost immediately upon ozone application, indicating very low ozone threshold values in NOM-free water. A summary of the *bromate yields* and correlation coefficients from linear regressions is presented in Table II. Bromate yields exhibit a wide range of values and good statistical fits for given sets of conditions.

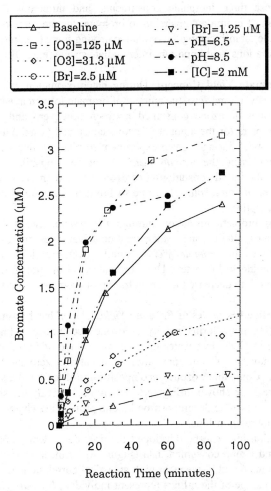

Figure 7 - Experimental results for bromate formation in NOM-free water from an orthogonal matrix design(Baseline condition: pH = 7.5, Ozone dose = 62.5 µM (3 mg/L) , [Br⁻] = 5 µM (400 µg/L))

Table I - Kinetic conversions of bromide to bromate in NOM-free water with different inorganic additives (pH=7.5, Br⁻= 5 μM, Ozone dose = 3 mg/L)

Reaction Time (minutes)	No Additives	Bicarbonate (2 mM)	Ammonia (5 μM)	Hydrogen Peroxide (31 μM)
0	0%	0%	0%	0%
1	1.7	2.5	1.9	30
5	7.0	7.2	3.0	33
15	19	21	3.1	36
30	30	33	3.3	--
60	43	48	--	--
90	48	55	--	--
24 hours	50	57	3.3	37

Table II - Bromate yields in NOM-free water
(Baseline Conditions: pH 7.5, Br⁻= 400 μg/L (5 μM), Ozone dose= 3 mg/L (62.5 μM))

Deviation from Baseline Condition	Bromate yield ($\mu M\ BrO_3^- / \mu M\ O_3$)	Average r^2
None	0.0423±0.0017	0.99
Ozone Dose = 125 μM	0.0295±0.0017	0.98
Ozone Dose = 31.3 μM	0.0351	0.99
Br⁻= 2.5 μM	0.0253	0.97
Br⁻= 1.25 μM	0.0101	0.98
pH = 6.5	0.0103±0.0006	0.95
pH = 8.5	0.0406±0.0005	0.96
Bicarbonate addition = 2 mM	0.05163	0.94

Table III - Bromate yields in NOM-free water and NOM isolates from SPW
(pH 7.5, Br⁻= 400 μg/L (5 μM), Ozone dose=6 mg/L (125 μM))

Solution Composition	Bromate yield ($\mu M\ BrO_3^- / \mu M\ O_3$)
Milli-Q	0.0295±0.0017
XAD8 Isolate	0.0053±0.0003
XAD4 Isolate	0.0049
RO Isolate	0.015
1K Isolate	0.016
30K Isolate	0.011

Note: Milli-Q contains less than 0.2 mg DOC/L, Solutions with NOM isolates contain 3 mg DOC/L.

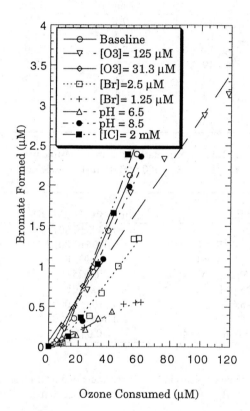

Figure 8 - *Bromate Yield* evaluation in NOM-free water from an orthogonal matrix design (Baseline condition: pH = 7.5, Ozone dose = 62.5 μM (3 mg/L) , [Br⁻] = 5 μM (400 μg/L))

A second example of *bromate yield* representations and computations are presented in Figure 9 and Table III for NOM-free water and five SPW isolates. Bromate formation data for this plot is also shown in Figure 4. While the highest *bromate yields* occur in the NOM-free water, the RO isolate from SPW exhibits the highest *bromate yield* among all the SPW isolates. The linear regressions illustrated in Figure 9 generally show a positive x-axis intercept. The ozone threshold values approach zero and represent less than 10% of the applied ozone dose. Even in the presence of NOM bromate formation begins rapidly, when a low bromate detection limit is available. The organic acid fraction isolates (i.e., XAD-8 and XAD-4) exhibit lower bromate yields than any of the UF/RO isolates. Overall, the conclusions drawn from bromate yield comparisons mirror those discussed for data in Figure 4. However, it requires just *one* bromate measurement, after all the ozone dissipated, to generate the bromate yield; whereas data interpretations similar to those drawn from Figure 4 require a complete kinetic time series of ozone and bromate measurements.

Numerical Representation of *Bromate Yields*. The bromate yield concept can serve as a powerful tool by substituting ozone consumed for reaction time as a critical component in bromate formation. Time only re-emerges as a parameter when bromate concentrations are calculated as a function of ozone decomposition. Bromate formation at any time during ozone consumption, or reactor hydraulic residence time (discussed in more detail later), relates to the *bromate yield* by the following relationship:

$$[BrO_3^-]_t = BY ([O_3]_0 - [O_3]_t) \qquad (7)$$

where $[BrO_3^-]_t$ is the bromate concentration (μM) at any time (t), BY is the *bromate yield* (μM BrO_3^- / μM O_3), $[O_3]_0$ is the initial transferred ozone dose, and $[O_3]_t$ is the ozone concentration at any time. If ozone decomposition follows first-order kinetics, represented by a first-order reaction rate coefficient k (time^{-1}), then the concentration of bromate at any time can be calculated by the following relationship:

$$[BrO_3^-]_t = BY \times [O_3]_0 (1 - e^{-kt}) \qquad (8)$$

Finally, the maximum bromate concentration ($[BrO_3^-]_{max}$) directly relates *bromate yield* to ozone dose by the following relationship:

$$[BrO_3^-]_{max} = BY \times (\text{Ozone dose } (\mu M)) \qquad (9)$$

Overall the concept of *bromate yield* provides a simple but powerful tool for interpreting experimental results since it represents an *apparent stoichiometry* between bromate formation and ozone decomposition. Comparisons of bromate yields for varying water quality and water treatment conditions can be used to quickly assess the rate and extent of bromate formation.

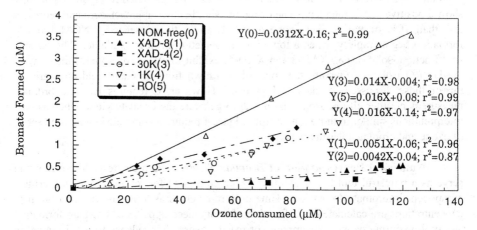

Figure 9 - *Bromate Yield* evaluation for NOM-free (MQW) water and several NOM fractions (DOC = 3 mg/L) from SPW (pH = 7.5, Ozone dose = 125 μM, [Br⁻] = 5 μM)

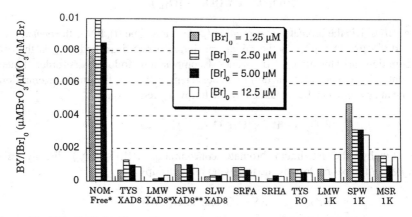

Figure 10 - Effect of initial bromide concentration on *normalized bromate yield* in NOM-free (MQW) water and several NOM fractions (DOC = 3 mg/L) from different sources, designated by acronyms (pH = 7.5, [Br⁻] = 5 μM, Ozone dose = 125 μM, *62.5 μM, or **94 μM)

Normalizing *Bromate Yield* to Initial Bromide Concentrations.   Previously data indicate that increasing bromide concentration results in an increase in the maximum bromate ($[BrO_3^-]_{max}$) concentration (Figure 8). In has been observed that the magnitude of initial bromide concentration variation equals the magnitude changes in $[BrO_3^-]_{max}$, all other conditions remaining constant. For example in the 1K isolate from SPW, doubling the initial bromide concentration from 2.5 μM to 5 μM approximately doubles the extent of bromate formation. Overall, an apparent linear correlation between initial bromide and bromate formation appears to exist over the range of bromide concentrations examined (100 to 1,000 μg/L). In NOM-free water, bromide may actually become limiting for bromate formation due to high conversions to bromate.

Comparison of *bromate yield* values (μM/μM) at varying initial bromide concentrations requires normalizing bromate yields to initial bromide concentrations. This is termed a *normalized bromate yield* and has the units of μM BrO_3^- per μM O_3 per μM Br^-. *Normalized bromate yields* for NOM-free water and solutions with NOM isolates are presented in Figure 10. These data do not suggest any strong trends in changes to the efficiency for conversion of bromide to bromate, supporting the observations that an approximately equal response in bromate formation occurs as initial bromide concentration varies.

Comparison of Bromate Control Strategies.   Both pH depression and ammonia addition can control bromate formation; hydrogen peroxide effectively reduces bromate formation at high pH (>8.0) levels but gives mixed results at intermediate (pH 7.5) and low (pH 6.5) levels. The effectiveness of applying bromate control strategies in NOM-free water or water containing organic acid fraction isolates indicates consistent trends (Figure 11). Since each of these six model solutions formed varying concentrations of bromate, the *bromate yields* were normalized to the *bromate yield* in each solution under baseline conditions (pH 7.5, $Br^- = 400$ μg/L, ozone dose = 3 to 6 mg/L, DOC = 3 mg/L except in Milli-Q water); the resulting values ($BY/BY_{baseline}$) permit comparison of different bromate control strategies with waters from different sources. A $BY/BY_{baseline}$ value of greater than 1.0 indicates higher *bromate yields* under a particular set of conditions than the *bromate yield* under baseline conditions. Therefore, a $BY/BY_{baseline}$ of less than 1.0 represents a reduction in bromate and constitutes a viable bromate control (minimization) strategy. Overall pH depression to approximately 6.5 emerges as the best bromate control strategy. Ammonia addition only slightly reduces bromate formation in the presence of NOM. Hydrogen peroxide addition can increase or decrease bromate formation.

Application of Bromate Yields to Continuous-Flow Ozone Reactors

The purpose of this section is three-fold: (i) to demonstrate bromate yield applications, (ii) to test our predictive tools at larger scales (pilot- and demonstration-scale), and (iii) to discuss how reactor design and operation may affect bromate formation. Both of these latter concepts will be addressed using the *bromate yield* approach.

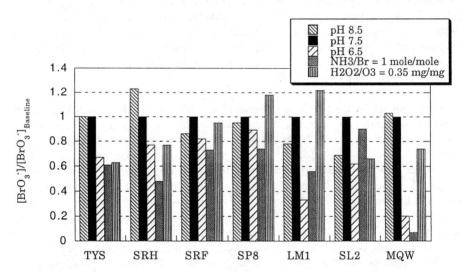

Figure 11 - Effect of bromate control strategies in several hydrophobic organic acid NOM fractions (represented by three letter acronyms) based on a *bromate yield* normalized to the *bromate yield* at pH 7.5

Laboratory work with batch reactors involves continuous mixing but no gas flow; ozone dosing occurs instantaneously at time-zero where the highest dissolved ozone concentration exists, and thereafter, ozone consumption progresses. Meanwhile, larger-scale contactors (pilot- and demonstration-scale) involve continuous-flow, mixing, and often more than one point of ozone application within a contactor, which is itself often compartmentalized.

Previous work suggests that bromate formation may be influenced by reactor/contactor type and hydrodynamics (mixing) (42). To examine these influences, the work presented herein incorporates laboratory first-order ozone decay constants (k), *bromate yield* estimates from batch experiments, and hydrodynamic descriptions of ozone contactors from pulse-input tracer studies.

Pilot-Scale Tests.   Four pilot-scale ozonation experiments were performed, with the matrix involving a transferred ozone dose of 4 mg/L applied to either the first reactor only, or split (2 mg/L) each to the first and second reactors (Figure 2); and two pH levels, an ambient (8.5) and reduced (6.5). Hereafter, Experiments 1, 2, 3, and 4 refer to the single-stage/high-pH, two-stage/high-pH, single-stage/low-pH, and two-stage/low-pH, respectively. At the time of the pilot tests, other water quality conditions were: DOC = 5.8 mg/L, $[Br]_0$ = 480 µg/L, and 12 °C.

Tracer tests were performed under both ozone application scenarios. For single-stage ozonation, simulations of the pulse-tracer input indicated equivalency to 2, 3, and 7 CSTRs in series (N) for C1, C2, and C3, respectively. For two-stage ozonation, the values of N were 2, 5, and 6. These values of N are *cumulative*; that is, the N specific for C2 was the composite effect of tracer flow through both C1 and C2.

Laboratory batch determinations of k, a first-order ozone decomposition rate constant, and *bromate yield* (BY) for a bulk Sacramento River Delta water sample obtained from CCWD at the time of pilot testing were made. Calculated batch k values equaled 1.14 and 0.14 min^{-1} for pH levels of 8.5 and 6.5, respectively. Corresponding *bromate yield* values were found to be 0.0101 and 0.0035 µM/µM.

Table IV summarizes ozone consumed predictions based on the application of a first-order ozone decay rate constant over the *cumulative* average residence time; 3, 6, and 9 minutes through C1, C2, and C3, respectively. Thus, these estimates correspond to plug-flow approximations (i.e., N=1) where the batch models are directly translated to the flow-through reactor results. The batch equation is as follows:

$$[O_3]_t = [O_3]_0 \exp(-kt) \tag{10}$$

and provides assessment for a steady state plug flow reactor if the time term t represents a *time of flow* through the reactor; thus, the equation provides an estimate of the time profile within a batch reactor or the spatial profile through a plug flow reactor. The cumulative concentration of ozone consumed relates to the difference between the ozone dose ($[O_3]_0$) and ozone concentration at any time ($[O_3]_t$):

$$[O_3]_{consumed} = [O_3]_0 (1 - e^{-kt}) \tag{11}$$

For tapered ozonation, estimates were assumed to be additive; that is, the decays of ozone applied to C1 and C2 were estimated independently with the two ozone demands, then added at either C2 or C3. As shown in Table IV, the predictions are fair, even through reactor hydrodynamics were not considered. An attempt was made to model ozone decay by taking into account the hydrodynamics of the system, based on:

$$[O_3]_t = [O_3]_0 / (1 + k(t/N))^N \qquad (12)$$

where N is equal to the number of CSTRs, t is the total hydraulic residence time, and t/N is equal to the residence time of each equal-sized CSTR. Predictions using this model were not much better than those shown in Table IV. It is important to note that, as N increases, the performance of a series of CSTRs approaches that of plug flow; thus, the plug flow approximations in Table IV have a reasonable basis. The CSTR-series approach would be expected to give better predictions for C1 than C2 and C3; for C1, N is only equal to 2, a value that clearly falls between plug flow and CSTR conditions. However, better predictions for ozone concentrations after C1 were not apparent.

Table V summarizes estimates of bromate concentrations, based on the *bromate yield* concept, where the measured values of ozone consumed in Table IV are coupled with:

$$[BrO_3^-]_{max} = BY \times [O_3]_{consumed} \qquad (13)$$

Predicted bromate concentration differ from measured values; a general pattern of underprediction can be seen. Once again, these estimates are made independent of the contactor hydrodynamics.

The experiments performed at CCWD allow a comparison of bromate formation under single-stage or two-stage ozonation. Concerns about minimizing bromate need to be balanced against the USEPA's CT requirements specified in the Surface Water Treatment Rule. Based on the tracer studies, t_{10} values were determined for the C1, C2, and C3 cells. The overall contactor CT was calculated as the summation of individual-cell CT's. These results are tabulated in Table VI, which also shows normalized bromate to CT values. Clearly, under high pH conditions, staged ozonation provides some benefit in (normalized) bromate reduction, while the converse is true under low pH conditions. Based on this one series of pilot experiments and the above discussions, chemical factors are likely to be more influential than hydrodynamic factors in bromate formation.

Demonstration-Scale Tests. The demonstration-scale system is located at MWD (Figure 3). Two tests were performed (Experiments 5 and 6, overall), with source water being the main variable: SPW and Colorado River Water (CRW). The water quality for SPW at the time of testing was: DOC = 3.8 mg/L, pH = 8.0, Br$^-$ = 320 µg/L, and T= 12° C. Corresponding conditions for CRW were 3.1 mg/L, 70 µg/L, pH=8.3, and T= 12° C, respectively.

Laboratory batch assessments were performed under identical conditions to those of demonstration-scale tests. The values of k were determined to be 0.63 and 0.15 min^{-1} for SPW and CRW, respectively. The *bromate yield* values for SPW and CRW were found to be 0.0042 and 0.0028 $\mu M/\mu M$ under ambient conditions.

Table VII summarizes ozone consumed predictions based on k and average residence times (i.e., Equation 11). The predictions for CRW are quite good whereas there is a significant disparity for SPW, possible due to a high estimate of k. Table VIII portrays results for bromate formation, based on the *bromate yield* concept (Equation 13), coupled with measured ozone consumption data from Table VII. The predictions shown for SPW are about 20% less than measured values; predictions for CRW were closer to measured values.

Recommended Bench-Scale Procedure for Determining *Bromate Yields*

The basis of the *bromate yield* concept is that an equal amount of bromate forms per unit concentration of ozone consumed. Therefore, knowing the batch transferred ozone dose and the final bromate concentration after all the ozone dissipates, the *bromate yield* can be estimated with just one pair of data points. The slope of a line between the experimental data pair (ozone dose and bromate formed) and x-y axis zero-intercept approximates the bromate yield. This analysis assumes that bromate forms immediately upon ozonation, which is a fair assessment based upon the low bromate detection capabilities and the low ozone thresholds found in this work.

The recommended procedure for obtaining *bromate yields* includes:

1. Divide water sample into 40 mL clean glass vials with Teflon septa
2. Adjust pH, inorganic matrix, and experimental conditions in each vial to prescribed conditions
3. Apply varying amounts of ozone stock solution to each vial, calculating the ozone dose
4. Measure bromate concentrations after 24 hours
5. Calculate *bromate yields* by dividing the measured bromate concentration by the ozone dose
6. Normalize *bromate yields* to initial bromide concentrations if necessary
7. Relate the water quality and experimental conditions to *bromate yields* with a spreadsheet or empirical multiple regression model as needed.
8. Bromate concentrations at intermediate ozone decomposition points can be estimated from Equation 13.

Conclusions

Based upon the linearity between ozone consumed and bromate formed during kinetic batch experiments, a technique, *bromate yield*, has been developed to assess the kinetics of bromate formation. Varying water quality and water treatment parameters will

Table IV - Measured and predicted values of ozone consumed for CCWD Pilot Plant, based on rate constant, k, and average hydraulic residence times (Ozone dose = 4 mg/L)

Exp. #	k (min^{-1})	Measured [O$_3$]$_{consumed}$, μM (mg/L)			Predicted [O$_3$]$_{consumed}$, μM (mg/L)		
		C1	C2	C3	C1	C2	C3
1	1.14	67 (3.2)	81 (3.9)	83 (4.0)	81 (3.9)	83 (4.0)	83 (4.0)
2	1.14	38 (1.8)	63 (3.0)	81 (3.9)	40 (1.9)	81 (3.9)	83 (4.0)
3	0.14	50 (2.4)	58 (2.8)	71 (3.4)	29 (1.4)	48 (2.3)	60 (2.9)
4	0.14	21 (1.0)	52 (2.5)	58 (2.8)	15 (0.7)	38 (1.8)	52 (2.5)

Table V- Measured and predicted values of bromate formed for CCWD Pilot Plant, based on *bromate yield* (BY) and [O$_3$]$_{consumed}$

Exp. #	BY (μM/μM)	Measured BrO$_3$ Formed , μM (μg/L)			Predicted BrO$_3$ Formed, μM (μg/L)		
		C1	C2	C3	C1	C2	C3
1	0.0101	0.81(104)	0.95 (121)	0.95 (121)	0.68 (87)	0.82(105)	0.84 (108)
2	0.0101	0.25 (32)	1.01 (129)	1.01 (129)	0.38 (49)	0.64 (82)	0.82 (105)
3	0.0035	0.24 (31)	0.29 (37)	0.29 (37)	0.18 (23)	0.20 (26)	0.25 (32)
4	0.0035	0.16 (21)	0.41 (52)	0.41 (52)	0.07 (9)	0.18 (23)	0.20 (26)

Table VI - CT evaluation of CCWD Pilot Experiments

Exp. #	pH	Ozone Application	CT (mg/L-min)	Bromate (μg/L)	Bromate/CT
1	8.5	one-stage	1.4	121	86
2	8.5	two-stage	2.3	129	56
3	6.5	one-stage	6.5	37	5.7
4	6.5	two-stage	6.7	52	7.8

Table VII - Measured and predicted values of ozone consumed for MWD Demonstration Plant, based on rate constant, k, and average hydraulic residence times (Ozone dose = 3.9 mg/L and 2.3 mg/L for SPW and CRW, respectively)

Exp. #	k (min^{-1})	Measured [O$_3$]$_{consumed}$, μM (mg/L)			Predicted [O$_3$]$_{consumed}$, μM (mg/L)		
		1B	2B	3B	1B	2B	3B
5	0.63	46 (2.2)	56 (2.7)	63 (3.0)	73 (3.5)	81 (3.9)	81 (3.9)
6	0.15	23 (1.1)	29 (1.4)	33 (1.6)	21 (1.0)	33 (1.6)	40 (1.9)

Table VIII - Measured and predicted values of bromate formed for MWD Demonstration
Plant, based on *bromate yield* (BY) and $[O_3]_{consumed}$

Exp. #	BY (μM/μM)	Measured BrO$_3$ Formed , μM (μg/L)			Predicted BrO$_3$ Formed, μM (μg/L)		
		1B	2B	3B	1B	2B	3B
5	0.0042	NA	NA	0.42 (54)	0.19 (24)	0.24 (31)	0.34 (43)
6	0.0028	0.05 (6)	0.07 (9)	0.09 (12)	0.06 (7)	0.08 (10)	0.09 (12)

influence *bromate yields* by changing the efficiency of bromide conversion to bromate. This type of technique appears to useful in assessing both chemical and hydrodynamic bromate control techniques.   The overall extent of bromate formation is critical to compliance with regulatory limits, but the kinetic rate of bromate formation may be crucial in ozone reactor design, ozone dose, point of application, and other operational considerations.

Acknowledgments

This research was funded by the American Water Works Research Foundation (Project Managers: Jeff Oxenford and Robert Allen), and the assistance of the Project Advisory Committee (Werner Haag, Robert Powell, Philip Singer, and James Symons) are greatly appreciated.   The authors also thank Mohammed Siddiqui, Kenan Ozekin, Larry McCollum (CCWD), Stuart Krasner (MWD), and Urs von Gunten for their assistance and input.

References

(1) Reckhow, D.A., Singer, P.C., and Malcolm, R.L. *Environ. Sci. and Tech.*, 1990, 24:11:1655-1664.
(2) McGuire, M.J. and Meadow, R.G. *J. Amer. Wat. Works Assoc.*, 1988, January, 61-68.
(3) Collins, M.R., Amy, G.L., King, P.H. *J. Environ. Eng.* 111:6:850-864.
(4) Weinberg, H.S., and Glaze, W.H., ., Disinfection By-Products in Water Treatment: The Chemistry of Their Formation and Control (eds. R.A. Minear and G.L. Amy), Lewis Publishers, New York, 1995, 165-186.
(5) *Ozone in Water Treatment: Application and Engineering*, Langlais, B., Reckhow, D.A., and Brink, D.R., Eds., Lewis Publishers, 1991.
(6) Haag, W.R. and David Yao, C.C., *Environ. Sci. and Tech.*, 1992, 26:5:1005-1013.
(7) Hoigne, J., and Bader, H., *Wat. Res.*, 1985, 17:173-183.
(8) Hoigne, J., and Bader, H., *Wat. Res.*, 1985, 17:185-194.
(9) Hoigne, J., Bader, H., Haag, W.R., and Staehelin, J., *Wat. Res.*, 1985, 19:8:993-1004.
(10) Reckhow, D.A., Singer, P.C., and Trussell, R.R., "Ozone as a Coagulant Aid", presented at the 1986 AWWA National Conference Sunday Seminar on Ozonation: Recent Advances and Reserach Needs, June.
(11) Ferguson, D.W., McGuire, M.J., Koch, B., Wolfe, R.L., and Aieta, E.M., *J. Amer. Wat. Works Assoc.*, 1990, April, 181-191.

(12) Westerhoff, P., Amy, G., Song, R., and Minear, R., Disinfection By-Products in Water Treatment: The Chemistry of Their Formation and Control (eds. R.A. Minear and G.L. Amy), Lewis Publishers, New York, 1995, 187-206.

(13) Haag, W.R. and Hoigne, J., Environ. Sci. and Tech., 17:5:261-267.

(14) von Gunten, U., and Hoigne, J., Environ. Sci. and Tech., 28:1234-1242.

(15) Siddiqui, M.S., and Amy, G.L., J. Amer. Wat. Works Assoc., 1993, 85:1:63-72.

(16) Schechter, D.S., and Singer, P.C., Ozone Sci. and Engin., 1995, 17:53-69.

(17) Siddiqui, M., Amy, G., Ozekin, K., Zhai, W., and Westerhoff, P., J. Am. Wat. Works Assoc., 1994, 86:10:81-96.

(18) Ono, U., Somiya, I., and Mohri, S., Ozone Sci. and Tech., 1994, 16:5:443-453.

(19) Wilbourn, J., Inter. Water Supply Assoc. Conference: Bromate and Water Treatment (Paris, France), 1993, November 22-24, pp. 3-12.

(20) Masschelein, W.J. Inter. Water Supply Assoc. Conference: Bromate and Water Treatment (Paris, France), 1993, November 22-24, pp. 13-23.

(21) Krasner, S.W., Gramith, J.T., Coffey, B.M., and Yates, R.S., J. Amer. Wat. Works Assoc., 1993, 85:1:73-81.

(22) Hoigne, J., and Bader, H., Wat. Res., 1976, 10:377-386.

(23) Peyton, G., and Bell, O., Ozone in Water and Wastewater Proceedings of the Eleventh Ozone World Congress, Volume 2., San Francisco, 1994, S-20-1-4.

(24) Haag, W.R., and David Yao, C.C., Ozone in Water and Wastewater Proceedings of the Eleventh Ozone World Congress, Volume 2., San Francisco, 1994, S-17-119-125.

(25) Buxton, G.V., Greenstock, C.L., Helman, W.P, and Ross, A.B., Critical Review of Rate Constants for Reactions of Hydrated Electrons, Hydrogen Atoms, and Hydroxyl Radicals in Aqueous Solution, Journal of Physical Reference Data, 1988, 17:2:513-851.

(26) Reckhow, D.A., Legube, B., and Singer, P.C., Wat. Res., 1986, 20:8:987-998.

(27) Chelkowska, K., Grasso, D., Fabian, I., and Gordon, R., Ozone Sci. and Eng., 1992, 14:33-49.

(28) Staehelin, J. and Hoigne, J., Environ. Sci. and Tech., 1985, 19:12:1206-1213.

(29) Westerhoff, P., Song, R., Amy, G., and Minear, R., "Applications of Ozone Decomposition Models", accepted for publication by Ozone Sci. and Eng., May 1996.

(30) Westerhoff, P., Aiken, G., Amy, G.L., and Debroux, J., "Reactivity of NOM from Aquatic Systems with Molecular Ozone and Hydroxyl Radicals", submitted to Environ. Sci. and Tech., 1996.

(31) Westerhoff, P., Ozone Oxidation of Bromide and Natural Organic Matter, Ph.D. dissertation, Depart. of Civil, Env., and Arch. Eng., Univ. of Colorado, Boulder, CO.

(32) Gurol, M. and Singer, P, Environ. Sci. and Tech., 1982, 16:7:377-383.

(33) Westerhoff, P., Ozekin, K., Siddiqui, M., and Amy, G., "Kinetic Modeling of Bromate Formation in Ozonated Waters: Molecular Ozone versus HO Radical Pathways", American Water Works Association Annual Conference Procedings, June 1994, New York City.

(34) Amy, G., Siddiqui, M., Debroux, J., Zhai. M., and Odem, W., Bromide Occurrence: national Bromide Survey, American Water Works Association Project, Denver, CO, 1995.

(35) Kruitohoff, J.C., and Meijers, R.T., Inter. Water Supply Assoc. Conference: Bromate and Water Treatment (Paris, France), 1993, November 22-24, pp. 125-133.

(36) Kisvirta, L. *Inter. Water Supply Assoc. Conference: Bromate and Water Treatment (Paris, France)*, 1993, November 22-24, pp. 151-156.
(37) Legube, B., Bourbigot, M.M., Bruchet, A., Deguin, A., Montiel, A., and Matia, L., *Inter. Water Supply Assoc. Conference: Bromate and Water Treatment (Paris, France)*, 1993, November 22-24, pp. 135-150.
(38) von Gunten, U., and Hoigne, J., ACS Div. of Env. Chem. National Meeting, Preprints of Papers, August 1995, Chicago, IL, 35:2:661-664.
(39) von Gunten, U., and Hoigne, J., <u>Disinfection By-Products in Water Treatment: The Chemistry of Their Formation and Control</u> (eds. R.A. Minear and G.L. Amy), Lewis Publishers, New York, 1995, 187-206.
(40) Haag, W.R., and Hoigne, J., *Inter. Ozone Assoc.*, 1984, 6:103-114.
(41) von Gunten, U., Hoigne, J., and Bruchet, A., 12th World Congress of IOA, May 15-18, Lille, France, 1995, Vol.1, 17-25.
(42) Siddiqui, M.S., Amy, G.L., Ozekin, K., and Westerhoff, P., *Ozone Sci. and Eng.*, 16:2:157-178.
(43) Amy, G., Siddiqui, M., Ozekin, K., and Westerhoff, P., International Water Supply Association Workshop: Bromate and Water Treatment, Paris, Nov. 22-24, 1993.
(44) Bader, H., and Hoigne, J., *Wat. Res.*, 1981, 15:449-456.
(45) Aiken, G.R., McKnight, D.M., Thorn, K.A., and Thurman, E.M., *Org. Geochem.*, 18:4:567-573.
(46) Westerhoff, P., Song, R., Amy, G., and Minear, R., "Affect of NOM During Ozonation of Bromide Containing Waters", in preparation for submission to *J. Amer. Wat. Works Assoc.*, 1996.

Chapter 19

Ion-Chromatographic Determination of Three Short-Chain Carboxylic Acids in Ozonated Drinking Water

Ching-Yuan Kuo, Hsiao-Chiu Wang, Stuart W. Krasner, and Marshall K. Davis

Water Quality Division, Metropolitan Water District of Southern California, 700 Moreno Avenue, La Verne, CA 91750–3399

A direct ion chromatography method was applied to the determination of short-chain carboxylic acids (acetic, formic, and oxalic) in drinking water treated with ozone. These organic acids were separated on an anion exchange column (Ionpac AS11; Dionex Corp., Sunnyvale, CA), using a 50-μL sample loop. Samples were collected from the ozone pilot plant of the Metropolitan Water District of Southern California to study the formation of carboxylic acids and their removal by two different types of biologically active filters. Preliminary results indicate that the carboxylic acid concentrations out of the ozone contactor were the greatest for oxalic acid (~380 μg/L), followed by formic (~130 μg/L) and acetic (~60 μg/L) acid, and that the sum of the three acids on a carbon basis was ~160 μg/L. Approximately 80 percent of these carboxylic acids were removed during biofiltration. Aldehydes and assimilable organic carbon (AOC) were also measured from the same batch of samples, and the removal percentages were similar to those obtained in the carboxylic acid analyses. The percent fractions of AOC at the ozone effluent for combined carboxylic acids and combined aldehydes were 36.9 and 3.7 percent, respectively.

Because of concerns about health effects and regulations governing chlorination by-products, the Metropolitan Water District of Southern California (MWDSC) and many other water agencies and utilities are actively investigating alternative disinfectants, such as ozone, to replace chlorine. The various by-products produced by the application of ozone are being extensively studied (1). Furthermore, the ozonation of natural organic matter in the source water can produce smaller and more biodegradable organic compounds, collectively referred to as assimilable organic carbon (AOC), which can be used as nutrients for bacteria and may cause microbial regrowth problems in a water distribution system (2-4).

0097–6156/96/0649–0350$15.00/0

AOC is measured by the maximum bacterial growth in the water sample and is compared with the previously established growth curves in an acetate or oxalate substrate (*2,4*). Although AOC is a good indicator for the biological regrowth potential of the water, nine days are required to determine biofiltration process removal efficiencies, and a more rapid test is needed. Schechter and Singer (*5*) indicated that there was a strong correlation between the AOC and aldehydes produced by ozone and that aldehydes could be utilized as potential surrogates for AOC. However, the total amount of aldehydes accounts for only a small percentage of the AOC. Other fractions identified were oxoacids and carboxylic acids (*6*), but the oxalic acid was not analyzed in the study. Thus, the development of a low-level analytical method for low-molecular-weight carboxylic acids, including oxalate, should provide additional information for the study of the formation and removal of ozonation DBPs and the biostability of the treated water.

Gas chromatography (GC), ion-exclusion chromatography (IEC), and ion chromatography (IC) have been used to determine organic acids. The GC method, however, involves time-consuming procedures such as extraction and derivatization (*6*), and the IEC method lacks the needed sensitivity to measure concentrations at the required µg/L level, as well as the ability to resolve oxalate from other components of the sample (Joyce, R., Dionex Corp., Sunnyvale, CA; personal communication, 1994). A few studies have utilized IC to measure carboxylic acids; however, they lack either the detailed analytical conditions (*7*) or the capacity to measure all three carboxylic acids simultaneously (Peldszus, S., Department of Civil Engineering, University of Waterloo; personal communication, 1995). The initial component of this research was the development of an IC method that (a) is a direct technique, with minimum sample pretreatment; (b) can measure acetate, formate, and oxalate in natural water in one single injection; and (c) achieves an acceptable low-µg/L reporting limit. The method used for low-level carboxylic acid determination is described below.

Following development of the IC method, parallel analyses of AOC (*4*), aldehydes (*8*), and carboxylic acids were conducted to establish baseline information on the formation of these components after ozonation and on their removal by biologically active filters.

Experimental Approach

Materials.
    Reagents. Ultrapure (double deionized) water (Super Q; Millipore Corp., Bedford, MA) was used to prepare eluent, stock solutions, and calibration standards. Standards were prepared using reagent-grade (>99 percent pure) carboxylic acids in the form of their sodium salts. Formate and acetate were obtained from Aldrich Chemical Co. (Milwaukee, WI). Oxalate was obtained from Spectrum Chemical Corp. (Gardena, CA). An anti-biodegradation agent was prepared by dissolving 1,000 mg of reagent-grade mercuric chloride ($HgCl_2$) in 100 mL of ultrapure water.
    Stock and Working Calibration Solutions. Individual stock carboxylic acid (acetate, formate, oxalate) solutions at 10,000 mg/L were prepared in ultrapure water and stored in amber glass bottles at 4°C. An intermediate stock standard solution of mixed acids (at concentrations of 10 mg/L each) was prepared weekly from the

individual stock solutions. Because of their unstable nature, working calibration standards were prepared daily at concentration levels of 20, 50, 100, 250, and 500 µg/L from the 10-mg/L intermediate stock solution.

IC and Data System. Analyses were performed on an ion chromatograph (model 2020; Dionex) equipped with an autosampler, an advanced gradient pump, a pulsed electrochemical detector operated in the conductivity mode, and an anion self-regeneration suppressor (ASRS). The separation of the carboxylic acids was achieved by using an Ionpac AG11 guard column and an Ionpac AS11 analytical column (both from Dionex). A 50-µL injection loop was installed on the injection port. A software package (AI-450; Dionex) was used to control all operations of the IC system modules and to integrate the data signal acquired by the interface module.

Eluent. A certified sodium hydroxide (NaOH) solution at 50 percent (on a weight basis), with low carbonate content, was obtained from Fisher Scientific (Tustin, CA). The working eluents were prepared as follows: The first eluent bottle contained a low concentration of NaOH (approximately 1.0 mM) prepared by pipetting and mixing 0.105 mL of the 50-percent NaOH solution into 2,000 mL of ultrapure water that had previously been purged with ultrapure helium gas at a head pressure of 8 psi for 10 min to remove dissolved carbon dioxide. A second NaOH eluent at approximately 100 mM was prepared by pipetting and mixing 10.5 mL of 50-percent stock NaOH solution into 1,984 mL (by removing 16 mL from 2,000 mL) of ultrapure water that had been purged with ultrapure helium gas for 10 min. A third eluent bottle filled with ultrapure water to be mixed in-line with the above two NaOH solutions to create a more diluted NaOH eluent during the isocratic and gradient analysis modes. An anion trap column (ATC-1; Dionex) was installed in-line to the gradient pump outlet and prior to the injection valve to remove anion impurities from the eluent before it entered the injection port and columns.

Anion Self-Regenerating Suppressor (ASRS). Baseline suppression was achieved by using an ASRS to yield a stable baseline with a low-conductivity output. The ASRS ran in the external water mode with the controller set at unit #1, which is equivalent to an electric current of 50 mA. The 4-L water reservoir was filled to capacity with ultrapure water and purged with helium gas before starting the overnight analysis to ensure retention of sufficient water to prevent damage to the ASRS.

Sample Preparation.

Ozone Quenching Agents. Residual ozone had to be removed from (or quenched out of) the sample from the ozone contactor to prevent further oxidation and formation of additional DBPs during storage. Preliminary testing results for diethylamine (9) and potassium iodide (considered as ozone-quenching agents) showed inconsistent recoveries on the spiked values of the three carboxylic acids, and the use of diethylamine or potassium iodide as a quenching agent was not pursued further.

Instead, an alternative procedure--very gentle and slow aeration of the sample bottle--was used in this study to remove residual ozone. This was done by leaving some headspace in the sample bottle and air-stripping the entire bottle with a capillary-tip glass pipette for only 1-2 min. The gentle aeration did not produce any apparent interference, nor did it strip out the carboxylic acid analytes of interest.

Anti-Biodegradation Agents. Based on early observations in this study, the carboxylic acids were not stable and degraded rapidly if not preserved. The spike recoveries of the samples without an anti-biodegradation agent were typically at low percentage levels (sometimes even 0) as a result of biological degradation. The IC manufacturer (Dionex Corp.) recommended using chromate at 10 mg/L to stabilize the sample and calibration standards from bacterial degradation at room temperature (*10*). However, the use of chromate at 10 mg/L severely interfered with the measurement of oxalate at low concentrations. The chromate peak overshadowed the tiny oxalate peak, which completely disappeared from the chromatogram.

Chloroform, chloramine, and $HgCl_2$ were also tested as anti-biodegradation agents during this study. $HgCl_2$ had been used successfully as an anti-biodegradation agent in the aldehydes method outlined by Sclimenti and colleagues (*8*). The $HgCl_2$ solution was added at 10 μL per 10 mL of sample. The chloramine was added to the sample at a final concentration of 1 mg/L. The chloroform (adding 50 μL at 0.1 percent to 50 mL of sample) was used by the University of Waterloo (Peldszus, S., Department of Civil Engineering; personal communication, 1995) in the analysis of organic acids by IC. The $HgCl_2$ and chloramines, but not the chloroform, showed satisfactory results based on spike recoveries and the sample stability test. Information recently received from the University of Waterloo (Peldszus, S., personal communication, July 3, 1996) indicates that chloroform should be added at 50 μL per 50 mL of sample. $HgCl_2$ was chosen for the IC method because of concern about possible reactions of chloramine with the organic matter in water. Because $HgCl_2$ is a very toxic compound, care must be taken in its handling and disposal.

Solid-Phase Sample Pretreatment. A hydrogen cartridge (OnGuard-H⁺, P/N 39596; Dionex) was attached in-line between the autosampler outlet and the sample injection loop. The purpose of using the hydrogen cartridge was to lower the sample pH (without adding acids) and to remove cations (primarily the mercury used as a preservative) from the sample matrix.

The cartridge has male and female Luer connectors on each end. Two adapters were needed to fit the Luer connectors to the pipe threads so that the hydrogen cartridge could be installed in-line to the autosampler outlet. A fresh cartridge was attached in-line for each daily analysis.

Analysis and Maintenance.

Analysis Conditions. All samples for regular analysis were collected in 60-mL brown glass bottles, with mercuric chloride preservative added before sample collection. The bottles were partially filled, but with some headspace left so the sample could be aerated with compressed laboratory air to remove residual ozone. The IC system setup and analytical conditions are summarized in Table I.

Acid-Washing of ASRS. Approximately 5 mL of 0.5 N sulfuric acid was injected through the eluent outlet of the ASRS and soaked for a few minutes to clean the ASRS. Both the eluent inlet and outlet lines were then connected back to the ASRS. Next, the system was stabilized by pumping the regular-strength eluent (0.1 mM NaOH) through the ASRS. This ASRS acid-washing step was performed before each daily analytical run.

Table I. IC System Conditions

IC system:	Dionex 2020 IC
Autosampler:	ASM-2
Analytical column:	Ionpac AS11
Guard column:	Ionpac AG11
Detector output range:	0.01 μS
Anion trap column:	Ionpac ATC-1
Gradient pump:	GPM-2 at 1 mL/min
Stock eluent:	Eluent 1--1 mM NaOH
	Eluent 2--100 mM NaOH
	Eluent 3--ultrapure water
Working eluents:	Isocratic mode--5 min, 0.1 mM NaOH

Eluent 1	10%	
Eluent 3	90%	

Gradient mode--30 min, linear increase
　　　from 0.1 mM to 10 mM NaOH

	Initial stage	Final stage
Eluent 1	10%	0%
Eluent 2	0%	10%
Eluent 3	90%	90%

Stabilizing mode--15 min, 0.1 mM

Eluent 1	10%
Eluent 3	90%

Suppression:	ASRS in external water mode @ 7 psi
	Controller current at setting #1; 50 mA
Detection:	Conductivity detector
Sample size:	50 μL
Sample pretreatment:	On-line H+ cartridge
Suppressor cleaning:	0.5N H$_2$SO$_4$

Cleaning of ATC-1. The ATC-1 column was cleaned with 100 mM NaOH when the system started to show high-conductivity output at the time of stabilization of the system baseline. The ATC-1 was disconnected from the system before cleaning to prevent this high-concentration NaOH eluent from being introduced into the columns and the suppressor.

Anion Interference. Fluoride is a potential interference in acetate analyses. During the initial period of method development, it was discovered that the fluoride and acetate peaks were located very close to each other, and both appeared at the early stage of the chromatogram (Figure 8). To ensure good separation of fluoride and acetate, the eluent concentration was lowered so that fluoride could be present in the water at up to 0.5 mg/L without interfering with acetate recovery under the current system configuration and analytical procedures.   As Figure 8 also indicates, the chloride peak appeared farther away from the acetate and formate, minimizing the interference of chloride.   The sulfate peak was too well separated from the oxalate peak to be considered a possible interference.   MWDSC's source waters contain fluoride in the range of 0.1 to 0.4 mg/L, chloride at 40 to 110 mg/L, and sulfate at 30 to 290 mg/L.

   Stability Tests.   Stability tests were conducted over a period of up to 3 weeks (testing was stopped when a component showed a significant decrease in concentration).   Three types of sample matrices (ultrapure water, raw water, and ozonated water) were prepared for the stability tests.   Each sample matrix was treated with and without $HgCl_2$.   Samples were stored in a refrigerator at 4°C when not being analyzed and were brought to room temperature before being prepared for analysis.

   The carboxylic acids, at concentrations of 50 µg/L each, were prepared in ultrapure water.   The raw-water sample for the stability test was spiked with 50 or 100 µg/L of the carboxylic acids to increase the background amounts of the three carboxylic acids.   In addition, an ozonated water (with ambient levels of the carboxylic acids) was evaluated with and without $HgCl_2$ preservation.

   Pilot-Plant Studies.   MWDSC has evaluated ozone for primary disinfection, followed by conventional treatment and biologically active filters to remove the AOC (*11*).   At the pilot plant, the ozonated water was diverted into two types of filter columns that utilized either granular activated carbon (GAC) or anthracite coal as the primary filter medium.   Both filter columns used 6-in. diameter glass pipe with 20 in. of filter media, 6 in. of sand, and 4 in. of gravel.   The flow was 1 gal/min per filter.

   The ozone dose was approximately 2 mg/L.   The total organic carbon concentration of the incoming raw water was approximately 3 mg/L.   The pH of the raw water was slightly below 8, and the water temperature was approximately 24°C.   The turbidity varied from 1.5 to 17.0 NTU during the testing period.

   Some limited and preliminary sampling of the pilot plant was performed for the carboxylic acids.   In addition, samples were collected for the determination of AOC and aldehydes.   The AOC was determined by a modification of the van der Kooij procedure (*4*), and the aldehydes were determined by a GC method (*8*).

Results and Discussion

Analytical Parameters.

   Stability Tests.   In ultrapure water, the carboxylic acids were stable throughout the 17-day testing period (either with or without $HgCl_2$) except for acetate in the unpreserved sample, which decreased some time after the fourth day of testing (Figures 1 and 2).   In the unpreserved raw water (without $HgCl_2$), the oxalate was not degraded throughout the 3-day test period (Figure 3), whereas the acetate and formate exhibited a significant decrease even on the first day of testing (Day 0).   The total analytical run time for a 10-sample analysis is typically 18 hours; obviously, the carboxylic acid reductions during this 18-hour period can be very significant.   This indicates that a preservative is a must for carboxylic acid analysis, even if the samples are analyzed

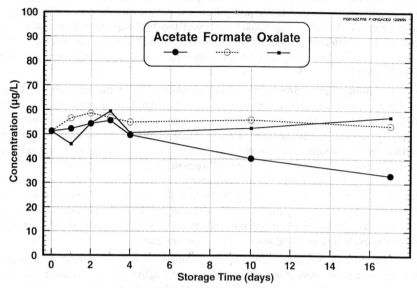

**Figure 1. Stability of Carboxylic Acids in
Ultrapure Water without Preservation**

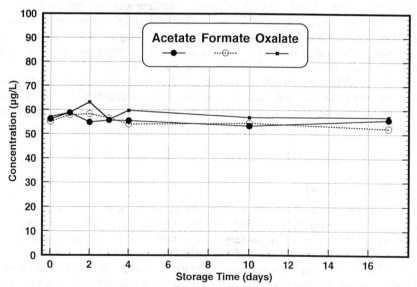

**Figure 2. Stability of Carboxylic Acids in
Ultrapure Water Preserved with HgCl₂**

immediately after collection. The raw-water sample preserved with HgCl$_2$ was stable for the full 17-day test period (Figure 4). The ozonated sample without preservation showed a decrease in acetate concentration on the third day of testing, whereas the other two carboxylic acids were not degraded during the 3-day test period (Figure 5). Again, the preserved ozonated sample showed excellent stability throughout the 17-day test period (Figure 6).

Solid-Phase Sample Pretreatment. Sample pretreatment with a hydrogen cartridge sharpened the peaks of the three carboxylic acids and allowed the separation of fluoride from acetate.   Figures 7 and 8 show two chromatograms of the same sample before and after hydrogen cartridge pretreatment.

Positive Analyte Area Value Drift. A positive analyte area value drift-- defined as a scenario in which the area values of the analytes gradually and positively increase over time (Figure 9)--was observed in the early methods development. The drift rate of the area values for acetate and formate was approximately twice as high as that for oxalate. This phenomenon was overcome by introducing the acid-washing step for the ASRS before an analytical run. Neither pre-stabilization of the IC system with the high-concentration NaOH eluent nor purging with high-concentration eluent after the oxalate peak was beneficial in diminishing the area value drift. In fact, the introduction of the high-concentration eluent to the ASRS might have been the primary cause of the positive analyte area value drift.

Quality Assurance.

Calibration Curves. Small amounts of acetic and oxalic acids were often observed in the reagent water; therefore, a blank was included in the daily run so that blank-adjusted area values of the standards could be used to construct the calibration curve. The calibration curves of the three carboxylic acids were linear over the range studied (i.e., 20 to 500 µg/L). The regression coefficient of the calibration curve was usually 0.999 or better.

Method Detection Limits (MDLs) and Minimum Reporting Limits (MRLs). MDLs were determined by measuring the 20-µg/L standard seven times under the same analytical conditions. The mean and the standard deviation (S.D.) of the seven replicates for acetate, formate, and oxalate were 20.9 ± 1.1, 23.3 ± 1.1, and 20.1 ± 0.7 µg/L, respectively. The MDLs were calculated at a 95-percent confidence level based on tripling the S.D. value. The calculated MDLs were 2-3 µg/L for the three carboxylic acids. Currently, the MRL is 15 µg/L for all three acids as a result of detection of the analytes in the blank. The MRL of 15 µg/L for formate is probably overconservative.

Accuracy and Precision. The study determined the accuracy and precision of seven ozonated samples spiked with 50 or 100 µg/L acetate, formate, and oxalate. The average accuracy for the three carboxylic acids ranged from 103 to 112 percent, and the S.D. of the accuracy ranged from 5.6 to 11.4 percent. The average precision of the replicate samples for the three analytes ranged from 2.0 to 5.8 µg/L; the S.D. of the precision ranged from 1.7 to 3.6 µg/L.

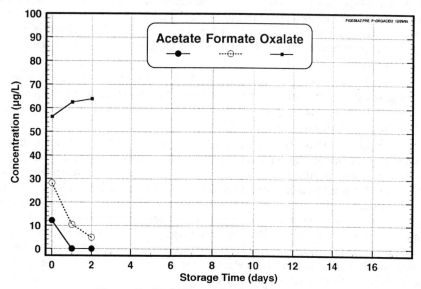

Figure 3. Stability of Carboxylic Acids in Fortified Raw Water without Preservation

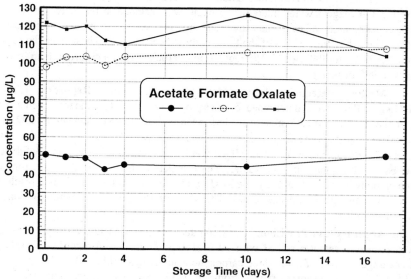

Figure 4. Stability of Carboxylic Acids in Fortified Raw Water Preserved with HgCl$_2$

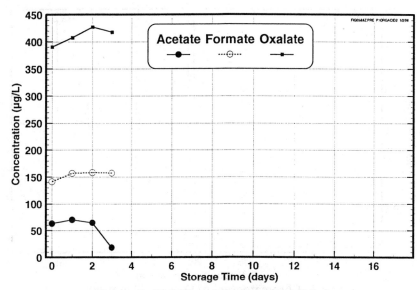

Figure 5. Stability of Carboxylic Acids in
Ozone Effluent without Preservation

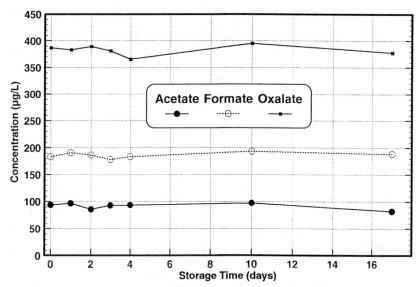

Figure 6. Stability of Carboxylic Acids in
Ozone Effluent Preserved with HgCl$_2$

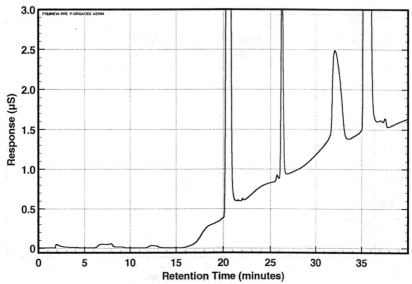

Figure 7. Chromatogram of Ozone Effluent without In-Line H⁺ Cartridge Pretreatment

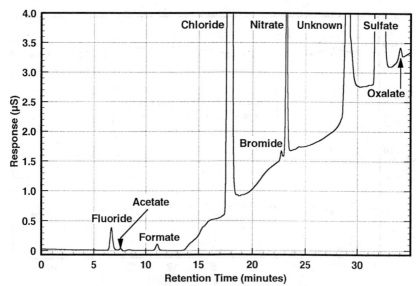

Figure 8. Chromatogram of Ozone Effluent with In-Line H⁺ Cartridge Pretreatment

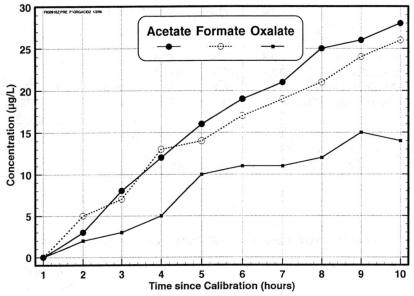

Figure 9. Analyte Area Value Drift

Pilot-Plant Studies.

Preliminary Carboxylic Acid Results. Preliminary results of the carboxylic acid measurements from the pilot-plant ozone contactor, filter influent, and filter effluents are presented in Table II. This testing was conducted after operating conditions in the pilot plant had stabilized and the filter beds were believed to be biologically active. In addition, the results of the analyses for AOC and aldehydes are presented in Tables III and IV, respectively.

Table II. Carboxylic Acids from Ozonated[a]/Biofiltered State Project Water[b]

	Concentration[c]			Reduction from Filter Influent		
Sample Location	Acetate (μg/L)	Formate (μg/L)	Oxalate (μg/L)	Acetate (%)	Formate (%)	Oxalate (%)
Ozone effluent	60.8	130.5	382.4			
Filter influent	37.5	103.8	283.1			
GAC filter effluent	7.2	20.8	57.8	80.4	80.0	79.5
Anthracite filter effluent	9.2	23.1	59.6	75.2	77.7	78.8

[a]Ozone dose = 2.1 - 2.2 mg/L, pH = 7.7, turbidity = 17.0 NTU.
[b]Sample preserved with $HgCl_2$.
[c]Average of three samples collected on consecutive days.

Table III. AOC from Ozonated[a]/Biofiltered State Project Water

	Concentration[b]			% Reduction from Filter Influent:
Sample Location	P17 (μg C/L)	NOX (μg C/L)	Total (μg C/L)	Total AOC
Ozone effluent	116.3	327.6	443.9	
Filter influent	106.5	320.3	426.9	
GAC filter effluent	33.4	84.1	117.5	72.5
Anthracite filter effluent	36.9	84.2	121.1	71.7

[a]Ozonated on same days as carboxylic acid samples in Table II.
[b]Average of three samples collected on consecutive days.

Table IV. Aldehydes from Ozonated[a]/Biofiltered State Project Water

	Concentration[b]			Reduction from Filter Influent		
	Formal-dehyde	Methyl Glyoxal	Glyoxal	Formal-dehyde	Methyl Glyoxal	Glyoxal
Sample Location	(µg/L)	(µg/L)	(µg/L)	(%)	(%)	(%)
Ozone effluent	19.1	10.7	8.3			
Filter influent	19.4	8.7	5.9			
GAC filter effluent	3.7	2.4	0.8	80.9	72.4	86.4
Anthracite filter effluent	0.3	2.6	0.6	98.5	70.1	89.8

[a]Ozonated on same days as carboxylic acid samples in Table II.
[b]Average of three samples collected on consecutive days.

The carboxylic acid that was produced in the largest amounts after the ozone contactor was oxalate (at approximately 380 µg/L), followed by formate (at approximately 130 µg/L) and acetate (at approximately 60 µg/L). The sum of the three carboxylic acids on a carbon basis was approximately 160 µg/L. The average percent removal of the three carboxylic acids from either type of biologically active filter was approximately 80 percent.

Comparison of Carboxylic Acid, Aldehyde, and AOC Results. Table V compares the levels of the sum of acetate, formate, and oxalate; the sum of formaldehyde, glyoxal, and methyl glyoxal; and the total AOC. All concentrations are presented on a carbon basis for easy comparison. Although only limited data are available at this time, some interesting observations can be made.

Table V. AOC and Proposed AOC Surrogates from Ozonated/Biofiltered California State Project Water

Sample Type	Sum of Acetate, Formate, and Oxalate (µg C/L)	Sum of Formaldehyde, Glyoxal, and Methyl Glyoxal (µg C/L)	Total AOC (µg C/L)
Ozone effluent	163.7	16.2	443.7
Filter influent	120.1	14.3	427.0
GAC filter effluent	24.2	2.9	117.7
Anthracite filter effluent	26.2	1.5	121.3

Based on the values after the ozone contactor, the sum of acetate, formate, and oxalate accounted for a greater percentage of the AOC values (36.9 percent on a carbon basis) than did the sum of the formaldehyde, glyoxal, and methyl glyoxal (3.7 percent on a carbon basis).

The removal percentages for the sum of the carboxylic acids, the sum of the aldehydes, and the total AOC through the biologically active filters are presented in Table VI. These removals are based on comparison of the filter effluents with the filter influent. In general, the removals were between 70 and 80 percent for all three analytical method groups and for both filter media. These preliminary data suggest that it may be possible to use the carboxylic acid IC method as a surrogate procedure for the measurement of AOC.

Table VI. Removal of AOC and Proposed AOC Surrogates Through Biologically Active Filters

Filter	Sum of Acetate, Formate, & Oxalate Removal (%)	Sum of Formaldehyde, Glyoxal, & Methyl Glyoxal Removal (%)	Total AOC Removal (%)
GAC	79.8	79.7	72.4
Anthracite	78.2	89.5	71.6

Summary and Conclusions

The quality assurance data indicate that IC is an excellent tool for the quantitative determination of low-molecular-weight carboxylic acids such as acetate, formate, and oxalate in water treated with ozone. In the results for the samples collected from the ozone pilot-plant studies, the Ionpac AS11 column (Dionex), in conjunction with a conductivity detector, showed acceptable accuracy and precision. All three analytes in the samples pretreated with an in-line H^+ cartridge (Dionex) showed reasonable peak shapes and sensitivities when a 50-μL injection loop was used. The stability tests indicated that acetate is the least stable analyte among the three carboxylic acids, whereas oxalate is the most stable. The preservation of samples with $HgCl_2$ eliminated biodegradation, allowing the samples to be stored for analysis at least as long as 17 days. Sulfuric acid cleaning of the ASRS before analysis was required to diminish positive analyte area drift. In addition, cleaning of the ATC-1 was a necessary step to reach a status of low baseline conductivity output during analysis. The sample turnaround time is approximately 24 hours for a batch of 10 to 15 samples (overnight analysis), and the results can be reported the next morning.

The pilot-plant data, though very preliminary, showed some interesting results. At an ozone dose of 2 mg/L, the ozone effluent samples of California state project water produced acetate, formate, and oxalate at approximately 160 μg C/L. The oxalate was the major product of the three carboxylic acids, accounting for approximately

60 to 65 percent of the sum of three carboxylic acids, followed by formate at 25 to 30 percent of the sum and acetate at 10 percent of the sum. There was approx-imately an 80-percent reduction of the three carboxylic acids through the biologically active filter. These carboxylic acids accounted for approximately 28 percent of the AOC (on a carbon basis) at the filter influent (37 percent at the ozone effluent) and may be useful surrogates for AOC in ozonation/biofiltration studies.

Currently, an alternative anti-biodegradation agent, additional carboxylic acids and ketoacids, and interference by other anions are being investigated. MWDSC is conducting biofiltration studies at its ozone demonstration plant. The method reported in this chapter is being used to collect information on carboxylic acids after ozonation and to evaluate their potential as surrogates for AOC.

Acknowledgments

The authors wish to thank the pilot-plant staff for the preparation of the ozonated samples; Patrick Hacker and Michael Sclimenti for the analyses of AOC and alde-hydes, respectively; and Peggy Kimball and Robert La Londe for technical editing of the manuscript and preparation of the graphics, respectively.

References

1. *Disinfection By-Products in Water Treatment: The Chemistry of Their Forma-tion and Control*; Minear, R.A.; Amy G.L., Eds.; CRC Press, Inc.: Boca Raton, FL, 1995.

2. van der Kooij, D.; Visser, A.; Hijnen, W.A.M. *J. AWWA* **1982**, *74*, 540-545.

3. Huck, P.M.; Fedorak, P.M.; Anderson, W.B. *J. AWWA* **1991**, *83*, 69-80.

4. Hacker, P.A.; Paszko-Kolva, C.; Stewart, M.H. *Ozone Sci. & Engrg.* **1994**, *16*, 197-212.

5. Schechter, D.S.; Singer, P.C. *Ozone Sci. & Engrg.* **1995**, *17*, 53-69.

6. Andrews, S.A.; Huck, P.M. *Ozone Sci. & Engrg.* **1994**, *16*, 1-12.

7. Le Lacheur, R.; Glaze, W. *Environ Sci. & Technol.* **1996**, *30*, 1072-1080.

8. Sclimenti, M.J.; Krasner, S.W.; Weinberg, H.S.; Glaze, W.H. In *American Water Works Association Proceedings: 1990 Water Quality Technology Conference;* AWWA: Denver, CO, 1991; pp 477-501.

9. Kuo, C-Y; Stalker, G; Weinberg, H.S. In *American Water Works Association Proceedings: 1990 Water Quality Technology Conference;* AWWA: Denver, CO, 1991; pp 503-525.

10. *Installation Instructions and Troubleshooting Guide for IONPAC AS11 Analyti-cal Column*; Dionex Document #034791; Dionex Corp.: Sunnyvale, CA, 1993.

11. Coffey, B.M.; Krasner, S.W.; Sclimenti, M.J.; Hacker, P.A.; Gramith, J.T. In *American Water Works Association Proceedings: 1995 Water Quality Tech-nology Conference;* AWWA: Denver, CO, 1996; pp 503-525.

Chapter 20

Bromate Ion Removal by Electric-Arc Discharge and High-Energy Electron Beam Processes

Mohamed S. Siddiqui[1], Gary L. Amy[2], and William J. Cooper[3]

[1]Department of Civil Engineering, University of South Alabama,
Mobile, AL 36688
[2]Department of Civil, Environmental, and Architectural Engineering,
University of Colorado, Boulder, CO 80309
[3]Drinking Water Research Center, Florida International University,
Miami, FL 33199

Proposed drinking water regulations in the U.S. will specify a maximum contaminant level (MCL) of 0.01 mg/L for bromate ion (BrO_3^-). The work reported herein involves removing BrO_3^- after its formation using electric arc discharge and high energy electron beam (HEEB) process, when other removal and minimization strategies are not effective. The electric arc discharge method destroyed 12-45% bromate for doses ranging from 130-1300 mW-s/cm^2 and a dose of 100 krads was sufficient to reduce 70% of BrO_3^- from an initial concentration of 100 μg/L using HEEB process in NOM-free water. The addition of hydrogen gas and the removal of dissolved oxygen enhanced BrO_3^- removal during electric arc discharge. During HEEB process the presence of electron scavengers such as hydrogen peroxide and nitrate significantly reduced BrO_3^- removal whereas the addition of OH radical scavenger such as t-butanol did not affect the removal of BrO_3^- indicating that aqueous electrons (e_{aq}^-) are mainly responsible for BrO_3^- destruction. The presence of natural organic matter (NOM) reduced BrO_3^- reduction efficiency.

INTRODUCTION

Bromate (BrO_3^-) has been shown to cause kidney, and possibly other tumors in laboratory animals. However, the mechanism of action and whether BrO_3^- is direct or indirect acting has yet to be resolved (1). BrO_3^- is also genotoxic *in vitro* and *in vivo* although this is primarily confined to causing physical damage to chromosomes (2). The World Health Organization (WHO) has proposed a provisional guideline of 25 μg/L for BrO_3^-; proposed drinking water regulations in the U.S. will specify a maximum contaminant level (MCL) of 10 μg/L for BrO_3^- and a Best Available Technology (BAT) of pH adjustment and ammonia addition (1).

Ozone is a strong oxidant and it oxidizes Br^- in water to BrO_3^- through two different pathways. BrO_3^- can form through both a molecular ozone (O_3) pathway and a hydroxyl radical $(OH\cdot)$ pathway, depending on the dissolved organic carbon (DOC), Br^- content, and pH of the source waters. By the molecular ozone pathway, Br^- is first oxidized by dissolved ozone (DO_3) to hypobromite ion (OBr^-) which is then further oxidized to BrO_3^- (3)(OH radicals are produced as a result of the decomposition of molecular ozone at pH > 8. Molecular ozone (O_3) is highly selective as an oxidant whereas OH radicals are non-selective. The reaction rates of OH radicals with bromine species are many times faster than with molecular ozone (18)). This reaction is pH-dependent since OBr^- is in equilibrium with hypobromous acid (HOBr). The molecular ozone theory suggests that BrO_3^- formation is directly driven by dissolved ozone (DO_3) and the OH radical theory indicates that dissolved ozone plays only an indirect role by decomposing O_3 to produce radicals which further react with bromine species to produce BrO_3^- (4).

BrO_3^- control strategies have been predicated upon minimization and removal strategies. BrO_3^- minimization (during ozonation) approaches have highlighted pH depression, ammonia addition, hydrogen peroxide addition, and modified ozone contactor design and operation (5, 6, 7, 8). Adjustment to pH 6.0 prior to ozonation will significantly reduce BrO_3^- formation; however, acid addition may not be viable or cost effective for high alkalinity waters. Ammonia addition can theoretically tie up bromine as monobromamine; however, the complexity of ammonia-ozone chemistry has yielded mixed BrO_3^- formation results. Removal approaches have focused on ferrous iron addition, UV irradiation, and activated carbon (9). If the proposed BrO_3^- MCL in the U.S. is lowered further, a combination of minimized production and subsequent removal may be required as a BrO_3^- control strategy. Recent work has shown that electron beam merits further consideration (9).

The HEEB process and electric arc system can potentially be used at various points in the process train, as either a pre-oxidant or a post-disinfectant. In one scenario, these processes could be used in place of pre-ozonation; here, they would provide CT credit, aid coagulation, destroys synthetic organic compounds and reduce DOC while not contributing to the formation of DBPs. In another scenario, HEEB would be used later in the process train following a pre-ozonation step; here it can be used to provide additional CT requirements while destroying ozonation by-products such as BrO_3^-. HEEB irradiation can destroy halogenated disinfection by-products such as trihalomethanes (THMs)(10) and may provide effective inactivation of microbes of present or future regulatory interest, including enteric viruses and *Giardia* .

This paper discusses the use of an innovative treatment processes, high energy electron beam irradiation and electric arc discharge, in drinking water treatment to remove BrO_3^-. The effect of dose, BrO_3^- concentration, pH, natural organic matter (NOM) and alkalinity on BrO_3^- removal is evaluated.

EXPERIMENTAL AND ANALYTICAL METHODS

SOURCE WATERS. Source waters evaluated included California State Project Water, CA (SPW), Colorado River Water, CA (CRW), Silver Lake Water, CO (SLW), Biscayne Aquifer Water, FL (BAW) and DOC-free (≤ 0.2 mg/L) Milli-Q water (MQW). A sample of SPW was taken from the Sacramento-San Joaquin delta, which is subject to saltwater intrusion. SLW is a sub-alpine lake in the Rocky mountains. CRW Br⁻ levels may be influenced by salt deposits (connate sea water) on the western slope of Colorado. Important characteristics of the source waters studied are summarized in Table 1.

Table 1. Raw Water Characteristics

Source Water	DOC mg/L	UV_{254} cm⁻¹	Specific Abs L/cm-mg	Alkalinity mg/L $CaCO_3$	pH	Br⁻ μg/L
CRW	3.3	0.08	0.024	120	8.0	70
SPW	3.5	0.14	0.041	87	8.0	150-400
SLW	3.1	0.07	0.023	15	7.0	10
BAW	4.5	0.101	0.023	90	8.0	115
MQW	0.15	<0.001	0.006	<10	6.0	<5

All samples were stored at 4 °C prior to analysis. All bench-scale experiments were performed at room temperature (20±3 °C). Reagent grade potassium bromate ($KBrO_3$) and potassium bromide (KBr) were used to prepare BrO_3^- and Br⁻ standards[2].

ARC DISCHARGE EXPERIMENTS. BrO_3^- destruction by arc discharge was evaluated in a lab-scale arc discharge apparatus (Figure 1). The intense electrical pulse is created by discharging a high voltage, low inductance capacitor by use of a very fast rise time switch, into a spark gap in the working fluid. The discharge is from 10 KV to 50 KV with a current from 10 to 100 amps and a pulse rate from 50 to 100 pulses/sec. The localized plasma pressure ranges between 100-1000 MPa, and the localized temperature is approximately 10,000 to 50,000 K. The working fluid in the cavity is quickly expands and then contracts as it cools very rapidly. This results in the creation of differential pressure waves, which are seen as shock waves within the cavity. The arc which produces the shock wave also produces a highly concentrated plasma. This plasma is a very powerful source of UV radiation. Production of free oxygen, OH radicals and H atoms is performed very efficiently by electric arc. H atoms are strong reducing species. An arc is generated between a point anode and a hollow cathode. The water being purified flows through the arc and out through the hollow cathode. Energy can be varied by pulse width or by pulse rate. Hydrogen, if used, was injected into the arc not directly into the water since hydrogen is sparingly soluble in water. Dissolved oxygen concentration was reduced by sparging with helium gas.

---

2 Aldrich Chemical Co., WI

HEEB PROCESS. The HEEB unit selected to process the samples in batch-mode was a 8 MeV Linatron machine[3] rated at 1kW (11). Electron beam equipment consists of a power supply unit, a control cabinet and electron accelerator with a beam scanner (see Figure 2). The samples were placed in a steel wire basket and irradiated. The conveyor speed and machine pulse rate were set to give the required dose to the samples. Radiochromic dosimeters[4] were placed above the samples to determine the doses. The dosimeters were read with a spectrophotometer. BrO_3^- varying in concentration from 100-320 μg/L was spiked into MQW and different source waters and transferred to 35 ml screw-cap vials.

The pilot-scale facility consists of an horizontal 1.5 MeV insulated core transformer electron accelerator whose beam current was varied from 0 to 50 mA to achieve absorbed doses of 0-800 krads at water flows of 120 gpm (12). The influent stream is presented to a scanned electron beam in a falling stream about 114 cm wide and 0.4 cm thick.

High energy electron are generally produced by electron accelerators or gamma ray sources such as ^{60}Co. In electron accelerators, the most widely used industrial source, a cathode is used to produce a stream of electrons which are accelerated by applying an electric field generated at a given voltage. The applied voltage determines the speed and thus the energy of the accelerated electrons.

ANALYTICAL METHODS. Br^- and BrO_3^- measurements were accomplished by ion chromatography (IC)[5] with an IonPac AS9-SC column. A 2.0 mM Na_2CO_3/0.75 mM $NaHCO_3$ eluent was used for Br^- determination and a 40 mM H_3BO_3/20 mM NaOH eluent was employed for BrO_3^- determination. Analytical minimum detection limit for BrO_3^- using borate eluent was 2 μg/L. For samples with high chloride ion (Cl^-) content, a silver cartridge was used to remove Cl^- prior to IC analysis to minimize its interference with BrO_3^- measurement. The source waters evaluated contained up to 150 mg/L as Cl^-.

DOC measurements were made using a TOC analyzer[6] after lowering the pH to 2.0 to strip inorganic carbon. UV_{254} measurements were made using a UV/VIS spectrophotometer[7].

In DOC-free water the concentration of hydrogen peroxide was monitored at 240 nm (ε = 40 $M^{-1}s^{-1}$) using a UV/VIS spectrophotometer[7].

---

[3] Nutek Corporation, Palo Alto, CA
[4] Far-West Technology, Palo Alto, CA
[5] Dionex Corporation, Sunnyvale, CA (4500i series)
[6] Shimadzu Corporation, CA (Model: TOC-5000)
[7] Shimadzu Corporation, CA (Model: UV-160)

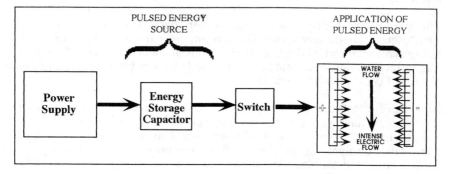

Figure 1. Electric Arc Discharge System

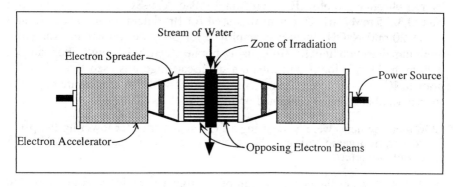

Figure 2. High Energy Electron Beam Irradiation

MECHANISM OF BROMATE DESTRUCTION

ELECTRIC ARC DISCHARGE. Two completely different pulse techniques were employed to reduce BrO_3^- from water using electric arc method. In both cases UV was also present due to the existence of a high energy arc, but it was not the major contributor to the reduction process. In one case, the arc was in water with no gas present. The arc breaks down the water into OH radicals (OH·) and free hydrogen atoms (H·). A small fraction of these hydrogen atoms can further react with hydroxide ions (OH⁻) in water to produce aqueous/solvated electrons (e^-_{aq}). These solvated electrons and hydrogen atoms (H·) are reducing species which can take part in BrO_3^- reduction. UV alone is not very effective in removal of BrO_3^-, howerver this may not be true in the presence of other reducing species such as H· and O_2·⁻ which is a conjugate base of HO_2·, the product of reaction between H· and O_2. The energy associated with UV is high enough to break down a Br-O bond in BrO_3^-. After this Br-O bond is broken down, it is very possible that the reducing radicals take over for further reduction of intermediate bromine species.

In the second case, the concentration of reducing species was enhanced by injecting hydrogen gas into the arc. The arc also reduces oxygen from the BrO_3^- into free oxygen atoms which are highly reactive towards other bromine species.

HEEB PROCESS. Electron beams are particulate radiation, consisting of charged electrons that exhibit wavelike properties when accelerated in an electric field. In contrast to other forms of radiation such as infrared and ultraviolet, ionizing radiation is absorbed almost totally by the target materials's electronic structure, which increases the energy level of its orbital electrons. The energy level of the radiation is high enough to produce changes in the molecular structure of materials, but sufficiently low to avoid inducing radioactivity. Electron beam uses that portion of the electromagnetic spectrum with energy levels between 10 electron volts (eV) and 10 million electron volts (MeV) (An eV is the amount of energy gained by an electron traversing a 1 volt potential field).

Irradiation of water by electron beam results in the formation of reducing species such as aqueous electrons (e^-_{aq}), hydrogen atoms (H·), and oxidizing species such as hydroxyl radicals (OH·). The yields of primary species in the radiolysis of pure water in millimoles per kilogram of material per Mrad are shown in Table 2 (14, 15). The formation of the reactive species is pH independent in the range 6-10 and therefore, any differences thatbexist in pH over time will not adveresely affect the treatment efficiency (provided there is no alkalinity present). The absorbed dose to the target material is measured in terms of energy absorbed per unit mass of material. An energy absorption of 100 ergs per gram is defined as one "rad". One million rads is commonly referred to as one Mrad (Million Radiation Absorbed Dose). Dosages delivered to source waters are also directly related to the nature of water constituents being addressed.

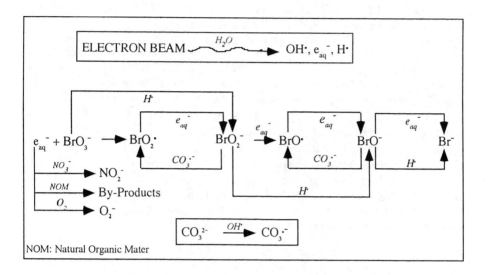

Figure 3. Mechanism of Bromate Destruction By HEEB

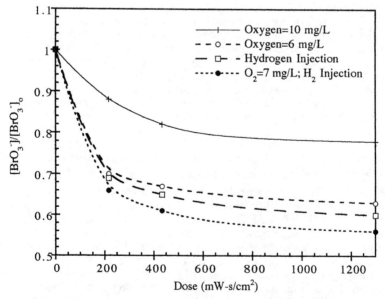

Figure 4. Effect of Electric Arc on Bromate Removal [pH=7.5; $(BrO_3^-)_o$=50 ug/L]

Table 2. Yields of Primary Species after HEEB Irradiation

Specie	G(X)	mM/kg water/Mrad
$e_{(aq)}^-$	2.6	2.7
H·	0.6	0.6
OH·	2.6	2.7
H_2O_2	0.75	0.8

The reaction of e_{aq}^- with BrO_3^- and other bromine intermediates leading to the formation of bromide ion are summarized in Figure 3. BrO_3^- is first converted to bromite ion (BrO_2^-) which is again reduced to hypobromite ion (BrO^-). Further reactions of e_{aq}^- and H· with BrO^- leads to the formation of terminal stable product Br^-. The e_{aq}^- is a powerful reducing agent with an E_o of -2.77. This is formed by solvation of the electron. It may react by neutralization reactions, or add to a BrO_3^- molecule and yield a dissociative electron-attachment reaction. The hydrogen radicals are the products of the solvated electrons reacting with hydrogen ions in water. H· can also abstract a halide or react with OH to produce e_{aq}^-.

RESULTS AND DISCUSSION

ELECTRIC ARC DISCHARGE EXPERIMENTS. Bromate destruction was evaluated at energy levels of 130-1300 mW-s/cm^2 ($\approx$ 50-300 Joules/L) and the percentage destruction varied from 12 to 45% as shown in Figure 4.

Experiments conducted by reducing the concentration of dissolved oxygen indicated increased destruction of BrO_3^-. Some of the aqueous electrons are scavenged by dissolved oxygen (O_2) and nitrate (NO_3^-). The addition of hydrogen gas into the arc increased BrO_3^- destruction. The removal was higher when dissolved oxygen was reduced before hydrogen gas addition. As shown in Figure 4 a combination of these techniques greatly enhances the reduction of BrO_3^-. Electric arc discharge represents a potential disinfection alternative and uses relatively less energy than the HEEB system.

HEEB PROCESS EXPERIMENTS. CRW, SPW, and MQW (NOM-free) source waters were irradiated with doses ranging from 0-750 krads with an initial BrO_3^- of 320 μg/L (Figure 5).

Br^- and free bromine (HOBr/OBr$^-$) concentrations were measured after irradiation and Br^- mass balances showed almost complete recovery of initial bromide (BrO_3^- is 63% Br by wt.) in MQW water. Complete recovery of initial Br^- in BrO_3^- was not observed at lower doses for CRW and SPW, ostensibly because of the formation of bromo-organic compounds. Reduction of BrO_3^- in MQW was higher than CRW and SPW because of the presence of carbonate alkalinity and DOC. The $CO_3\cdot^-/HCO_3\cdot$ radicals are produced by the reaction of e$^-$

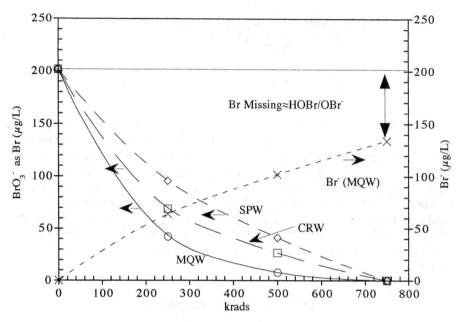

Figure 5. Effect of Irradiation on Bromide Formation
(pH=7.5, DO=7.0 mg/L, $[BrO_3^-]_o$=320 ug/L)

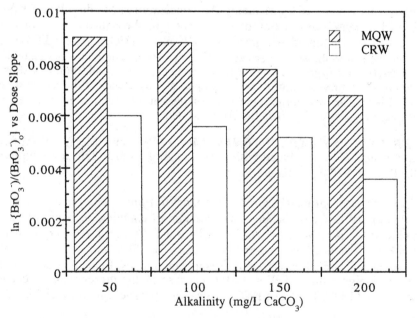

Figure 6. Effects of Alkalinity and Source Water on Bromate Removal
$[pH=7.5; (BrO_3^-)_o$=100 ug/L]

$_{aq}$ and OH· with bicarbonate ions and dissolved carbon dioxide in water (Figure 6). The carbonate radicals act predominantly as electron acceptors; hydrogen abstraction by the carbonate radicals is generally very slow. Increasing alkalinity from 50 to 200 mg/L as $CaCO_3$ in MQW, BrO_3^- reduction by HEEB irradiation decreased by 25%. On ultrafiltration analysis, SPW was found to contain significantly higher fraction of DOC in the higher molecular weight range than CRW and this higher molecular weight fraction is more amenable to lower molecular weight transformation by HEEB process. In other words some of the e^-_{aq} are being scavenged by higher molecular weight DOC resulting in decreased BrO_3^- destruction.

The effect of background water characteristics on BrO_3^- removal is shown in Figure 7. The H· undergoes two general types of reactions with DOC and BrO_3^-, hydrogen addition and hydrogen abstraction whereas the OH· radical can undergo several types of oxidation reactions with chemicals in aqueous solution.

The addition of hydrogen peroxide had a mixed effect on BrO_3^- removal. Experiments performed by spiking H_2O_2 showed a relatively strong inhibition effect on BrO_3^- reduction indicating that H_2O_2 is acting as a scavenger of e^-_{aqs} at concentrations greater than 2 mg/L. BrO_3^- reduction was enhanced when H_2O_2 concentration was less than 2 mg/L indicating that there is a threshold level of H_2O_2 above which it simply acts as a sink for e^-_{aqs} and OH radicals. This threshold level of H_2O_2 is dependent on dose, pH, alkalinity and DOC. The relevant reactions are shown below (13, 16):

$$OH· + OH· \rightarrow H_2O_2 \qquad\qquad k = 5E9\ M^{-1}s^{-1} \qquad\qquad [1]$$

$$e^-_{aq} + H_2O_2 \rightarrow OH^- + OH· \qquad\qquad k = 2E10\ M^{-1}s^{-1} \qquad\qquad [2]$$

$$OH· + H_2O_2 \rightarrow H_2O + HO_2· \qquad\qquad k = 3E7\ M^{-1}s^{-1} \qquad\qquad [3]$$

Reactions involving e^-_{aq} and H· produced during HEEB irradiation can also lead to the formation of hydrogen peroxide (H_2O_2) in the presence of DOC and dissolved oxygen, depending upon the concentration of natural organic matter Thus H_2O_2 acts as e^-_{aq} scavenger or promoter of excited state species.

Several experiments were conducted to show the effect of radical scavengers on the removal of BrO_3^-. In the presence of t-butanol which is a strong OH· radical scavenger, reduction of BrO_3^- was not impaired suggesting that e^-_{aq} is mainly responsible for BrO_3^- reduction. The OH· radicals cannot reduce BrO_3^- since they are oxidizing species and possibly cannot participate in Br⁻ oxidation since they are consumed by e^-_{aq} faster than they can react with bromine species.

To further confirm that only e^-_{aq} are responsible for BrO_3^- reduction, solutions containing BrO_3^- were irradiated in the presence of electron scavengers such as nitrate, dissolved oxygen and chloroacetate; the destruction of BrO_3^- was found

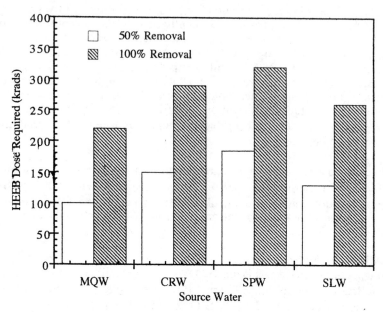

Figure 7. Effect of Source Waters on Bromate Removal [$(BrO_3^-)_o$=100 ug/L, pH=7.0]

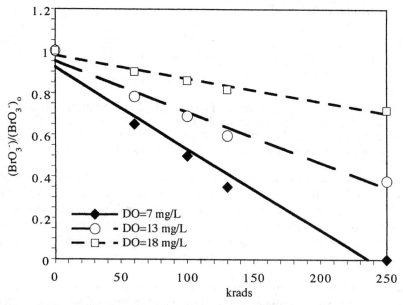

Figure 8. Effects of Dissolved Oxygen on Bromate Removal
[MQW, pH=7.5, $(BrO_3^-)_o$=100 ug/L]

to decrease significantly indicating that the e^-_{aqs} are in fact responsible for BrO_3^- reduction during HEEB irradiation. A concentration of 10 mg/L as NO_3^- was sufficient to completely inhibit the reduction of BrO_3^- with an initial BrO_3^- concentration of 100 μg/l (HEEB dose=200 krads). However, nitrate has some rather complex radiation chemistry associated with it. That is, the reaction with e^-_{aq} will form nitrite (NO_2^-), which can further react with hydroxyl radicals and in the presence of an aromatic ring can actually form compounds such as nitrobenzene. The reaction between e^-_{aq} and NO_3^- is given below (13).

$$e^-_{aq} + NO_3^- \rightarrow O^{\cdot-} + NO_2^- \qquad k = 9.7E9 \ M^{-1} \ s^{-1} \qquad [4]$$

To evaluate the effect of dissolved oxygen on BrO_3^- destruction, MQW was evaluated with dissolved oxygen concentration ranging from 7-18 mg/L (Figure 8). The ambient dissolved oxygen level of source waters evaluated was approximately 7 mg/L. However, after ozonation, DO levels could be as high as 15-20 mg/L. In source waters with high DO, e^-_{aq} and H· will be both effectively scavenged by DO, and the HEEB then becomes really an oxidation process in practice. For a given drinking water source, irradiation induced oxygen uptake rate may be significant and may reach upto 12 mg/L per 100 krads. BrO_3^- destruction decreased on increasing dissolved oxygen concentration since e^-_{aq} are strong scavengers of oxygen (13).

$$e^-_{aq} + O_2 \rightarrow O_2^- \qquad k = 2E10 \ M^{-1} \ s^{-1} \qquad [5]$$

If the the oxygen concentration is more than 8 mg/L, the use of HEEB process can become prohibitively expensive and oxygen concentration may have to be decreased. In water treatment, e^-_{aq} and H· will be both effectively scavenged by DO, and the HEEB then becomes really an oxidation process in practice.

Full-scale HEEB irradiation experiments were conducted using BAW water (flow≈100 gpm) by spiking BrO_3^- at concentrations ranging from 180 to 700 μg/L at two different pH levels (pH 4.5 and 9.0) and the results are shown in Figures 9. Better removal of BrO_3^- was observed at pH 9.0 than at pH 4.5 because the scavenging efficiency of protons (H^+) by e^-_{aqs} decreases (13).

$$e^-_{aq} + H^+ \rightarrow H\cdot \qquad k = 1E10 \ M^{-1}s^{-1} \qquad [6]$$

DOC concentration and UV_{254} absorbance levels decreased by 26% and 42% at pH=4.5 and 25% and 32% at pH 9.0 respectively on increasing HEEB doses from 50 to 200 krads. Better removal of DOC at lower pH was observed because of increased scavenging efficiency for OH radicals (13).

$$e^-_{aq} + OH^{\cdot} \rightarrow OH^- \qquad k = 3E10 \ M^{-1}s^{-1} \qquad [7]$$

Also the presence of alkalinity in natural source waters may decrease the concentration of OH radicals and e^-_{aq} available for DOC oxidation and some

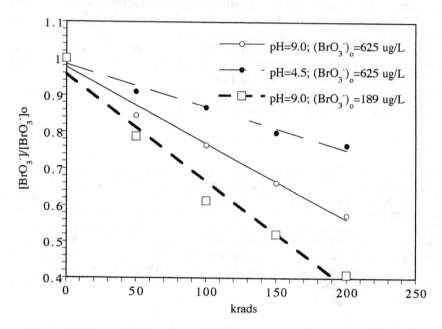

Figure 9. Effects of pH and Bromate on Bromate Removal [Source: BAW]

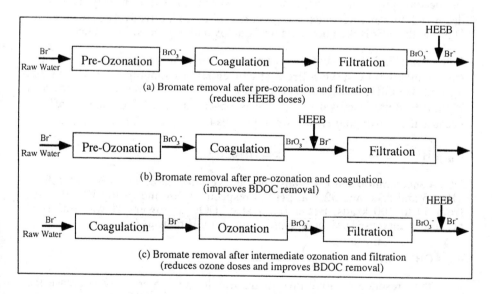

(a) Bromate removal after pre-ozonation and filtration
(reduces HEEB doses)

(b) Bromate removal after pre-ozonation and coagulation
(improves BDOC removal)

(c) Bromate removal after intermediate ozonation and filtration
(reduces ozone doses and improves BDOC removal)

Figure 10. Removal of Bromate After Ozonation During Drinking Water Treatment

humic/fulvic material might change in reactivity when the pH is increased. Also e^-_{aq} acts as a nucleophile in its reactions with natural organic matter, and its reactivity is greatly enhanced by electron-withdrawing substituents adjacent to alkene double bonds or attached to aromatic rings (13, 16).

The reaction of OH radicals with DOC may also lead to the formation of oxidized products. The DOC can be converted to CO_2 or secondary elimination processes may occur. The yields achieved by hydroxylation reactions is dependent on the yield of OH radicals whose efficiency in turn is dependent on the pH of source water. The pH of water slightly decreased after irradiation as expected because OH^- are consumed to produce OH· radicals which subsequently react with DOC and other solutes in water. HEEB process can oxidize larger molecules to smaller compounds such as lower molecular weight organic acids resulting in increased negative charge density of water.

Figure 10 presents the schematic for different scenarios for destroying BrO_3^- after ozonation during drinking water treatment. Figure 10a shows the reduction of BrO_3^- to Br^- in finished water with the added advantage of getting extra CT credit and enhanced coagulation. The reactions of radiolytic products in water are responsible for changes in the stability of colloids through oxidations and secondary reactions. Also dissolved macromolecules could be oxidized to polyelectrolytes, which can adsorb and flocculate dispersed particles (11). Figure 10b uses HEEB irradiation to destroy BrO_3^- after coagulation to enhance the formation of biodegradable organic compounds through organic matter transformation which are subsequently removed at a later stage during filtration thereby reducing DOC and DBP formation potential.

The comparative assessment of bromate removal by a medium pressure lamp, electric arc discharge and HEEB process are shown in Figure 11.

Electron beam systems have been in commercial use since the 1950s. Early applications involved the cross-linking of polyethylene film and wire insulation. The number of applications has since grown to include sterilization of medical supplies, rubber vulcanization, disinfection of wastewater, food preservation, curing of coatings, etc. Today there are several hundred electron processing systems installed for industrial applications in over 25 countries. Polymerization of cable insulation and cross-linking of plastic film still account for the bulk of the applications. More than half of the total installed world capacity of 15 MW of electron beam power is devoted for sterilization of medical products. Only a small amount of the installed capacity is used for biological disinfection and detoxification. One of the most promising uses for this evolving technology is for the treatment of hazardous wastes prior to discharge to the environment and for the removal of micropollutants from drinking water sources. The dosages required for BrO_3^- removal and other applications is illustrated in Figure 12. This Figure shows that electron beam irradiation for disinfection and BrO_3^- removal requires relatively lower dosages than treatment of stack gases and detoxification of chlorinated hydrocarbons.

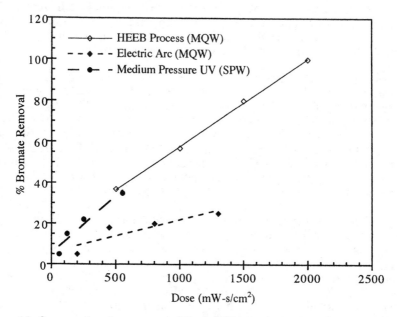

Figure 11. Comparative Assessment of Various Technologies for Bromate Removal
[pH=7.5; $(BrO_3^-)_o$=100 ug/L)

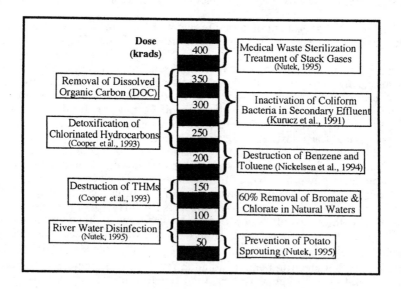

Figure 12. Electron Beam Dose Requirements for Bromate Removal
and Selected Applications

CONCLUSIONS

Given the assumption that ozonation is one of the best methods for disinfection and micropollutants removal and that it will continue to be used, the concentration of bromate in drinking water needs to be controlled by either removing the chemical precursor, i.e., bromide ion, before ozonation or by removing the bromate ion after ozonation. Because there are few economically acceptable methods for the removal of bromide ion from water, the removal of bromate by electric arc and HEEB process may serve as potential alternatives.

Solutions irradiated with HEEB and electric arc reduced BrO_3^- to Br^- with OBr^- as an intermediate species. A HEEB dose of 100 krads is sufficient to reduce 50% of BrO_3^- from an initial concentration of 100 $\mu g/L$, the highest concentration normally formed upon oxidation of Br^- during ozonation. During HHEB irradiation, some metal ions if present could in principle be dissolved after the irradiation-induced oxidation or reduction, and some chelating properties of organic molecules may be lost, while others will be formed. Generally, organic solutes will become more hydrophilic and hence possibly less toxic. The dose will be low to yield aromatizations of organic solutes and no carcinogenic compounds will be formed.

The presence of hydrogen peroxide, NO_3^- and dissolved oxygen (strong electron scavengers) significantly decreased reduction of BrO_3^- whereas addition of radical scavenger t-butanol did not affect the reduction of BrO_3^- indicating that e_{aq}^- are stronger reducing species than other excited species. Hence in source waters with significant levels of NO_3^- and dissolved oxygen, HEEB use may not be feasible nor cost effective. Also the removal of BrO_3^- can be enhanced by the addition of hydrogen gas by electric arc method.

The presence of dissolved organic matter (DOC) decreased BrO_3^- reduction efficiency. HEEB irradiation reduced DOC concentration by 15-26% for a HEEB dose of 100-200 krads. This reduction efficiency is dependent on the type of DOC in source water.

REFERENCES

(1) US Environmental Protection Agency (USEPA), "National Primary Drinking Water Regulations; Disinfectants and Disinfection By-Products: Proposed Rule", *Federal Registery*, 59(145):38668-38829 (1994).

(2) Melnick, R. L., "An Alternative Hypothesis on the Role of Chemically Induced Protein Droplet Nephropathy in Renal Carcinogenesis", *Regulatory Toxicology and Pharmacology*, 16:111-125 (1992).

(3) Haag, W. R. and J. Hoigne, "Ozonation of Bromide Containing Waters: Kinetics of Formation of Hypobromous Acid and Bromate", *Envi. Sci. & Technol.*, 17:261-267, 1983.

(4) Richardson, L. B.; D. T. Burton; G. R. Helz; J. C. Rhoderick, "Residual Oxidant Decay and Bromate Formation in Chlorinated and Ozonated Sea-Water", *Water Research*, 15:1067-1074, 1981.

(5) Siddiqui, M.; Amy, G. and Rice, R. G., "Bromate Ion Formation: A Critical Review", *JAWWA*, 87(10): 58, 1995.

(6) Krasner, S. W.; W. Glaze; H. Weinberg; P. Daniel, and I. Najm, "Formation and Control of Bromate During Ozonation of Waters Containing Bromide", *JAWWA*, 85(1):73, 1993.

(7) Siddiqui, M. and G. Amy "Factors Affecting DBP Formation During Ozone-Bromide Reactions" *JAWWA*, 85(1):63, 1993.

(8) Siddiqui, M. and G. Amy, "Formation of Bromate: Effect of NH_3/Br Ratio", *IOA World Congress*, Lillee, France, 1995.

(9) Siddiqui, M. et al., "Alternative Strategies for Removing Bromate", *JAWWA*, 86(10):81, 1994.

(10) Cooper, W.; E. Cadavid; M. Nichelsen; K. Lin; C. Kurucz; and T. Waite, "Removing THMs From Drinking Water Using High-Energy Electron-Beam Irradiation", *JAWWA*, 85(9):106, 1993

(11) Nutek Corporation, "Electron Beam Process Technical Overview", Research Report, Palo Alto, CA, 1994.

(12) Kuruez C. et. al. "High-Energy Electron-Beam Irradiation of Water, Wastewater and Sludge", *Advances in Nuclear Science and Technology*, Plenum Press, New York, 1991.

(13) Buxton, G. and C. Greenstock, "Critical Review of Rate Constants for Reactions of e^-_{aq}, H· and HO· in Aqueous Solutions" *J. Phy. Chem. Ref. Data*, 17(2), 1988.

(14) Hoigne, J., "Aqueous Radiation Chemistry in Relation to Waste Treatment", in *Radiation for a Clean Environment*, International Atomic Energy Agency, Vienna, 1975.

(15) Allen, A. O., "The Radiation Chemistry of Water and Aqueous Solutions", Van Nostrand, Princeton, 1961.

(16) Neta, P.; R. Huie; A. Ross, "Rate Constants for Reactions of Inorganic Radicals in Aqueous Solutions", *J. Phys. Chem. Ref. Data*, 17:3, 1988.

(17) Eliassen, R.; J. Trump, "High Energy Electrons Offer Alternative to Chlorine", *Calif. Water Pollt. Control Assoc. Bull.*, 10(3), 1974.

INDEXES

Author Index

Affiliation Index

Subject Index

A

B

Highlights from ACS Books

Good Laboratory Practice Standards: Applications for Field and Laboratory Studies
Edited by Willa Y. Garner, Maureen S. Barge, and James P. Ussary
ACS Professional Reference Book; 572 pp; clothbound ISBN 0–8412–2192–8

Silent Spring Revisited
Edited by Gino J. Marco, Robert M. Hollingworth, and William Durham
214 pp; clothbound ISBN 0–8412–0980–4; paperback ISBN 0–8412–0981–2

The Microkinetics of Heterogeneous Catalysis
By James A. Dumesic, Dale F. Rudd, Luis M. Aparicio, James E. Rekoske,
and Andrés A. Treviño
ACS Professional Reference Book; 316 pp; clothbound ISBN 0–8412–2214–2

Helping Your Child Learn Science
By Nancy Paulu with Margery Martin; Illustrated by Margaret Scott
58 pp; paperback ISBN 0–8412–2626–1

Handbook of Chemical Property Estimation Methods
By Warren J. Lyman, William F. Reehl, and David H. Rosenblatt
960 pp; clothbound ISBN 0–8412–1761–0

Understanding Chemical Patents: A Guide for the Inventor
By John T. Maynard and Howard M. Peters
184 pp; clothbound ISBN 0–8412–1997–4; paperback ISBN 0–8412–1998–2

Spectroscopy of Polymers
By Jack L. Koenig
ACS Professional Reference Book; 328 pp;
clothbound ISBN 0–8412–1904–4; paperback ISBN 0–8412–1924–9

Harnessing Biotechnology for the 21st Century
Edited by Michael R. Ladisch and Arindam Bose
Conference Proceedings Series; 612 pp;
clothbound ISBN 0–8412–2477–3

From Caveman to Chemist: Circumstances and Achievements
By Hugh W. Salzberg
300 pp; clothbound ISBN 0–8412–1786–6; paperback ISBN 0–8412–1787–4

The Green Flame: Surviving Government Secrecy
By Andrew Dequasie
300 pp; clothbound ISBN 0–8412–1857–9

For further information and a free catalog of ACS books, contact:
American Chemical Society
Customer Service & Sales
1155 16th Street, NW
Washington, DC 20036
Telephone 800–227–5558

Bestsellers from ACS Books

The ACS Style Guide: A Manual for Authors and Editors
Edited by Janet S. Dodd
264 pp; clothbound ISBN 0–8412–0917–0; paperback ISBN 0–8412–0943–X

Understanding Chemical Patents: A Guide for the Inventor
By John T. Maynard and Howard M. Peters
184 pp; clothbound ISBN 0–8412–1997–4; paperback ISBN 0–8412–1998–2

Chemical Activities (student and teacher editions)
By Christie L. Borgford and Lee R. Summerlin
330 pp; spiralbound ISBN 0–8412–1417–4; teacher ed. ISBN 0–8412–1416–6

Chemical Demonstrations: A Sourcebook for Teachers,
Volumes 1 and 2, Second Edition
Volume 1 by Lee R. Summerlin and James L. Ealy, Jr.;
Vol. 1, 198 pp; spiralbound ISBN 0–8412–1481–6;
Volume 2 by Lee R. Summerlin, Christie L. Borgford, and Julie B. Ealy
Vol. 2, 234 pp; spiralbound ISBN 0–8412–1535–9

Chemistry and Crime: From Sherlock Holmes to Today's Courtroom
Edited by Samuel M. Gerber
135 pp; clothbound ISBN 0–8412–0784–4; paperback ISBN 0–8412–0785–2

Writing the Laboratory Notebook
By Howard M. Kanare
145 pp; clothbound ISBN 0–8412–0906–5; paperback ISBN 0–8412–0933–2

Developing a Chemical Hygiene Plan
By Jay A. Young, Warren K. Kingsley, and George H. Wahl, Jr.
paperback ISBN 0–8412–1876–5

Introduction to Microwave Sample Preparation: Theory and Practice
Edited by H. M. Kingston and Lois B. Jassie
263 pp; clothbound ISBN 0–8412–1450–6

Principles of Environmental Sampling
Edited by Lawrence H. Keith
ACS Professional Reference Book; 458 pp;
clothbound ISBN 0–8412–1173–6; paperback ISBN 0–8412–1437–9

Biotechnology and Materials Science: Chemistry for the Future
Edited by Mary L. Good (Jacqueline K. Barton, Associate Editor)
135 pp; clothbound ISBN 0–8412–1472–7; paperback ISBN 0–8412–1473–5

For further information and a free catalog of ACS books, contact:
American Chemical Society
Customer Service & Sales
1155 16th Street, NW, Washington, DC 20036
Telephone 800–227–5558